WP3—Innovation in Agriculture and Forestry Sector for Energetic Sustainability

WP3—Innovation in Agriculture and Forestry Sector for Energetic Sustainability

Editors

Andrea Colantoni
Massimo Cecchini
Alvaro Marucci
Fabio Recanatesi
Elena Di Mattia
Rodolfo Picchio
Mauro Villarini
Valerio Cristofori

MDPI • Basel • Beijing • Wuhan • Barcelona • Belgrade • Manchester • Tokyo • Cluj • Tianjin

Editors

Andrea Colantoni
Department of Agricultural and Forestry Sciences (DAFNE), Tuscia University
Italy

Massimo Cecchini
Department of Agricultural and Forestry Sciences (DAFNE), University of Tuscia
Italy

Alvaro Marucci
Department of Agricultural and Forestry Sciences (DAFNE), University of Tuscia
Italy

Fabio Recanatesi
Department of Agricultural and Forestry Sciences (DAFNE), University of Tuscia
Italy

Elena Di Mattia
Department of Agricultural and Forestry Sciences (DAFNE), University of Tuscia
Italy

Rodolfo Picchio
Department of Agricultural and Forestry Sciences (DAFNE), University of Tuscia
Italy

Mauro Villarini
Dipartimento di Scienze Agrarie e Forestali, Università degli Studi della Tuscia
Italy

Valerio Cristofori
Department of Agricultural and Forestry Sciences (DAFNE), University of Tuscia
Italy

Editorial Office
MDPI
St. Alban-Anlage 66
4052 Basel, Switzerland

This is a reprint of articles from the Special Issue published online in the open access journal *Energies* (ISSN 1996-1073) (available at: https://www.mdpi.com/journal/energies/special_issues/WP3).

For citation purposes, cite each article independently as indicated on the article page online and as indicated below:

LastName, A.A.; LastName, B.B.; LastName, C.C. Article Title. *Journal Name* **Year**, *Volume Number*, Page Range.

ISBN 978-3-0365-0226-7 (Hbk)
ISBN 978-3-0365-0227-4 (PDF)

Contents

About the Editors

Andrea Colantoni is an Associate Professor in Agricultural Mechanics full time at Tuscia University. Deputy director of the Experimental Farm 'Nello Lupori' of Tuscia University. Member of the Italian Association of Agricultural Engineers which is part of the European Society of Agricultural Engineers (EurAgEng), of the International Commission of Agricultural Engineering (CIGR), of the Italian Association of Scientific Agricultural Societies (AISSA) for the 3rd edition on "Mechanization and technologies for agriculture" and for the 5th edition on "Ergonomics and work organization". Author of more than 200 publications in excellent journals with high impact factors.

Massimo Cecchini is an Associate Professor in Agricultural Mechanics full time at the University of Tuscia. He obtained his degree in Agricultural Sciences in 1993 with 110/110 cum laude. From 2012 to 2016, he was the coordinator of the PhD program in Agricultural and Forestry Systems Engineering at the University of Tuscia. Since 2016, he has been a member of the PhD course in Engineering for Energy and Environment of the University of Tuscia. Since 2001, he is has been a member of AIIA, the Italian Society of Agricultural Engineering, member of EurAgEng—European Society of Agricultural Engineers, of CIGR—International Commission of Agricultural Engineering, and AISSA—Italian Association of Agricultural Scientific Societies.

Alvaro Marucci is a Full Professor in rural buildings and agro-forest land planning. The main issues addressed in his research include the following:

(A) Construction and energetic aspects of structures for protected crops;

(B) Functional and energetic aspects of soil solarization;

(C) Rural buildings;

(D) Recovery and reuse of rural buildings;

(E) Environmental modeling, planning of rural areas, and wastewater management.

Fabio Recanatesi is an Associate Professor in Land Use Planning and Geographic Information System (GIS) at the University of Tuscia (Italy). Professor Recanatesi is an expert in geomatics and remote sensing for the environmental risk assessment of natural and semi-natural environments and a scientific consultant at the Site of Community Importance (SCI) of the state nature reserve of Castelporziano (Rome).

Elena Di Mattia received her PhD degree in 1999 on "Energetics and Environmental Technologies to Development" at the University of Roma "La Sapienza" and, since 2002, has been a researcher (AGR/16) at DAFNE, in Tuscia University, working on Agro-Environmental Microbiology and Soil Microbiology. Her main research works focus on microbial ecology and physiology of soil and rhizosphere: microbial activities of inoculants and plant biostimulation. Moreover, she is involved in some studies concerning waste recycle valorization (i.e., biochar, wastewater, poly-caprolactone, and biodegradable polymer) using bacterial strains and microbial consortia also managed by nanotechnology.

Rodolfo Picchio obtained a Forest Science degree and a PhD degree in Forest Mechanizationfrom Tuscia University in Viterbo. From 2004 to 2014, he worked as a university researcher and, since 2014, has been an Associate Professor in Forest Logging and Wood Technologies. Since 2014, he has also led the 'Forest Utilization Work Group' at Tuscia University (DAFNE). From 2015 to 2017, he served as the president of the Master program of Forest Science "Conservation and Restoration of the Forest Environment and Soil Defense" at Tuscia University (DAFNE). He has served on the board of the PhD programs of "Science and Technologies for Forest and Environmental Management" (2005–2013) and "Engineering for Energy and Environment" (since 2018) at Tuscia University. His fields of expertise include forest logging, forest mechanization, precision forestry, wood technologies, and the environmental impacts of wood harvesting operations. He has worked as the Italian representative in two EU COST Actions on silviculture and biomass. Furthermore, he has co-authored 85 peer-reviewed papers in international journals indexed by WoS and Scopus, 25 technical books, and over 180 papers in technical magazines. He is also on the editorial boards of four international scientific journals covering forestry and forest engineering subjects. He is a member of the scientific Society of Silviculture and Forest Ecology.

Mauro Villarini is a mechanical engineer. He attained his PhD in Energy Systems from 2007 to 2012 and also served as a Research Assistant at Sapienza University. Since 2012, he has been an Assistant Professor at the DAFNE Department of Tuscia University of Viterbo teaching the following courses: Energy and Environment, Renewable Energies, Energy Systems, Fluid Dynamics, Management of Energy and Industrial Plants within the Master's Degree course of Mechanical Engineering (DEIM Department) and the course of Forest and Natural Sciences (DAFNE Department) of Tuscia University.

Valerio Cristofori is an Associate Professor in Horticulture at the Department of Agriculture and Forest Sciences—Tuscia University. His research skills are focused on the evaluation of qualitative traits in nuts and temperate fruit species. Other research activities treat the individuation, characterization, and conservation of local fruit germplasm, including minor fruit species of nutraceutical interest. He is also involved in the investigation of innovation in cultivar choice of nut species and orchard management. He is currently the Principal Investigator of the Local Unit in the European Project Horizon 2020 "Pantheon—Precision Farming of Hazelnut Orchards (Grant agreement n. 774571)". His research activity is manifested in more than 50 publications in international journals (indexed in Scopus), several book chapters, and in many other publications in national journals and proceedings of International Congresses.

Preface to "WP3—Innovation in Agriculture and Forestry Sector for Energetic Sustainability"

This Special Issue was founded after the creation of a department of excellence at the University of Tuscia (Viterbo, Italy). In particular, in the context of the WP3 objective, there is a need for innovation in the agricultural and forestry sectors for energy sustainability. Renewable energy sources and the rational use of energy represent an important agricultural and forestry resource in a local context against climate change. The first topic that this Special Issue will address is identifying the energy potential from agroforestry biomass, also dealing with the production of agricultural and forestry biomass in terms of supply and logistics; short rotation forestry (SRF); agricultural and forestry residues and their valorization; the main techniques of cultivation, the mechanization of biomass production, harvesting, and pre-treatment; the energy and environmental balance of biomass production; further forms of by-product valorization; evaluation of the potential of land use and planning tools. The second topic encompasses the performance analysis of prototypal systems for energy conversion, including hydrogen, power, and/or heat production plants and the pertinent thermodynamic cycles; sustainable renewable energy technologies (RETs); biomass thermochemical energy conversion technologies; system analysis and the integration of production and conversion and integrated bioenergy systems, including economic, environmental, and management aspects. The third topic focuses on energy sustainability for environmental mitigation, such as the impact of large-scale/small-scale bioenergy systems.

Andrea Colantoni, Massimo Cecchini, Alvaro Marucci, Fabio Recanatesi, Elena Di Mattia, Rodolfo Picchio, Mauro Villarini, Valerio Cristofori

Editors

Editorial

WP3—Innovation in Agriculture and Forestry Sector for Energetic Sustainability

Andrea Colantoni *, Rodolfo Picchio *, Alvaro Marucci, Elena Di Mattia, Valerio Cristofori, Fabio Recanatesi, Mauro Villarini, Danilo Monarca and Massimo Cecchini

Department of Agricultural and Forestry Sciences (DAFNE), University of Tuscia, via S. Camillo De Lellis, 01100 Viterbo, Italy; marucci@unitus.it (A.M.); dimattia@unitus.it (E.D.M.); VALERIO75@UNITUS.IT (V.C.); fabio.rec@unitus.it (F.R.); mauro.villarini@unitus.it (M.V.); monarca@unitus.it (D.M.); cecchini@unitus.it (M.C.)

* Correspondence: colantoni@unitus.it (A.C.); r.picchio@unitus.it (R.P.); Tel.: +39-0761357356 (A.C.)

Received: 15 September 2020; Accepted: 22 October 2020; Published: 16 November 2020

1. Documents in This Special Issue

Papers submitted and published in this Special Issue "WP3—Innovation in Agriculture and Forestry Sector for Energetic Sustainability" bring together some of the latest research results in the field of biomass valorization and the process of energy production and climate change and other items about energetic sustainability [1–20]. Moreover it is very important to evaluate the safety aspects for energy plant use [21–24].

Responses to our call generated the following statistics:

- Submissions (21);
- Publications (15);
- Rejections (6);
- Article types: research articles (13), reviews (2).

Published submissions are related to 15 published articles.

We found the edition and selections of papers for this exercise very inspiring and rewarding. We also thank the editorial staff and reviewers for their efforts and help during the process.

For better comprehension, the contributions to this special issue are split in parts, as follows.

1.1. Research Articles

The first contribution in this section explores the Performance Assessment of Front-Mounted Beet Topper Machine for Biomass Harvesting by Volodymyr Bulgakov, Simone Pascuzzi, Semjons Ivanovs, Francesco Santoro, Alexandros Sotirios Anifantis and Ievhen Ihnatiev. In this article, the authors focused their attention on the analysis of performance related to the operation of a new high-quality prototype of sugar beet top harvester that they built in Ukraine. Sugar beet is an extensive crop of of great agronomic value with significant productive and economic returns and Ukraine's sugar beet accounts for about 5.1% of the overall world production. Sugar beet and the by-products resulting from its manufacturing transformation are a significant renewable energy resource; a top sugar beet harvester, front-mounted on a tractor, was built by the authors in Ukraine. After the description of the beet topper machine features, the field tests, which took place in the Kiev region, were presented with reference to the tractor traction power using sensors able to measure torque and angular speeds. The experimental data were processed and showed that the energy costs related to the single work row of the prototype beet top harvesting machine were significantly lower than the corresponding performance parameter values of beet top-harvesting machines currently in use in Ukrainian farms.

The second paper is entitled Environmental and Economic Analysis of an Anaerobic Co-Digestion Power Plant Integrated with a Compost Plant and has been written by Sara Rajabi Hamedani,

Mauro Villarini, Andrea Colantoni, Maurizio Carlini, Massimo Cecchini, Francesco Santoro and Antonio Pantaleo. The article presents an analysis of economic and environmental issues of anaerobic digestion power-generation plants considering the Italian market evolution and basedon the scenario of a real case study. The spread of anaerobic digestion power plants slowed down after the remarkable growth that occurred between 2009 and 2016. The specific analysis considers also the benefits of use of digestate as fertilizer and of cogeneration heat in order to exploit all the resources potentially coming from these systems. Furthermore, a life-cycle analysis (LCA) has been carried out and the unreleased environmental emissions were converted into economic benefits by means of a stepwise approach. The final results of the analysis showed that the integration of a compost plant within the biogas plant integrated in the Italian electricity grid regulation allowed it to reach good financial performance in terms of IRR (internal rate of return) and NPV (net present value).

The third article entitled "The techno-economic modelling of biomass pellet routes: feasibility in Italy" was written by Antonio Pantaleo, Mauro Villarini, Andrea Colantoni, Maurizio Carlini, Francesco Santoro and Sara Rajabi Hamedani. Wood and agricultural biomass pellets boost the potential of bio-fuels for power production in tertiary and residential sectors. The production of pellets, however, is a multi-stage process where the supply-processing phases and the overall energy input strongly depend on the characteristics of the input biomass. In this paper, a model to evaluate the economic issues of agro-pellet was developed. The breakdown structure of costs has been represented after having described the manufacturing process from raw biomass to finished pellet. CAPEX and OPEX have been determined by means of a mathematical model. In the CAPEX, pelletizing, drying, pretreatment storage plants in addition to installation and engineering costs have been considered. The annual operation and maintenance costs have been calculated considering the following expense items: raw biomass supply and transport, biomass drying, electricity consumption, plants maintenance and personnel. After the model implementation, the analysis was applied to the case study of an Italian firm producing doors and windows in laminated wood. The analysis was then carried out within 4 different scenarios where the organization and, especially, the origin of the biomass used for the pellet production changes. The best scenario is the second one in which all the biomass treated comes from the production process of the company. Finally, the result showed that use of forestry residues with high moisture and high ash content, high costs of collection/transport, and high costs of pre-treatment and drying is not financially competitive.

The fourth papers entitled Evaluation of Compressor Heat Pump for Root Zone Heating as an Alternative Heating Source for Leafy Vegetable Cultivation has been written by Chiara Terrosi, Sonia Cacini, Gianluca Burchi, Maurizio Cutini, Massimo Brambilla, Carlo Bisaglia, Daniele Massa and Marco Fedrizzi. It aims to investigate the best value heating system for protected horticulture represented by leafy vegetable cultivation with a focus on the performance of the heat pump. It used 840 sweet basil seedlings during the experimental tests within the case study. The three systems examined were a condensing boiler, a heating pump and a air-to-air heater. The purpose was to verify the applicability considering the moderate temperature achieved by this system and the energy efficiency considering the absence of electricity, produced from renewable energy resources, to feed the heat pump itself. The heating distribution was described and the results presented showing that an optimization of the system can be achieved by reducing energy needs, using energy more efficiently and using less expensive energy sources. In particular, the adoption of an electric heat pump for greenhouse heating allows remarkable energy savings to be obtained and, especially, with respect to the condensing boiler and the air heater, the energy reduction amounts to 45%.

The fifth article entitled Influence of Oxidant Agent on Syngas Composition: Gasification of Hazelnut Shells through an Updraft Reactor has been written by Francesco Gallucci, Raffaele Liberatore, Luca Sapegno, Edoardo Volponi, Paolo Venturini, Franco Rispoli, Enrico Paris, Monica Carnevale and Andrea Colantoni. This work concerns a laboratory test at Sapienza University of Rome of an updraft gasifier reactor fed with hazelnut shells, profitably used as fuel in thermo-chemical processes, such as direct combustion or gasification. The tests were aimed to study the effect of an oxidant agent on syngas quality (namely its lower heating value), composition, producible energy, and cold gas efficiency. Then, temperature distribution, syngas composition and heating value, and producible energy were measured. The syngas flow produced by the two different oxidant agent, air and steam, was roughly the same but its quality was considerably different. This different performance can be notice by the following results: syngas produced by steam gasification had a lower heating value of 13.1 MJ/Nm^3 with an energy flow of 5.4 MJ/s. On the contrary, the syngas produced by air gasification had a lower heating value of less than 6 MJ/Nm^3 with a 3.3 MJ/s energy flow.

The sixth article entitled Sustainability Assessment of Alternative Strip Clear Cutting Operations for Wood Chip Production in Renaturalization Management of Pine Stands was written by Janine Schweier, Boško Blagojević, Rachele Venanzi, Francesco Latteriniand Rodolfo Picchio. The object of this paper is a sustainability impact assessment approach was applied to understand how to modify forest operation planning in order to minimize environmental impact.

The work highlights the urgent exigence to apply silvicultural management strategies in order to support vegetation dynamics and enhance stand ecology, like the renaturalization concept. The forest operations are essential with respect to the environmental issues of this topic and sustainable forest management should be implemented. The objective of the forest operations presented by this work was that the forest wood chains could support the aforementioned strategy of renaturalization in typical afforested pine plantations in the Mediterranean basin. Considering the plethora of factors and criteria, often conflicting, to be considered, a multi-criteria decision analysis was applied. Three different forest wood chains were applied in pine plantations, all differing in the extraction system (animal, forestry-fitted farm tractor with winch, and double drum cable yarder). Twelve economic, environmental and socio-ecological indicators were selected and calculated in order to address the sustainability assessment. After that a multi-criteria decision analysis has been implemented. Results showed that first ranked alternative was case 2, in which extraction was conducted by a tractor with a winch. The main reason was that this alternative had best performance for 80% of the analyzed criteria.

The seventh paper is entitled Production of Wood Pellets from Poplar Trees Managed as Coppices with Different Harvesting Cycles and was written by Vincenzo Civitarese, Andrea Acampora, Giulio Sperandio, Alberto Assirelli and Rodolfo Picchio. This article aims to study high-density biomass plantations and the particular case short rotation wood coppice of poplar has been focused on and, by means of the CREA farm, was exploited to develop the experimental activity using different treatments with harvesting cycles of 3, 6 and 9 years. The objective of the study was to identify the best raw material suitable for pellet production from trees or stems. The crops were subdivided by crop cycle and type of product in six groups separately stored in six bins after having been chipped and refined. Then, dehydration and pelletization the moisture content was measured in three different times. During the monitoring process, the crops were sized and the dehydration process controlled. The pelletizing process using high density poplar plantation as a raw material highlights the possibility of obtaining a product that meets many of the quality standards required on the market.

The eighth article is entitled Optimizing the 3D Distributed Climate inside Greenhouses Using Multi-Objective Optimization Algorithms and Computer Fluid Dynamics and was written by Kangji Li, Wenping Xue, Hanping Mao, Xu Chen, Hui Jiang and Gang Tan. This work focuses on greenhouses considered important for densely-populated regions. Then, the modelling of the micro-climate of greenhouses by means of a hybrid computational fluid dynamics -evolutionary algorithm was implemented. The objective was to determine the optimal combination of parameters in order to make the crops grow and corresponding to the Pareto frontier. To that aim, a commercial greenhouse located in China was used for the validation of the aforementioned model. Then, after the model construction, it was validated with a field experiment. The temperature was measured and compared with values determined by the model. Afterwards, the optimization process was based on the following two problems: first, how to find out multiple variables' optimal setting points according to multiple environmental requirements; and second, for local planting areas, how to adjust environmental variables with high spatial resolution to find the balance of energy saving and environment suitability. Twenty five pairs of control variables from a total of 250 chromosomes were identified belonging to the optimum set point basis. This way, the optimal tradeoff between energy efficiency and environmental suitability was determined. A detailed analysis may be provided that helps find the potential of the crop yield and energy conservation.

Moreover, other articles published in this special issue are:

Oscillations Analysis of Front-Mounted Beet Topper Machine for Biomass Harvesting by Volodymyr Bulgakov, Simone Pascuzzi, Alexandros Sotirios Anifantis and Francesco Santoro. The goal of this study was to assess the opportunity to use beet leaves and tops for the production of renewable energy. In this regard, one of the main issues is related to harvesting operation and waste recovery. In particular, considering the mechanization applied and the natural soil roughness, the machines are affected by angular oscillations in a longitudinal–vertical plane that strongly affect the cutting uniformity. Using Lagrange II-type equations some simulations were performed to assess the design and kinematic parameters of a front-mounted beet topper. According to the main findings, in order to improve the efficiency of this harvesting machine, soil preparation is first needed, while the influence of the stiffness and damping parameters of the feeler wheels pneumatic tires is not so clear.

Life Cycle Assessment and Environmental Valuation of Biochar Production: Two Case Studies in Belgium by Sara Rajabi Hamedani, Tom Kuppens, Robert Malina, Enrico Bocci, Andrea Colantoni and Mauro Villarini. This paper aimed to understand the economic feasibility of biochar production. Currently, the biochar production process may not be considered very cheap, hence it is difficult to find lenders and business owners. In this initial phase, other aspects need to be carefully assessed, mainly related to two of the sustainability pillars, social and environmental. These issues were assessed through life-cycle analysis (LCA) performed for two potential biochar production systems and two different feedstocks: willow and pig manure. The functional unit was one ton of biochar and the LCA database was SimaPro. The findings showed that the biochar production from willow achieves better results for all environmental impact categories surveyed in comparison to biochar from pig manure. Also, a monetary valuation was applied in order to weigh environmental benefits against environmental costs using the Ecotax, Ecovalue, and Stepwise approach. The final remarks highlight once again that willow biochar is preferable to biochar production from pig manure from the environmental point of view.

The last two research articles are:

Wood Chip Drying through the use of a Mobile Rotary Dryer by Angelo Del Giudice, Andrea Acampora, Enrico Santangelo, Luigi Pari, Simone Bergonzoli, Ettore Guerriero, Francesco Petracchini, Marco Torre, Valerio Paolini and Francesco Gallucci. One of the main problems related to biomass use for energy production is the moisture content. Generally, moisture content negatively affects the energy conversion efficiency and the feedstock storage. For a number of issues, biomass drying is a crucial operation in which technology has tried in various ways to solve the most critical aspects. Currently, rotary dryers seem to be the best solution, concerning low cost of maintenance and consume of 15% and 30% less in terms of specific energy. This paper focused on the use of a new prototype of mobile rotary dryer concurrent flow on wood chips and three wooden biomass typologies have been assessed (*Populus* spp., *Robiniapseudoacacia* L. and *Vitis vinifera* L.). The drying process was affected by the initial moisture content: poplar in 8 h from 50% reached 41%; black locust in 6 h from 30% reached 21%; grapevine in 6 h from 30% reached 21%. Moreover, this study showed that other biomass characteristics (particle size distribution and bulk density) have an influence on drying operation and as a consequence on the process energy consumption. Findings showed that the three-biomass needed 1.61 (poplar), 0.86 (grapevine), and 1.12 MJ kg dry solids^{-1} (black locust), with an efficiency of thermal drying (η) respectively of 37%, 12%, and 27%. From this work some suggestions were formulated in order to improve the sustainability of the process: the need to increase the efficiency of the thermal insulation of mobile dryer; the application of the mobile dryer in small-farms, and using exhaust gases from thermal power plants.

The final article is entitled: Sensitivity Analysis of Different Parameters on the Performance of a CHP Internal Combustion Engine System Fed by a Biomass Waste Gasifier by Mauro Villarini, Vera Marcantonio, Andrea Colantoni and Enrico Bocci. This paper presented a study on the energetic valorization of residues from biomasses production. The energy production or proper conversion was undertaken through gasification and then by using an internal combustion engine with a generator. Sampling was undertaken selecting the most representative types of biomass waste from agricultural productions, and the most suitable one to be used in the gasification process. Generally, good quality syngas with up to 16.1% CO–4.3% CH_4–23.1% H_2 can be produced. The syngas's lower heating value may vary from 1.86 MJ/Nm^3 to 4.5 MJ/Nm^3 in the gasification with air and from 5.2 MJ/Nm^3 to 7.5 MJ/Nm^3 in the gasification with steam. The cold gas efficiency may vary from 16% to 41% in the gasification with air and from 37% to 60% in the gasification with steam. A sensitivity analysis was used considering the cold gas efficiency and the LHV, in order to select the best configuration process for the best quality syngas. The syngas quality was also assessed through the electrical efficiency and the cogeneration efficiency.

1.2. Review Articles

In addition to the original articles, two reviews were published in this special issue. Briefly, the first review treats Pellet Production from Woody and Non-Woody Feedstocks: A Review on Biomass Quality Evaluation by Rodolfo Picchio, Francesco Latterini, Rachele Venanzi, Walter Stefanoni, Alessandro Suardi, Damiano Tocci and Luigi Pari.

Forest and agricultural biomasses represent a notable fuel source and they are renewable and sustainable feedstock for energy production. Nowadays, many factors (economic, social and environmental) have greatly contributed to the increase of their consumption. Among these, pellet has a substantial importance with an increase in production and innovations, concerning both woody and non-woody biomass. This form of densified biomass can be composed of a broad spectrum of possible raw materials, and for this reason the assessment of its quality may be considered an important issue.

In this regard, worldwide research in the last decade produced a consistent number of scientific papers, and this review work is aimed to highlight the most interesting ones and to give the readers an overall view of the most current knowledge about this large and interesting topic. The authors focused on pellets from agricultural and forestry origin with a selection of papers from the last five years (2016–2020) and grouped them in four main topics: influence of different agro-forest management systems on pellet quality; analysis of pellets from pure feedstocks; influence of blending and binders on pellet quality; influence of pre- and post- treatments. A critical discussion on research that is missing, on future developments, and trends closed this work.

The other review is Revolutionizing towards Sustainable Agricultural Systems: The Role of Energy by Ilaria Zambon, Massimo Cecchini, Enrico Maria Mosconi and Andrea Colantoni. The purpose of this work was to increase the performances of primary sector focusing on bioeconomy and sustainability. The application of innovations is a progressive and integrated process. Knowing the governance and opening a dialogue with stakeholders is a fundamental step for innovation and development at national or international scale. However, when opposing normative guidelines for alternative systems of agriculture that arise, modernizations in agricultural and forestry may contribute to outlining more sustainable systems. Currently the primary sector, except for industrial agriculture, does not seem to develop adequately in terms of innovation. This work highlights the main innovations of recent years in the primary sector, including agriculture and forestry. In this sector, one of the main aids for pursuing adequate sustainable development is undoubtedly represented by energy. In this emerging framework, adequate technologies for concrete energy efficiency are needed. Moreover, energy sustainability itself is one of the most discussed issues currently. With this review we try to understand which innovations have actually been received by the primary sector, highlighting their limits and opportunities.

2. Conclusions

In summary, the papers of the special issue represent some of the latest and most promising research results in this new and exciting field, which continues to make significant impact on real-world applications. We are confident that this special issue will stimulate further research in this area.

We thank all the authors for their contributions to this special issue.

Author Contributions: Conceptualization, A.C., M.C., D.M., A.M., R.P.; methodology, V.C., F.R., E.D.M., M.C., D.M., A.M., R.P., A.C.; data curation, M.V., A.C., V.C., R.P.; supervision, A.C. All authors have read and agreed to the published version of the manuscript.

Funding: Ministry for Education, University and Research (MIUR) according to the Italian Law 232/2016.

Acknowledgments: The special issue was partially supported by the Italian Ministry for Education, University and Research (MIUR) according to the Italian Law 232/2016 within the fund for the financing of universities "Departments of Excellence".

Conflicts of Interest: The authors declare no conflict of interest.

References

1. Villarini, M.; Marcantonio, V.; Colantoni, A.; Bocci, E. Sensitivity Analysis of Different Parameters on the Performance of a CHP Internal Combustion Engine System Fed by a Biomass Waste Gasifier. *Energies* **2019**, *12*, 688. [CrossRef]
2. Del Giudice, A.; Acampora, A.; Santangelo, E.; Pari, L.; Bergonzoli, S.; Guerriero, E.; Petracchini, F.; Torre, M.; Paolini, V.; Gallucci, F. Wood Chip Dryingthrough the Using of a Mobile Rotary Dryer. *Energies* **2019**, *12*, 1590. [CrossRef]
3. RajabiHamedani, S.; Kuppens, T.; Malina, R.; Bocci, E.; Colantoni, A.; Villarini, M. Life Cycle Assessment and Environmental Valuation of Biochar Production: Two Case Studies in Belgium. *Energies* **2019**, *12*, 2166. [CrossRef]
4. Colantoni, A.; Villarini, M.; Marcantonio, V.; Gallucci, F.; Cecchini, M. Performance Analysis of a Small-Scale ORC Trigeneration System Powered by the Combustion of Olive Pomace. *Energies* **2019**, *12*, 2279. [CrossRef]
5. Bulgakov, V.; Pascuzzi, S.; Anifantis, A.; Santoro, F. Oscillations Analysis of Front-Mounted Beet Topper Machine for Biomass Harvesting. *Energies* **2019**, *12*, 2774. [CrossRef]
6. Li, K.; Xue, W.; Mao, H.; Chen, X.; Jiang, H.; Tan, G. Optimizing the 3D Distributed Climate inside Greenhouses Using Multi-Objective Optimization Algorithms and Computer Fluid Dynamics. *Energies* **2019**, *12*, 2873. [CrossRef]
7. Civitarese, V.; Acampora, A.; Sperandio, G.; Assirelli, A.; Picchio, R. Production of Wood Pellets from Poplar Trees Managed as Coppices with Different Harvesting Cycles. *Energies* **2019**, *12*, 2973. [CrossRef]
8. Schweier, J.; Blagojević, B.; Venanzi, R.; Latterini, F.; Picchio, R. Sustainability Assessment of Alternative Strip Clear Cutting Operations for Wood Chip Production in Renaturalization Management of Pine Stands. *Energies* **2019**, *12*, 3306. [CrossRef]
9. Zambon, I.; Cecchini, M.; Mosconi, E.; Colantoni, A. Revolutionizing Towards Sustainable Agricultural Systems: The Role of Energy. *Energies* **2019**, *12*, 3659. [CrossRef]
10. Gallucci, F.; Liberatore, R.; Sapegno, L.; Volponi, E.; Venturini, P.; Rispoli, F.; Paris, E.; Carnevale, M.; Colantoni, A. Influence of Oxidant Agent on Syngas Composition: Gasification of Hazelnut Shells through an Updraft Reactor. *Energies* **2020**, *13*, 102. [CrossRef]
11. Terrosi, C.; Cacini, S.; Burchi, G.; Cutini, M.; Brambilla, M.; Bisaglia, C.; Massa, D.; Fedrizzi, M. Evaluation of Compressor Heat Pump for Root Zone Heating as an Alternative Heating Source for Leafy Vegetable Cultivation. *Energies* **2020**, *13*, 745. [CrossRef]
12. Pantaleo, A.; Villarini, M.; Colantoni, A.; Carlini, M.; Santoro, F.; RajabiHamedani, S. Techno-Economic Modeling of Biomass Pellet Routes: Feasibility in Italy. *Energies* **2020**, *13*, 1636. [CrossRef]
13. RajabiHamedani, S.; Villarini, M.; Colantoni, A.; Carlini, M.; Cecchini, M.; Santoro, F.; Pantaleo, A. Environmental and Economic Analysis of an Anaerobic Co-Digestion Power Plant Integrated with a Compost Plant. *Energies* **2020**, *13*, 2724. [CrossRef]
14. Picchio, R.; Latterini, F.; Venanzi, R.; Stefanoni, W.; Suardi, A.; Tocci, D.; Pari, L. Pellet Production from Woody and Non-Woody Feedstocks: A Review on Biomass Quality Evaluation. *Energies* **2020**, *13*, 2937. [CrossRef]
15. Bulgakov, V.; Pascuzzi, S.; Ivanovs, S.; Santoro, F.; Anifantis, A.; Ihnatiev, I. Performance Assessment of Front-Mounted Beet Topper Machine for Biomass Harvesting. *Energies* **2020**, *13*, 3524. [CrossRef]

16. Zambon, I.; Colantoni, A.; Carlucci, M.; Morrow, N.; Sateriano, A.; Salvati, L. Land quality, sustainable development and environmental degradation in agricultural districts: A computational approach based on entropy indexes. *Environ. Impact Assess. Rev.* **2017**, *64*, 3746. [CrossRef]
17. Colantoni, A.; Marucci, A.; Monarca, D.; Pagniello, B.; Cecchini, M.; Bedini, R. The risk of musculoskeletal disorders due to repetitive movements of upper limbs for workers employed to vegetable grafting. *J. Food Agric. Environ.* **2012**, *10*, 14–18.
18. Monarca, D.; Colantoni, A.; Cecchini, M.; Longo, L.; Vecchione, L.; Carlini, M.; Manzo, A. Energy characterization and gasification of biomass derived by hazelnut cultivation: Analysis of produced syngas by gas chromatography. *Math. Probl. Eng.* **2012**, *2012*, 102914. [CrossRef]
19. Anifantis, A.S.; Colantoni, A.; Pascuzzi, S.; Santoro, F. Photovoltaic and hydrogen plant integrated with a gas heat pump for greenhouse heating: A mathematical study. *Sustainability* **2018**, *10*, 378. [CrossRef]
20. Marucci, A.; Monarca, D.; Cecchini, M.; Colantoni, A.; Cappuccini, A. The heat stress for workers employed in laying hens houses. *J. Food Agric. Environ.* **2013**, *11*, 20–24.
21. Di Giacinto, S.; Colantoni, A.; Cecchini, M.; Monarca, D.; Moscetti, R.; Massantini, R. Dairy production in restricted environment and safety for the workers. *Ind. Aliment.* **2012**, *530*, 5–12.
22. Marucci, A.; Pagniello, B.; Monarca, D.; Cecchini, M.; Colantoni, A.; Biondi, P. Heat stress suffered by workers employed in vegetable grafting in greenhouses. *J. Food Agric. Environ.* **2012**, *2*, 1117–1121.
23. Monarca, D.; Cecchini, M.; Colantoni, A. *Plant for the Production of Chips and Pellet: Technical and Economic Aspects of an Case Study in the Central Italy*; Springer: Berlin/Heidelberg, Germany, 2011; pp. 296–306.
24. Monarca, D.; Cecchini, M.; Colantoni, A.; Marucci, A. *Feasibility of the Electric Energy Production through Gasification Processes of Biomass: Technical and Economic Aspects*; Springer: Berlin, Germany, 2011; pp. 307–315.

Publisher's Note: MDPI stays neutral with regard to jurisdictional claims in published maps and institutional affiliations.

Article

Life Cycle Assessment and Environmental Valuation of Biochar Production: Two Case Studies in Belgium

Sara Rajabi Hamedani [1,*], Tom Kuppens [2], Robert Malina [2,3], Enrico Bocci [4], Andrea Colantoni [1] and Mauro Villarini [1,*]

1 Department of Agriculture and Forest Sciences, Tuscia University, 01100 Viterbo, Italy; colantoni@unitus.it
2 Environmental Economics Research Group, Centre for Environmental Sciences (CMK), Hasselt University, 3590 Diepenbeek, Belgium; tom.kuppens@uhasselt.be (T.K.); robert.malina@uhasselt.be (R.M.)
3 Laboratory for Aviation and the Environment, Massachusetts Institute of Technology, 77 Massachusetts Avenue, Cambridge, MA 02139, USA
4 Department of Innovation and Information Engineering, Marconi University, 00193 Rome, Italy; e.bocci@unimarconi.it
* Correspondence: sara.rajabi1322@gmail.com (S.R.H.); mauro.villarini@unitus.it (M.V.)

Received: 10 April 2019; Accepted: 3 June 2019; Published: 6 June 2019

Abstract: It is unclear whether the production of biochar is economically feasible. As a consequence, firms do not often invest in biochar production plants. However, biochar production and application might be desirable from a societal perspective as it might entail net environmental benefits. Hence, the aim of this work has been to assess and monetize the environmental impacts of biochar production systems so that the environmental aspects can be integrated with the economic and social ones later on to quantify the total return for society. Therefore, a life cycle analysis (LCA) has been performed for two potential biochar production systems in Belgium based on two different feedstocks: (i) willow and (ii) pig manure. First, the environmental impacts of the two biochar production systems are assessed from a life cycle perspective, assuming one ton of biochar as the functional unit. Therefore, LCA using SimaPro software has been performed both on the midpoint and endpoint level. Biochar production from willow achieves better results compared to biochar from pig manure for all environmental impact categories considered. In a second step, monetary valuation has been applied to the LCA results in order to weigh environmental benefits against environmental costs using the Ecotax, Ecovalue, and Stepwise approach. Consequently, sensitivity analysis investigates the impact of variation in NPK savings and byproducts of the biochar production process on monetized life cycle assessment results. As a result, it is suggested that biochar production from willow is preferred to biochar production from pig manure from an environmental point of view. In future research, those monetized environmental impacts will be integrated within existing techno-economic models that calculate the financial viability from an investor's point of view, so that the total return for society can be quantified and the preferred biochar production system from a societal point of view can be identified.

Keywords: life cycle analysis; environmental valuation; biochar; willow; pig manure

1. Introduction

Biochar is the stable, carbon-rich substance obtained from the pyrolysis of biomass materials such as wood, manure, or leaves [1]. The application of this pyrogenic black carbon can have substantial advantages from a social, economic, and environmental point of view, such as (1) job creation (social), (2) soil improvement for higher biomass yields and possible cost savings (economic), and (3) climate change mitigation and water or air pollutant absorption due to its porous form (environmental) [2–7]. Since sustainable biochar systems are essential to the future of biochar, these systems need to address a wide range of potential environmental, social, and economic impacts [8].

The economic features of biochar production have been reflected through many techno-economic assessments (TEA) [9–15], where the production cost is estimated based on the investment and operation costs of conversion technologies. In addition, life cycle assessment (LCA) has been applied several times to quantify the environmental impacts of biochar production systems. The majority of the research is focused on calculating potential savings in greenhouse gas (GHG) emissions, which is the most quoted benefit of biochar production and application [16–22]. It has also been illustrated that agricultural land occupation might become an issue when dedicated crops are grown specifically for biochar production [23]. However, in the sustainability framework, a comprehensive assessment involves not just the quantification of the financial impact of biochar technology and issues such as global warming, but also the broader, societal, cultural, political, and environmental impacts. While an understanding of societal impact is important for decision making and product project design, collecting and analyzing data on societal impacts is difficult and requires considerable time and interpersonal skills.

One way to solve this issue is to use the optional weighting approaches in LCA for converting and aggregating the results into a single indicator. Weights can be determined in a quantitative or qualitative way [24], or can be expressed in monetary units, both for midpoints and endpoints [25]. Biophysical impacts are then translated into monetary values by means of shadow prices reflecting the societal value of non-market goods, such as environmental quality, for which no prices exist. The advantage of using shadow prices is that they make environmental impacts comparable, so that all impacts can be aggregated and integrated in a techno-economic assessment containing private costs and benefits related to the production of market goods such as biochar. In fact, disregarding external costs imposed on society over the entire life cycle of biochar can lead to inefficient market pricing of this product, which results in non-sustainable biochar systems.

To the best of our knowledge, an assessment of biochar production systems that includes a wide range of environmental impact categories and that integrates the environmental aspects with the economic and social aspects is still missing.

The use of monetary valuation is recognized in LCA [26] and is easy to understand by communicating with a wide range of decision-makers [27,28].

However, the use of monetary values in LCA is controversial as the choice of valuation method is subjective and mirrors underlying social, ethical, and political values [29,30]. Therefore, we apply and compare three monetary valuation methods to LCA results for a case study in Belgium in order to answer the following research question: "What is the monetary value of the environmental impact of biochar production and application?". In other words, the aim of this study is to perform an LCA of two biochar production systems in Belgium, and to monetize the environmental impacts via applying and comparing three environmental valuation methods.

2. Methodology

2.1. *Life Cycle Assessment (LCA)*

The life cycle assessment methodology was used for the evaluation of the environmental impacts associated with the production and application of biochar in soil. The impacts were calculated in SimaPro software (version 8.3.0), according to the ISO 14040:2006 [31] requirements.

2.1.1. Goal Definition

The goal of this study was to compare the positive and negative environmental impacts of two potential cases in Belgium for biochar production and its use for soil amendment in drought-sensitive agricultural soils in the Campine region situated in the province of Limburg, Belgium. Some, but not all, of the soils in this vast region have been moderately polluted with cadmium (Cd) as a consequence of the pyrometallurgical processing of zinc until the seventies [12]. Hence, several opportunities for biochar application exist within the region as all soils are drought-sensitive and sandy, some of which are contaminated with heavy metals:

a. Metal-contaminated soils benefit from biochar application thanks to its capacity to immobilize the heavy metals;
b. Non-contaminated soils benefit from biochar application thanks to its capacity to better retain nutrients and water, especially within the context of climate change.

We propose the use of locally available feedstock to avoid the fact that biochar needs to be produced at distant locations, so that CO_2 emissions from transport are prevented. Additionally, no (indirect) land use change will result from crop cultivation specifically for (unsustainable) biochar production from, e.g., tropical woods. Examples of local feedstock for biochar production include:

i. Pig manure, which is abundantly available and needs to be processed anyhow to avoid a local oversupply of nutrients. Moreover, the processing of pig manure in three steps seems to result in a positive business case [32]. First, the pig manure is separated in water (44%), a thick fraction (17.5%), and a thin fraction (38.5%). The nutrients nitrogen (N) and potassium (K) end up in the thin fraction or concentrate, whereas phosphorus (P) is concentrated in the thick fraction. Second, the thick fraction is dried to a dry matter content of 95%. Third, the dried thick fraction is pyrolyzed for the production of biochar and energy. In full operation, it is expected that 60,000 tons of wet thick fraction can be processed annually;
ii. Willow can be cultivated on marginal soils that remain largely unused for agricultural production as they are not fertile and might even be contaminated with heavy metals. Hence, willow can be cultivated in short rotation to either produce energy and biochar from marginal, non-contaminated soils or to extract the cadmium from 2400 ha [33] of contaminated soil to produce energy and "activated" biochar.

So far, those local opportunities for biochar production and application have been investigated at Hasselt University by conducting pyrolysis experiments for willow and pig manure feedstock within the research group Applied and Analytical Pyrolysis, and by building techno-economic models within the research group of Environmental Economics at Hasselt University [12,32]. However, the societal value of its environmental impact remains unknown.

As the application of biochar in metal-contaminated soils has not been tested yet within this region, the monetary value of the environmental impact of biochar production and application has been determined for the following two case studies:

- Case 1: production and application of biochar from willow cultivated on non-contaminated marginal soils;
- Case 2: production and application of biochar from the dried thick fraction of pig manure.

In the future, other feedstocks will also be investigated, but no experimental data or traditional techno-economic models are available yet. Hence, we calculated and compared the environmental benefit/cost of both feedstocks to augment the available techno-economic models. The paper is to be considered as a first iteration that needs to be refined in the future based on the questions raised by the results after the first iteration. The functional unit is defined as 1 t of produced biochar because the main function of the system is biochar production [19,21]. The system boundary is shown in Figure 1. Either pig manure (case study 1) or willow woodchips (case study 2) are used as a feedstock for the pyrolysis process. Two system boundary expansions are included to represent additional functions of biochar [21]: (i) the pyrolysis process generates excess energy as a co-product in the form of bio-oil and syngas, avoiding some consumption and production of electricity and natural gas; and (ii) the application of biochar for soil amendment reduces the use of NPK fertilizer.

The syngas is burnt to provide the internal energy requirements for heat and electricity of the pyrolysis process. Excess energy from burning syngas on top of internal energy requirements is offset to the market as a substitute for natural gas and electricity. Additionally, the bio-oil co-product is sold on the market as a replacement for natural gas.

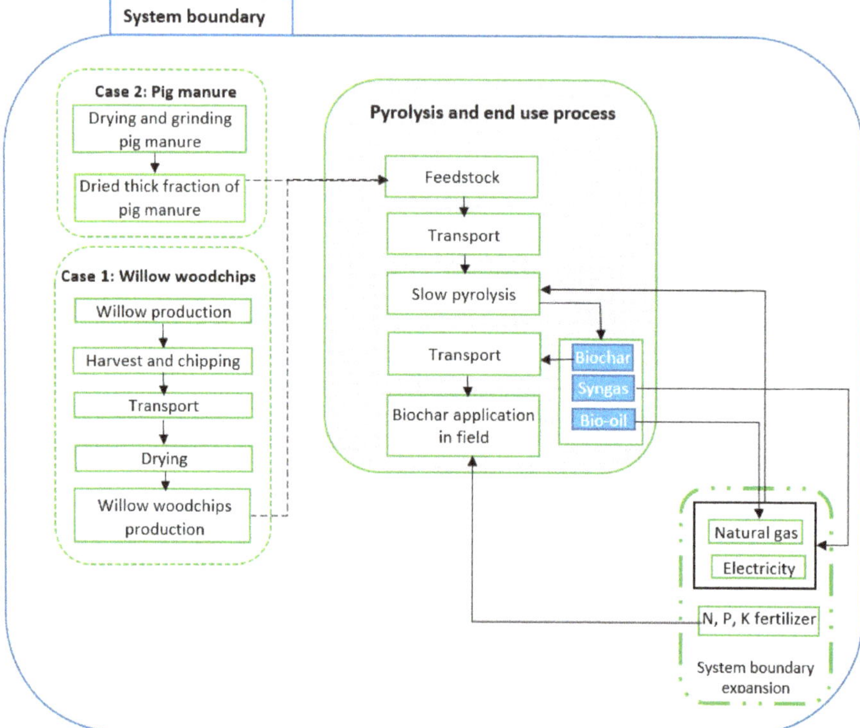

Figure 1. System boundaries for life cycle assessment (LCA) of biochar (and bioenergy) production.

2.1.2. Life Cycle Inventory (LCI)

Data were collected through laboratory tests, and scientific and technical literature. Some processes, such as willow woodchips production, transportation, electricity, and fuel production, were modeled using the available databases (Table A1) from the SimaPro 8.3 software. Fuel consumption emissions were estimated by IPCC guidelines [34]. Emissions released from burning syngas in combined heat and power (CHP) were estimated based on reported emission limits [35].

According to a unit process selected from the Ecoinvent 3.0 database (Table A1), activities for the production of willow woodchips comprise establishment, operation, and clearing of the plantation. The inputs of seed, mineral fertilizers, and pesticides are considered. It is assumed that no organic fertilizers are applied. In addition, the database includes all machine operations, namely soil cultivation, planting, fertilization, weed control, pest and pathogen control, harvest and chipping of willow stems, transport from the field to farm (2 km), drying of wood chips under a roof (air drying; no electricity input), and clearing of the plantation by a rotary tiller including the growing of oil radish (not harvested). Corresponding machine infrastructure and sheds are also covered in this database. Further, direct field emissions are included. This activity ends after mulching of the oil radish and with the provision of willow wood chips at the farm gate.

It is assumed that the pyrolysis plant will be operational for 20 years with 7000 working hours per year. The residence time of the feedstock is 60 min and the process temperature is set at 500 °C, allowing the volatile components to escape while a charred solid is left behind.

In the first case study (Table 1), wood chips were transported to the pyrolysis plant. According to pyrolysis experiments conducted at Hasselt University, the measured product yields for biochar, syngas, and bio-oil were 33.5%, 31.9%, and 34.6% of $W_{\text{dry feedstock}}$, respectively. Moreover, the calorific values of the syngas and the bio-oil were calculated to be 11 and 16 MJ/kg, respectively. The carbon sequestration potential of biochar application to the soil was calculated based on the total carbon content of biochar as 75% of $W_{\text{dry biochar}}$ for willow, of which a conservative share of 80% consists of stable carbon.

Table 1. Inventory data for 1 ton of biochar obtained via willow pyrolysis.

Inputs		Outputs	
Energy and transport		*Products*	
Willow woodchips	3.73 ton	Biochar	1 ton
Transport feedstock to pyrolysis plant	149.2 tkm	*Avoided products*	
Heat (pyrolysis)	1.92 GJ	Natural gas	0.37 ton
Transport biochar to filed	40 tkm	Electricity	1.01 GJ
		N fertilizer	0.66 kg
		K fertilizer	0.13 kg
		P fertilizer	0.1 kg
		Emissions	
		Syngas combustion in CHP	
		SO_2	0.015 kg
		NO_x	0.2 kg
		Biochar application in soil	
		CO_2 avoided	−2.2 ton
		N_2O avoided	−2.6 kg

In the second case study (Table 2), first, the pig manure is separated in water (44%), a thick fraction (17.5%), and a thin fraction (38.5%). Second, the thick fraction is dried to a dry matter content of 95%, and this dried thick fraction, after grinding, is the feedstock of the pyrolysis plant. This implies that additional pretreatment (drying and grinding) is required after the reception of the separated thick fraction of pig manure at the farm. The management of pig farms itself is not included in the system, as it is not expected that choosing pyrolysis instead of anaerobic digestion as the preferred manure processing technology will influence the farm's operations. According to the experimental results that have been obtained from the Cleantech business case for pig manure (second case study), pyrolysis of the dried thick fraction resulted in 48.8% (in terms of $W_{\text{dry feedstock}}$) biochar, 23.3% bio-oil, and 27.9% of syngas. Based on Cleantech estimation, one ton of biochar was produced by 2.9 tons of dried tick fraction. Furthermore, the calorific values of the syngas and the bio-oil were calculated to be 4 and 17.5 MJ/kg, respectively. The carbon content in the biochar from the dried thick fraction of pig manure that is used in calculating its carbon sequestration potential was estimated to be 33.7% of $W_{\text{dry biochar}}$.

Syngas in both cases was assumed to burn in a CHP with an electric efficiency of 25.6% and thermal efficiency of 54.4% [36]. In willow biochar, 12% of electricity generated entirely met the need of pyrolysis and the excess electricity (88%) was considered as an avoided product. However, heat from syngas burning covered only 73% of pyrolysis requirements. Comparatively, in the pig manure case, burning syngas could only provide 30% and 37% of heat and electricity demands of pyrolysis, respectively. Therefore, the avoided product in this case was only connected to bio-oil production.

As part of the application to soil, the biochar not only sequesters C, but also improves crop performance [37,38], which is a result of the enhancement in fertilizer use efficiency. This improvement can therefore reduce the amount of commercial chemical fertilizers applied. The dose of biochar applied to the soil as a main factor affects the results [39,40]. According to [17], 30 t ha^{-1} application of biochar for winter wheat crops can lead to a 10%, 5%, 5%, and 25% decrease in N, P, and K, fertilizers and N_2O emissions, respectively. Therefore, the total amount of N, P, and K fertilizers avoided and reduction of N_2O under normal management conditions of winter wheat [41] were calculated as 20, 3,

4, and 78 kg ha^{-1}, respectively. Since these assumptions can be affected by the area's climate, type of biochar, and agricultural plant, uncertainty of these assumptions is considered as a defined range for sensitivity analysis. The transportation distance of biomass feedstock to the pyrolysis facility and biochar to the field was also considered to be 40 km.

Table 2. The inventory data for 1 ton of biochar obtained via pig manure pyrolysis.

Inputs		Outputs	
Energy and transport		*Products*	
Heat (dried and ground pig manure production)	3.28 GJ	Biochar	1 ton
Electricity (dried and ground pig manure production)	0.7 MWh	*Avoided products*	
Transport feedstock to pyrolysis plant	116 tkm	Natural gas	0.14 ton
Heat (pyrolysis)	3.63 GJ	N fertilizer	0.66 kg
Electricity (pyrolysis)	0.05 MWh	K fertilizer	0.13 kg
Transport biochar to field	40 tkm	P fertilizer	0.1 kg
		Emissions	
		Syngas combustion in CHP	
		SO_2	0.003 kg
		NO_x	0.04 kg
		Biochar application in soil	
		CO_2 avoided	−0.98 ton
		N_2O avoided	−2.6 kg

2.1.3. Impact Assessment

In LCA studies on biochar, impact methods such as ReCipe midpoint [21] and Eco indicator 99 [20] have been developed for biochar systems. In this study, the life cycle impact assessment was performed using IMPACT 2002+ and CML-baseline methods in SimaPro 8.3. The former was selected since the IMPACT 2002+ [42] model is one of the main applied models in LCA analysis [43,44] and it enables researchers to consider environmental impacts on both a midpoint and endpoint level. However, the latter was chosen as a basis for the quantification of monetary values.

Monetizing Environmental Impacts

Monetization of environmental impacts can be carried out by means of benefit transfer using shadow prices that represent the value of those environmental aspects [45]. So far, there is no consensus in the scientific community on the most appropriate monetization method for weighting environmental impacts in LCA [46]. Therefore, three monetary valuation methods were employed: Ecotax02 [47], Ecovalue08 [48], and Stepwise2006 [49]. The Ecotax method is based on taxes and fees that are paid in Sweden for emissions and resource use and hence are an expression of the revealed value society attributes to the environmental effects. Stepwise2006 is based on a relatively new method [37] that takes into account the budget constraint, i.e., the annual income an average person can pay for an additional life year [35]. The use of a budget constraint reduces the uncertainty or bias that is associated with stated preference methods for the economic valuation of environmental impacts as respondents may not adequately consider their real income when answering questions related to their willingness to pay for environmental goods and services. The Ecovalue08 method, on the other hand, is based on the value individuals (rather than society) place on environmental goods and services. The Ecovalue08 method has been specifically developed in order to have a consistent weighting set which is based on the same valuation principle for all environmental impact categories considered [34]. The three methods hence represent different approaches (revealed versus stated preference, whether or not taking into account budget constraints) that can be used for a monetary valuation of environmental impacts and thus give an indication of the range within which the true value of the environmental impact will fall. As the existing techno-economic models for the two case studies are expressed in Euro2012 terms, the monetary values from the weighting methods have been converted into Euro2012 using European inflation rates between 2002 and 2012 [50].

The three methods also differ in terms of the characterization method and impact category levels (midpoint versus endpoint) for which they have been designed. Ecotax02 and Ecovalue08 have been designed for weighting at the midpoint level using CML midpoint categories [47,48], whereas Stepwise2006 provides the option of expressing results in both midpoints and endpoints through combining monetarization values with midpoint impact categories of IMPACT2002+ and EDIP 2003 [49]. In the present study, the results of this method are expressed at the midpoint level. To compare these methods, Table 3 presents the relevant weighting factors connected to each method. Since these factors are defined for the CML method's impact categories, first, the characterization of impact categories was conducted according to the CML life cycle impact assessment (LCIA) method. Next, the quantified environmental impacts were multiplied by the weighting factors presented in Table 3.

Table 3. Shadow prices used in different monetary valuation methods [51].

LCA Application Euryear	STEPWISE2006 EUR2003	ECOTAX02 EUR2002	ECOVALUE08 EUR2010
Global warming [eur/kgCO_2eq]	0.08	0.07	0.23
Ozone depletion [eur/kgCFC11eq]	100	139.56	-
Acidification [eur/kgSO_2eq]	0.00015	2.09	3.49
Eutrophication [eur/kgPO_4eq]	1.2	3.32	25.35
Photochemical oxidation [eur/kgC_2H_4eq]	0.00056	55.82	4.65
Abiotic resources [eur/MJ]	0.004	0.02	0.00047
Human toxicity [eur/kg1.4DBeq]	0.00154	0.17	1.4

2.1.4. Uncertainty Analysis

Uncertainty issues are always relevant in LCA studies. However, they are deemed to be more critical when developing comparative models. Therefore, uncertainty analysis of the main assumptions is necessary to support the results of comparative studies [52].

In this study, data uncertainty is assessed and quantified for NPK savings, as well as by products of the process, namely, syngas and bio-oil. As we do not have access to empirically-based data related to those uncertainties, one can apply the same arbitrary variation to the uncertainties [53], e.g., a coefficient of 10% from the nominal value of the uncertain variable [54,55]. Therefore, there is no specific rational for using ±10% variation, except for applying a conventional way of conducting sensitivity analysis when true ranges are missing.

3. Results and Discussion

3.1. Interpretation of LCA Midpoints

The characterization results of the life cycle impact assessment for the two case studies are reported in Table 4 in terms of Impact 2002+ midpoint categories. Case 1 and 2 represent biochar production from willow and pig manure, respectively. Negative values mean that environmental savings are generated by avoiding the use of products during biochar production and its application in soil, while positive values represent a burden for the environment. The results show that ionizing radiation, non-renewable energy, and global warming impacts were reduced in willow biochar production compared with pig manure biochar production ((−9500 vs. 22,392 Bq C-14 eq t^{-1}), (−16,830 vs. 6100 MJ t^{-1}), and (−2063 vs. −472 kg CO_2 eq t^{-1}), respectively). This is explained by a high contribution from natural gas and electricity production processes to ionizing radiation and non-renewable energy categories. Since the willow biochar process results in a higher amount of natural gas and electricity being avoided, related impacts are greatly reduced compared with pig manure biochar. In terms of global warming, the difference mainly refers to the higher potential of willow biochar with regards to CO_2 emission saving in soil compared with pig manure biochar.

On the contrary, pig manure biochar represents lower impacts than those of willow biochar in other impact categories. Particularly, differences are highlighted in terms of aquatic ecotoxicity, terrestrial ecotoxicity, and land use categories due to agricultural machinery, nitrogen fertilizer application, and land occupation in the cultivation and chipping phase of willow.

Table 4. IMPACT2002+ mid-point results (per ton of biochar).

Impact Category	Units	Case 1 (Willow)	Case 2 (Pig Manure)
Carcinogens	kg C_2H_3Cl eq	4.40	10.09
Non-carcinogens	kg C_2H_3Cl eq	11	3.08
Respiratory inorganics	kg $PM_{2.5}$ eq	0.31	0.17
Ionizing radiation	Bq C-14 eq	−9500	22,392
Ozone layer depletion	kg CFC-11 eq	2.14×10^{-5}	1.12×10^{-4}
Respiratory organics	kg C_2H_4 eq	0.01	0.06
Aquatic ecotoxicity	kg TEG water	31,000	23,400
Terrestrial ecotoxicity	kg TEG soil	48,200	6654
Terrestrial acid/nutri	kg SO_2 eq	6.69	2.68
Land occupation	m^2org.arable	3693	8.84
Aquatic acidification	kg SO_2 eq	0.91	0.68
Aquatic eutrophication	kg PO_4 P-lim	0.58	−0.005
Global warming	kg CO_2 eq	−2063	−472
Non-renewable energy	MJ primary	−16,830	6100
Mineral extraction	MJ surplus	15.37	4.29

In the next step, normalization is used to solve the incompatibility of units and simplify the interpretation of the results. In fact, normalization shows the relevant share of each impact category to the overall impacts through the application of the normalization factor. The normalization factor is defined as the impacts of all substances in their specific categories per person per year. The normalized values are obtained through dividing the characterization results by normalization factors, so the unit of all normalized values is [pers year/$unit_{emission}$], i.e., the number of equivalent persons affected during one year per unit of emission [42]. The Impact 2002+ normalization set defined for the European zone was employed.

According to the obtained results (Figure 2), it can be inferred that the most affected categories are terrestrial ecotoxicity, land occupation, global warming, and non-renewable energy. These categories are analyzed in detail below.

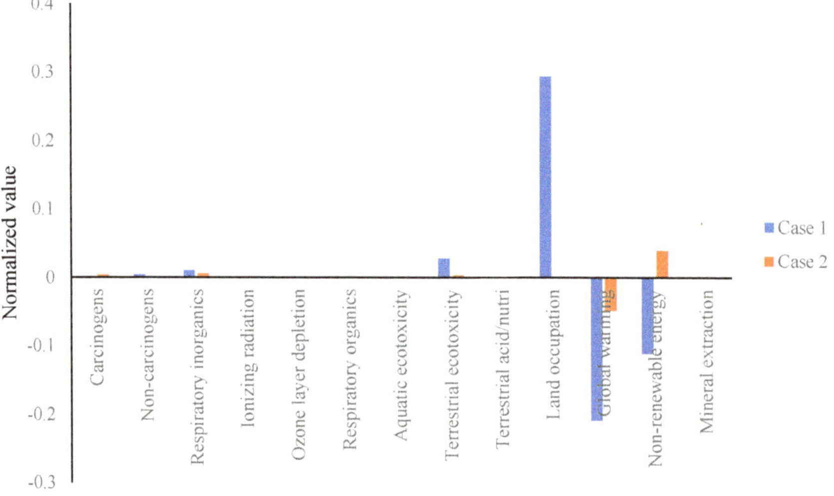

Figure 2. Normalized impact categories in each case.

3.1.1. Land Occupation

Case 1 (willow) has the most adverse impact on land occupation. This is due to land use for the willow production process. Hence, if case study 1 is implemented, the willow should be cultivated on marginal land (though not polluted with Cd). Case 2 (pig manure) has the lowest impact as the dried thick fraction is considered as waste from a pig farm.

3.1.2. Global Warming

Both case studies result in net savings of CO_2 emissions and thus can be considered as a measure to fight global warming. The expected savings in CO_2 emissions can be explained by the substituted amount of heat and electricity production and reduced fertilizer production, amongst other factors, but the highest share in total CO_2 savings is attributable to the application of biochar in soils. The difference in savings of CO_2 emissions can be explained by the different stable carbon content of the produced biochars. The biochar produced from willow can reduce GHG emissions more than pig manure biochar (2.2 t CO_2 vs 0.98 t CO_2 t^{-1} of biochar) because the stable carbon content of willow biochar is higher than that of pig manure biochar. The value obtained for savings of CO_2 emissions as a consequence of the application of willow biochar is close to those reported by Hammond et al. [17], being between 2.1 and 2.7 t CO_2 t^{-1} biochar.

3.1.3. Non-Renewable Energy

Case 1 (willow) reduces the amount of primary energy consumed, whereas case 2 (pig manure) results in a net increase of primary energy consumption. The reduction of 18,109 MJ of primary energy per ton biochar in case 1 (willow) can be explained by the substitution of natural gas and electricity resulting from the use of the pyrolysis byproducts (syngas and bio-oil). The increase of 10,820 MJ primary energy per ton biochar in case 2 (pig manure) is the result of the energy needed during the pretreatment process (especially drying) for pig manure.

3.1.4. Terrestrial Ecotoxicity

Additionally, in the impact category of terrestrial ecotoxicity, case 1 (willow) results in a more intensive impact than case 2 (pig manure). The main contribution to emissions in case 1 comes from fertilizer and agricultural machinery application during the production of willow wood chips, whereas the main contribution to emissions in case 2 comes from high electricity and heat consumption in the pretreatment of pig manure.

3.2. *Interpretation of LCA Endpoints*

Table 5 shows the damage endpoint categories and total impact single scores for each case per ton of biochar production. Case 1 (willow) resulted in reduced impacts on all categories, except ecosystem quality, due to land occupation during willow production. Case 2 (pig manure), on the other hand, results in increased impacts on all categories except climate change.

Table 5. IMPACT 2002+ endpoint results (per ton of biochar).

Damage Category	Unit	Case 1	Case 2
Human health	DALY	-8.68×10^{-8}	1.65×10^{-7}
Ecosystem quality	PDF·m^2·yr	2.63	0.06
Climate change	kg CO_2 eq	−2.22	−0.47
Resources	MJ primary	−23.59	6.11
Total points	μPt	−199.38	20.65

Figure 3 can be used to analyze the contribution of the process steps to the total damage. For case 1 (willow), the net reduction of resource consumption is caused by the avoidance of electricity and fuel production during the biochar production process. In both cases, the use of heat in the biochar production process and additionally in case 2 (pig manure) for drying the feedstock is the hotspot in the human health impact category.

According to the single score in the last line of Table 5, which represents a weighted score of overall impact categories that is not based on monetization, one can conclude that biochar production from willow is preferred over biochar production from pig manure from a life cycle perspective based on the aforementioned assumptions. In addition, according to the single score, one can even say that the production and application of biochar from willow is beneficial for the environment. Another important take-home message from Figure 3 is that one should look for more sustainable solutions for the pretreatment of pig manure. If these can be found, another iteration of the life cycle analysis should provide better insight into the environmental balance for both biochar production pathways as a basis for selecting the preferred biochar production pathway.

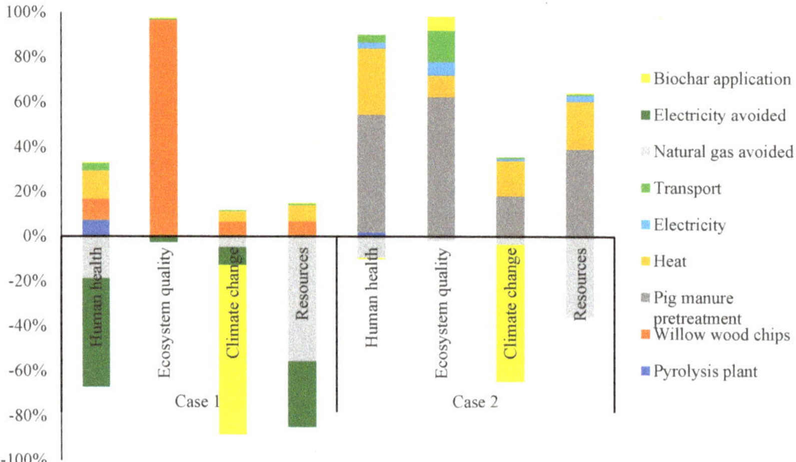

Figure 3. Process contribution to the damage categories.

3.3. *Interpretation of the Monetized Value of the Environmental Impacts*

The characterization results via the CML method (Table 6) were multiplied by the weighting factors presented in Table 3. The summary of results for the environmental valuation of willow and pig manure biochar production are reported in Table 7 for each of the three monetary valuation methods (Stepwise2006, Ecotax02, and Ecovalue08). Tables A2 and A3 present results in detail.

Table 6. CML characterization results (per ton of biochar).

Impact Category	Unit	Case 1 (Willow)	Case 2 (Pig Manure)
Global warming (GWP100a)	kg CO_2 eq	−2089.65	−466.56
Ozone layer depletion	kg CFC-11 eq	2.37×10^{-5}	1×10^{-4}
Human toxicity	kg 1,4-DB eq	97.42	69.63
Abiotic depletion (fossil fuels)	MJ	−15,085.11	2106.38
Eutrophication	kg PO_4 eq	2.66	0.20
Photochemical oxidation	kg C_2H_4 eq	−0.03	0.02
Acidification	kg SO_2 eq	0.89	0.66

All values in Table 7 are aggregated and visually represented in Figure 4. As prices cannot be negative, the signs in Table 7 reflect the sign of the environmental impact, i.e., negative values reflect avoided environmental impacts, whereas positive values represent processes that release emissions and hence are an environmental burden. After multiplying the avoided or additional emissions by their shadow price, the negative values can be interpreted as a benefit for society, whereas the positive values represent the external cost for each impact category considered.

Table 7. Environmental valuation per 1 ton of biochar produced from willow woodchips and pig manure.

Impact Category	Case 1			Case 2		
	Ecovalue08	Stepwise2006	Ecotax02	Ecovalue08	Stepwise2006	Ecotax02
Abiotic resources	−7.09 €	−60.32 €	−301.59 €	0.99 €	8.42 €	42.11 €
Global warming	−480.62 €	−167.17 €	−146.28 €	−107.31 €	−37.33 €	−32.66 €
Ozone depletion	0	0	0	0	0.01 €	0.02 €
Human toxicity	136.36 €	0.15 €	16.56 €	97.49 €	0.11 €	11.84 €
Photochemical oxidation	−0.18 €	0	−2.14 €	0.11 €	0	1.30 €
Acidification	3.11 €	0	1.86 €	2.32 €	0	1.39 €
Eutrophication	67.65 €	3.20 €	8.86 €	5.32 €	0.25 €	0.70 €
Net balance	**−280.77 €**	**−224.14 €**	**−422.72 €**	**−1.08 €**	**−28.53 €**	**24.69 €**

In terms of the net balance of external benefits and costs, all three valuation methods lead to the same conclusion for the first case on biochar production and application with willow feedstock. Using shadow prices as weights for the environmental impacts does not lead to a different conclusion compared to the single score of Table 5, in which non-monetized weights are used: all of the three methods indicate that the external benefits of biochar production and application with willow are higher than the external costs.

For the second case study, it was concluded from the single score (using non-monetized weights) in Table 5 that biochar production and application from pig manure was rather detrimental for the environment, which was mainly due to the high energy demand in the pretreatment step for drying the thick fraction. Applying the Ecotax02 method gives the same conclusion: biochar production and application from pig manure results in a net external cost and again, the pretreatment step is the largest contributor to the external cost. However, the distance between the external benefits and external costs, which corresponds to the value of the net external cost/benefit, is not as large as the distance or net benefit in the case study for willow. Moreover, according to the Ecovalue08 and Stepwise2006 methods, the external benefits are even higher than the external costs of the pig manure biochar system. Therefore, if sustainable solutions can be found for the pretreatment step of pig manure, the sign of the net result might be reversed. Hence it is important to investigate the effect of alternative pretreatment pathways on the net external cost/benefit in the pig manure case.

If we look at the results in more detail (see Figure 4), according to the Ecovalue08 method, in both cases, the application of biochar to soils is the main contributor to the external benefits from reduced global warming, which again can be traced back to the stable carbon content of the biochar. For either method and either case, it is also clear that the production of energy from the pyrolysis byproducts results in external benefits as a consequence of the avoided use of natural gas and electricity. However, the external benefits from avoided energy use are smaller for the pig manure case, because more biochar and less byproducts are produced in the latter case study.

When we compare the three methods, the Ecovalue08 and Ecotax02 methods indicate a different system component as the main contributor to the total external environmental benefit. For the Ecovalue08 method, it is concluded that the application of biochar contributes the most to the total external benefits. The Ecotax02 method, however, indicates that the reduced demand for primary energy or abiotic resources, i.e., the avoided energy use because of the valorization of the pyrolysis byproducts, is the most important contributor to external benefits. Another difference can be found in the relatively high value attached to human toxicity according to the Ecovalue08 method for both the production of willow woodchips and the pretreatment of pig manure. This can be partly explained by the relatively higher price the Ecovalue08 method attaches to this environmental impact category (see Table 3).

Comparing the willow and manure biochar system, it can be concluded that the external benefits for the willow biochar system are double the external benefits for the manure biochar system, which is explained by (i) the higher amount of saved energy consumption thanks to the pyrolysis byproducts in the willow biochar production system and (ii) the higher carbon content of the willow biochar (Figure 4).

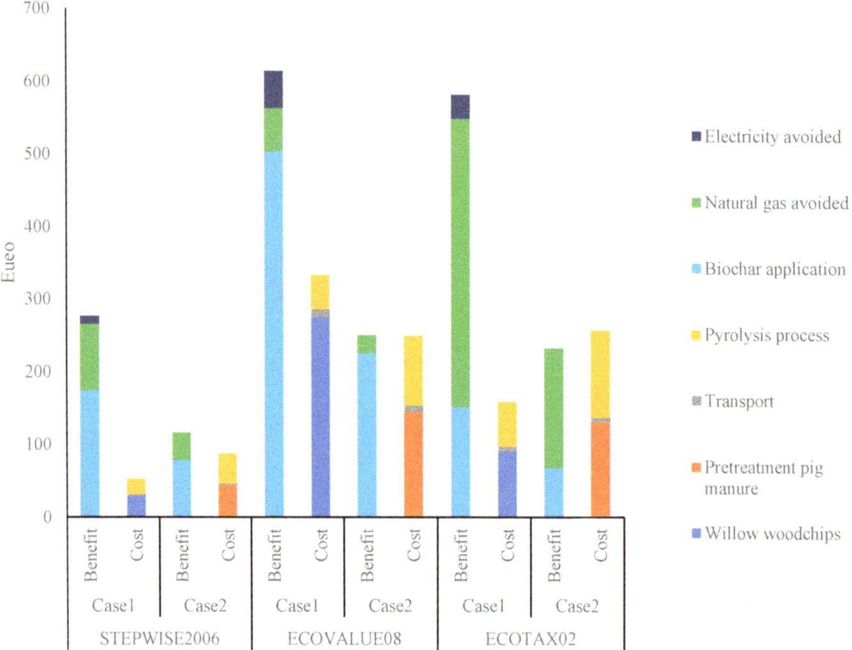

Figure 4. Environmental benefits and costs of two biochar production systems.

3.4. Sensitivity Analysis

This analysis assesses the impact of variation in NPK savings and byproducts of the process on monetized life cycle assessment results. Stepwise 2006 is selected as representative of valuation methods. Uncertainty analyses of the effect of different emission sources on the environmental value of impact categories for willow and pig manure biochar are presented as tornado diagrams in Figure 5. These diagrams show the variables which have the greatest effects on each of the impact categories and associated environmental values. The results present the impact of a 10% increase and 10% decrease in the average quantities of bio-oil and syngas and NPK savings on the outcome of the model. As it is seen, in willow biochar production, the most sensitive variable for human toxicity and eutrophication was willow woodchip production. However, human toxicity and eutrophication in pig manure biochar production showed the highest sensitivity for the pretreatment of pig manure. This is interpreted as a result of the high dependency of the pretreatment process of pig manure on heat and electricity consumption partly supplied by pyrolysis gas. Avoiding natural gas contributed the most to abiotic resources in both willow and pig manure biochar production. The results of the variation in NPK savings for all impact categories in both cases are negligible. Overall, the results of uncertainty analysis in W5 and P5 signified that feedstock provision and avoided products had high impacts on environmental values of biochar production in both cases. The total net balance in Figure 5. indicates that a ±10% variation in the quantity of byproducts produced can result in a range of environmental benefits from 159.9 euro to 367.6 euro and from 26.6 euro to 31.9 euro per 1 ton of biochar produced from willow and pig manure, respectively.

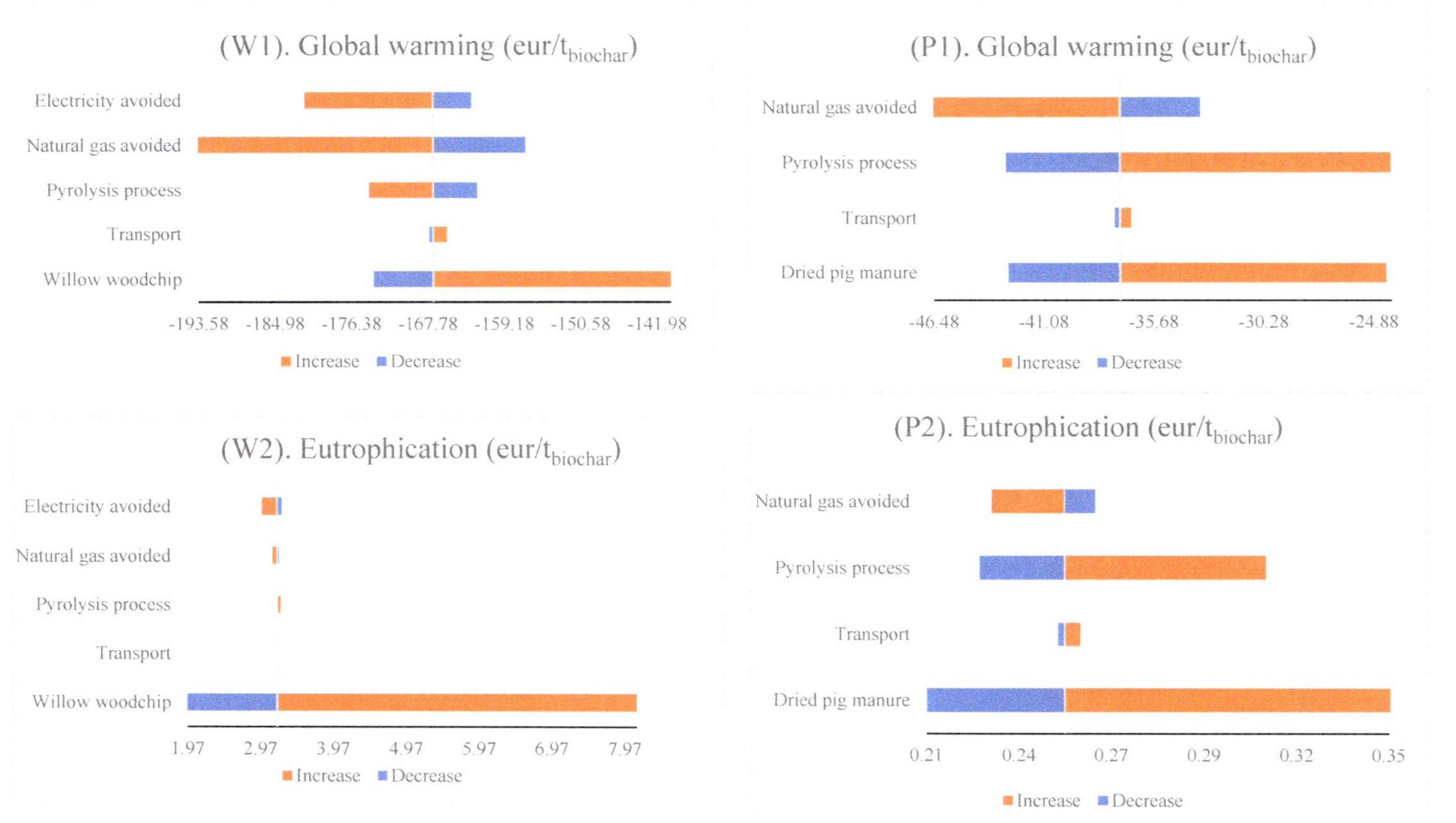

Figure 5. *Cont.*

Figure 5. *Cont.*

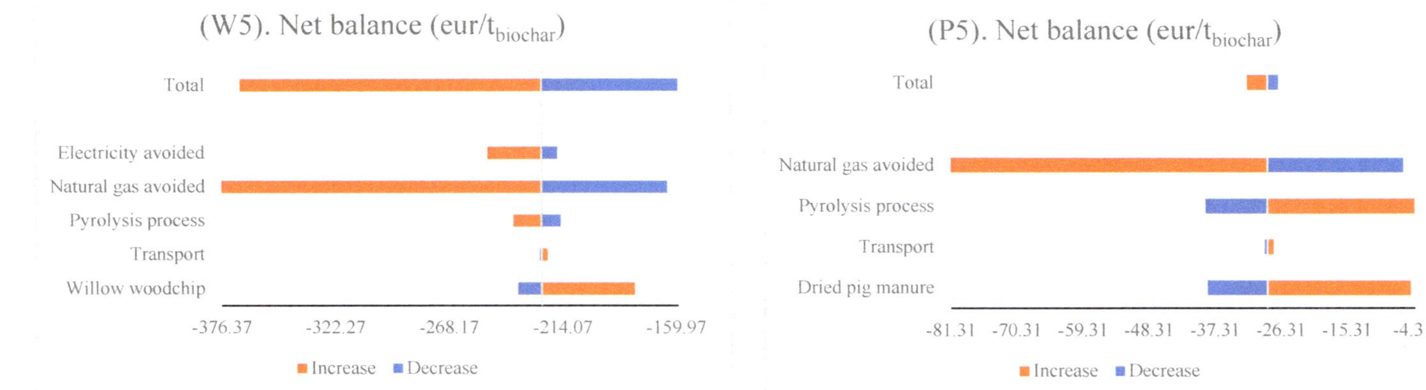

Figure 5. Sensitivity analysis of byproducts and NPK savings for each impact category-based environmental value for willow (**W1–W5**) and pig manure (**P1–P5**) biochar. (Note: vertical lines in the middle of the graph show the average environmental value of impact categories. Deviations from the average value show the changes in average environmental values by a 10% decrease or 10% increase in the average byproduct and NPK saving quantities).

4. Conclusions

This study investigated the environmental impact of biochar production from two feedstocks and its use for soil amendment by means of LCA. The novelty of the work consists of a comprehensive assessment of biochar production systems comprising a wide range of environmental impact categories and quantifying the environmental aspects as monetary values. This can be done by a weighting step in which monetary and non-monetary weights can be used. The advantage of monetary weights is that they reflect the values society or individuals attach to environmental goods or impacts, and they can be integrated with private costs and benefits from a traditional techno-economic assessment (TEA). For this purpose of integration, in future research, social impacts will also be quantified and monetized, and those monetized environmental and social impacts will be complemented by existing techno-economic models in order to develop a new methodological framework, i.e., a "societal techno-economic assessment", that takes into account both private and external costs and benefits and ultimately covers economic, social, and environmental aspects. This paper contributed to the calculation of the monetized value of the environmental aspects. Governments can use this information to devise policies for new technologies and to determine the right amount of taxes and subsidies to correct market failures.

The novel and main contribution of this paper, with respect to the literature, was the determination of the monetary value of the environmental impact of biochar production and its application for two case studies in Belgium using (i) willow and (ii) pig manure as a feedstock, for which techno-economic models were available. For the relevant (normalized) impact categories (global warming and non-renewable energy use), the willow biochar pathway outperforms the manure pathway. There are problems for land occupation and terrestrial ecotoxicity in the willow case, but these can be solved by restricting the growth of willow crops to marginal soils and the efficient application of fertilizers during willow growing. One of the main reasons why producing biochar from pig manure seems less beneficial is due to the high energy cost in the pretreatment step. Therefore, it is advised that researchers take a closer look at more sustainable ways of handling pig manure before it enters the pyrolysis reaction. If these can be found, another iteration of the life cycle analysis should provide better insight into the environmental balance for both biochar production pathways as a basis for selecting the preferred biochar production pathway.

In both cases, applying monetary weights resulted in the same conclusion as the one from using a single score environmental impact using non-monetary weights: under current assumptions, the willow biochar pathway appears to be better for the environment compared to the manure biochar pathway. Hence, a potential successful application of a willow-biochar system can consist of growing willow trees on non-contaminated marginal land for the production of biochar, and subsequently using the biochar for soil amendment within contaminated land to immobilize the metals.

However, the applied shadow prices differ and, as a next step, it should be investigated which method reflects the biochar production systems under investigation the best. For instance, the geographical scope might explain divergence: Ecotax2002 and Ecovalue08 are based on Swedish conditions, whereas Stepwise2006 has a more global scope. The annual income can be easily adjusted to the regional context when applying the Stepwise2006 approach, though Ecotax2002 and Ecovalue08 require more extensive adjustment steps that are beyond the scope of the current paper. Sensitivity analysis results also revealed that avoided products and feedstock provision had great impacts on the environmental values of biochar production in both cases.

Author Contributions: Conceptualization, S.R.H. and T.K.; methodology, S.R.H. and M.V.; software, S.R.H.; validation, E.B., A.C. and M.V.; formal analysis, S.R.H. and M.V.; investigation, S.R.H.; resources, R.M., M.V. and E.B.; data curation, S.R.H. and T.K.; writing—original draft preparation, S.R.H.; writing—review and editing, S.R.H., M.V. and E.B.; visualization, S.R.H.; supervision, M.V.; project administration, A.C. and M.V.; funding acquisition, A.C., E.B. and M.V.

Funding: The research was partially supported by MIUR (Ministry for education, University and Research), Law 232/2016, "Department of excellence" and partially supported by MISE (Italian Ministry of the Economic Development), "HBF2.0" Project, Grant number: CCSEB_00224. Additionally, the research was partially supported by the Italian SME company Enertecna.

Conflicts of Interest: The authors declare no conflict of interest.

Appendix A

Table A1. List of the processes extracted from SimaPro in this study.

Input	Process	Project
Willow woodchip	Wood chips and particles, willow {RoW}| willow production, short rotation coppice | Alloc Def, U	Ecoinvent 3
Transport willow woodchip to pyrolysis plant	Transport, freight, lorry >32 metric ton, EURO6 {GLO}| market for | Alloc Def, U	Ecoinvent 3
Heat	Heat, district or industrial, natural gas {Europe without Switzerland}| heat production, natural gas, at boiler modulating >100 kW | Alloc Def, U	Ecoinvent 3
Transport Willow biochar to field	Transport, freight, lorry 16–32 metric ton, EURO6 {GLO}| market for | Alloc Def, U	Ecoinvent 3
Natural gas	Natural Gas Mix, technology mix, consumption mix, at consumer, onshore and offshore production incl. pipeline and LNG transport EU-27 S	ELCD
Electricity	Electricity, medium voltage {BE}| market for | Alloc Def, U	Ecoinvent 3
N fertilizer	Nitrogen fertilizer, production mix, at plant/US	USLCI
K fertilizer	Potassium chloride (NPK 0-0-60), at plant/RER Economic	Agri-footprint
P fertilizer	Phosphorous fertilizer, production mix, at plant/US	USLCI
Transport pig manure to pyrolysis plant	Transport, freight, lorry >32 metric ton, EURO6 {GLO}| market for | Alloc Def, U	Ecoinvent 3
Transport pig manure biochar to field	Transport, freight, lorry 16–32 metric ton, EURO6 {GLO}| market for | Alloc Def, U	Ecoinvent 3

Table A2. Environmental valuation per 1 ton biochar produced from willow.

Impact Category	Willow Woodchips	Transport	Pyrolysis Plant	Biochar Application	Natural Gas Avoided	Electricity Avoided	Total
				Ecovalue08			
Abiotic resources	0.99 €	0.10 €	1.13 €	0.03 €	−8.82 €	−0.53 €	−7.09 €
Global warming	51.14 €	2.98 €	32.66 €	−504.73 €	−45.31 €	−17.36 €	-480.62 €
Ozone depletion	0	0	0	0	0	0	0
Human toxicity	148.76 €	7.31 €	10.89 €	2.10 €	−2.44 €	−30.25 €	136.36 €
Photochemical oxidation	0.28 €	0.01 €	0.06 €	0	−0.47 €	−0.06 €	−0.18 €
Acidification	4.85 €	0.12 €	1.00 €	−0.11 €	−2.08 €	−0.67 €	3.11 €
Eutrophication	69.40 €	0.18 €	1.52 €	−0.83 €	−0.80 €	−1.83 €	67.65 €
Net balance	275.41 €	10.71 €	47.27 €	−503.54 €	−59.91 €	−50.71 €	**−280.77 €**
				Stepwise2006			
Abiotic resources	8.42 €	0.89 €	9.65 €	0.28 €	−75.05 €	−4.50 €	−60.32 €
Global warming	17.79 €	1.04 €	11.36 €	−175.56 €	−15.76 €	−6.04 €	−167.17 €
Ozone depletion	0	0	0	0	0	0	0
Human toxicity	0.16 €	0.01 €	0.01 €	0	0	−0.03 €	0.15 €
Photochemical oxidation	0	0	0	0	0	0	0
Acidification	0	0	0	0	0	0	0
Eutrophication	3.29 €	0.01 €	0.07 €	−0.04 €	−0.04 €	−0.09 €	3.20 €
Net balance	29.65 €	1.94 €	21.10 €	−175.32 €	−90.85 €	−10.66 €	**−224.14 €**

Table A2. *Cont.*

Impact Category	Willow Woodchips	Transport	Pyrolysis Plant	Biochar Application	Natural Gas Avoided	Electricity Avoided	Total
				Ecotax02			
Abiotic resources	42.08 €	4.44 €	48.27 €	1.38 €	−375.24 €	−22.50 €	−301.59 €
Global warming	15.56 €	0.91 €	9.94 €	−153.61 €	−13.79 €	−5.28 €	−146.28 €
Ozone depletion	0	0	0	0	0	0	0
Human toxicity	18.06 €	0.89 €	1.32 €	0.25 €	−0.30 €	−3.67 €	16.56 €
Photochemical oxidation	3.34 €	0.12 €	0.76 €	−0.03 €	−5.60 €	−0.71 €	−2.14 €
Acidification	2.91 €	0.07 €	0.60 €	−0.07 €	−1.24 €	−0.40 €	1.86 €
Eutrophication	9.09 €	0.02 €	0.20 €	−0.11 €	−0.10 €	−0.24 €	8.86 €
Net balance	91.04 €	6.44 €	61.09 €	−152.19 €	−396.28 €	−32.82 €	**−422.72 €**

Table A3. Environmental valuation per 1 ton of biochar produced from pig manure.

Impact Category	Pig Manure Pretreatment	Transport	Pyrolysis Plant	Biochar Application	Natural Gas Avoided	Total
		Ecovalue08				
Abiotic resources	2.31 €	0.08 €	2.23 €	0.02 €	−3.66 €	0.99 €
Global warming	70.35 €	2.34 €	65.24 €	−226.44 €	−18.80 €	−107.31 €
Ozone depletion	0	0	0	0	0	0
Human toxicity	66.25 €	5.73 €	25.06 €	1.45 €	−1.01 €	97.49 €
Photochemical oxidation	0.18 €	0.01 €	0.12 €	0	−0.19 €	0.11 €
Acidification	1.91 €	0.10 €	1.30 €	−0.13 €	−0.86 €	2.32 €
Eutrophication	4.30 €	0.14 €	2.07 €	−0.86 €	−0.33 €	5.32 €
Net balance	145.31 €	8.40 €	96.03 €	−225.97 €	−24.86 €	**−1.08 €**
			Stepwise2006			
Abiotic resources	19.70 €	0.70 €	19.01 €	0.15 €	−31.14 €	8.42 €
Global warming	24.47 €	0.81 €	22.69 €	−78.76 €	−6.54 €	−37.33 €
Ozone depletion	0.01 €	0	0	0	0	0.01 €
Human toxicity	0.07 €	0.01 €	0.03 €	0	0	0.11 €
Photochemical oxidation	0	0	0	0	0	0
Acidification	0	0	0	0	0	0
Eutrophication	0.20 €	0.01 €	0.10 €	-0.04 €	-0.02 €	0.25 €
Net balance	44.45 €	1.52 €	41.84 €	−78.65 €	−37.70 €	**−28.53 €**
			Ecotax02			
Abiotic resources	98.50 €	3.48 €	95.07 €	0.77 €	−155.71 €	42.11 €
Global warming	21.41 €	0.71 €	19.86 €	−68.92 €	−5.72 €	−32.66 €
Ozone depletion	0.01 €	0	0.01 €	0	0	0.02 €
Human toxicity	8.05 €	0.70 €	3.04 €	0.18 €	−0.12 €	11.84 €
Photochemical oxidation	2.13 €	0.09 €	1.46 €	−0.05 €	−2.32 €	1.30 €
Acidification	1.15 €	0.06 €	0.78 €	−0.08 €	−0.52 €	1.39 €
Eutrophication	0.56 €	0.02 €	0.27 €	−0.11 €	−0.04 €	0.70 €
Net balance	131.81 €	5.06 €	120.48 €	−68.21 €	−164.44 €	**24.69 €**

References

1. Gaunt, J.L.; Lehmann, J. Energy balance and emissions associated with biochar sequestration and pyrolysis bioenergy production. *Environ. Sci. Technol.* **2008**, *42*, 4152–4158. [CrossRef] [PubMed]
2. Lehmann, J.; Joseph, S. *Biochar for Environmental Management: An Introduction*, 1st ed.; Lehmann, J., Joseph, S., Eds.; Routledge: Abington, UK, 2009; ISBN 9781844076581.
3. Verheijen, F.; Jeffery, S.; Bastos, A.C.; Van Der Velde, M.; Diafas, I. *Biochar Application to Soils: A Critical Review of Effects on Soil Properties, Processes and Functions*; European Commission: Brussel, Belgium, 2010; ISBN 9789279142932.
4. Kauffman, N.; Dumortier, J.; Hayes, D.J.; Brown, R.C.; Laird, D.A. Producing energy while sequestering carbon? The relationship between biochar and agricultural productivity. *Biomass Bioenergy* **2014**, *63*, 167–176. [CrossRef]

5. Houben, D.; Evrard, L.; Sonnet, P. Beneficial effects of biochar application to contaminated soils on the bioavailability of Cd, Pb and Zn and the biomass production of rapeseed (*Brassica napus* L.). *Biomass Bioenergy* **2013**, *57*, 196–204. [CrossRef]
6. Hilioti, Z.; Michailof, C.M.; Valasiadis, D.; Iliopoulou, E.F.; Koidou, V.; Lappas, A.A. Characterization of castor plant-derived biochars and their effects as soil amendments on seedlings. *Biomass Bioenergy* **2017**, *105*, 96–106. [CrossRef]
7. Mohammadi, A.; Cowie, A.L.; Kristiansen, P.; Cacho, O.; Lan, T.; Mai, A. Biochar can improve the sustainable use of rice residues in rice production systems. In Proceedings of the EGU General Assembly Conference, Vienna, Austria, 8–13 April 2018; Volume 20, p. 2652.
8. Lehmann, J.; Joseph, S. (Eds.) *Biochar for Environmental Management: Science, Technology and Implementation*, 2nd ed.; Routledge: Abington, UK, 2015.
9. Sahoo, K.; Bilek, E.; Bergman, R.; Mani, S. Techno-economic analysis of producing solid biofuels and biochar from forest residues using portable systems. *Appl. Energy* **2019**, *235*, 578–590. [CrossRef]
10. Jaroenkhasemmeesuk, C.; Tippayawong, N. Technical and Economic Analysis of A Biomass Pyrolysis Plant. *Energy Procedia* **2015**, *79*, 950–955. [CrossRef]
11. Brown, T.R.; Wright, M.M.; Brown, R.C. Estimating profitability of two biochar production scenarios: slow pyrolysis vs fast pyrolysis. *Biofuels Bioprod. Biorefining* **2011**, *5*, 54–68. [CrossRef]
12. Kuppens, T.; Van Dael, M.; Vanreppelen, K.; Thewys, T.; Yperman, J.; Carleer, R.; Schreurs, S.; Van Passel, S. Techno-economic assessment of fast pyrolysis for the valorization of short rotation coppice cultivated for phytoextraction. *J. Clean. Prod.* **2015**, *88*, 336–344. [CrossRef]
13. Homagain, K.; Shahi, C.; Luckai, N.; Sharma, M. Life cycle cost and economic assessment of biochar-based bioenergy production and biochar land application in Northwestern Ontario, Canada. *Ecosystems* **2016**, *3*, 1–10. [CrossRef]
14. Huang, Y.; Anderson, M.; McIlveen-Wright, D.; Lyons, G.A.; McRoberts, W.C.; Wang, Y.D.; Roskilly, A.P.; Hewitt, N.J. Biochar and renewable energy generation from poultry litter waste: A technical and economic analysis based on computational simulations. *Appl. Energy* **2015**, *160*, 656–663. [CrossRef]
15. Patel, M.; Zhang, X.; Kumar, A. Techno-economic and life cycle assessment on lignocellulosic biomass thermochemical conversion technologies: A review. *Renew. Sustain. Energy Rev.* **2016**, *53*, 1486–1499. [CrossRef]
16. Roberts, K.G.; Gloy, B.A.; Joseph, S.; Scott, N.R.; Lehmann, J. Life Cycle Assessment of Biochar Systems: Estimating the Energetic, Economic, and Climate Change Potential. *Environ. Sci. Technol.* **2010**, *44*, 827–833. [CrossRef] [PubMed]
17. Hammond, J.; Shackley, S.; Sohi, S.; Brownsort, P. Prospective life cycle carbon abatement for pyrolysis biochar systems in the UK. *Energy Policy* **2011**, *39*, 2646–2655. [CrossRef]
18. Ibarrola, R.; Shackley, S.; Hammond, J. Pyrolysis biochar systems for recovering biodegradable materials: A life cycle carbon assessment. *Waste Manag.* **2012**, *32*, 859–868. [CrossRef] [PubMed]
19. Harsono, S.S.; Grundman, P.; Lau, L.H.; Hansen, A.; Salleh, M.A.M.; Meyer-Aurich, A.; Idris, A.; Ghazi, T.I.M. Energy balances, greenhouse gas emissions and economics of biochar production from palm oil empty fruit bunches. *Resour. Conserv. Recycl.* **2013**, *77*, 108–115. [CrossRef]
20. Homagain, K.; Shahi, C.; Luckai, N.; Sharma, M. Life cycle environmental impact assessment of biochar-based bioenergy production and utilization in Northwestern Ontario, Canada. *J. For. Res.* **2015**, *26*, 799–809. [CrossRef]
21. Muñoz, E.; Curaqueo, G.; Cea, M.; Vera, L.; Navia, R. Environmental hotspots in the life cycle of a biochar-soil system. *J. Clean. Prod.* **2017**, *158*, 1–7. [CrossRef]
22. Brassard, P.; Godbout, S.; Pelletier, F.; Raghavan, V.; Palacios, J.H. Pyrolysis of switchgrass in an auger reactor for biochar production: A greenhouse gas and energy impacts assessment. *Biomass Bioenergy* **2018**, *116*, 99–105. [CrossRef]
23. Frazier, R.S.; Jin, E.; Kumar, A. Life cycle assessment of biochar versus metal catalysts used in syngas cleaning. *Energies* **2015**, *8*, 621–644. [CrossRef]
24. Huppes, G.; Van Oers, L. *Background Review of Existing Weighting Approaches in Life Cycle Impact Assessment (LCIA)*; European Union: Luxembourg, 2011.

25. Weidema, B.; Brandão, M.; Pizzol, M. *The Use of Monetary Valuation of Environmental Impacts in Life Cycle Assessment: State of the Art, Strengths and Weaknesses*; 2.-0 LCA Consultant: Aalborg, Denmark, 2013; Volume 33.
26. Itsubo, N.; Sakagami, M.; Kuriyama, K.; Inaba, A. Statistical analysis for the development of national average weighting factors-visualization of the variability between each individual's environmental thought. *Int. J. Life Cycle Assess.* **2012**, *17*, 488–498. [CrossRef]
27. Wu, X.; Zhang, Z.; Chen, Y. Study of the environmental impacts based on the "green tax"—Applied to several types of building materials. *Build. Environ.* **2005**, *40*, 227–237. [CrossRef]
28. van Harmelen, T.; van Korenromp, R.; Deutekom, C.; Ligthart, T.N.; van Leeuwen, S.; van Gijlswijk, R. *The Price of Toxicity. Methodology for the Assessment of Shadow Prices for Human Toxicity, Ecotoxicity and Abiotic Depletion*; Springer: Dordrecht, The Netherlands, 2006.
29. Johnsen, F.M.; Løkke, S. Review of criteria for evaluating LCA weighting methods. *Int. J. Life Cycle Assess.* **2013**, *18*, 840–849. [CrossRef]
30. Finnveden, G. *A Critical Review of Operational Valuation/Weighting Methods for Life Cycle Assessment*; Swedish Environmental Protection Agency: Stockholm, Swedish, 1999.
31. International Organization for Standardization. *Environmental Management-Life Cycle Assessment-Principles and Framework*; ISO: Geneva, Switzerland, 1947.
32. Kuppens, T.; Maggen, J.; Carleer, R.; Yperman, J.; Van Dael, M.; Van Passel, S. *Valorisatie van Varkensmest Door Pyrolyse—Cleantech Business Case in Limburg*; University of Hasselt: Hasselt, Belgium, 2014.
33. Kuppens, T.; Rafiaani, P.; Vanreppelen, K.; Yperman, J.; Carleer, R.; Schreurs, S.; Thewys, T.; Van Passel, S. Combining Monte Carlo simulations and experimental design for incorporating risk and uncertainty in investment decisions for cleantech: a fast pyrolysis case study. *Clean Technol. Environ. Policy* **2018**, *20*, 1195–1206. [CrossRef]
34. IPCC. *Guidelines for National Greenhouse Gas Inventories*; IPCC: Geneva, Switzerland, 2006.
35. European Union. *PE-CONS 42/15 DGE 1 EB/vm*; European Union: Brussels, Belgium, 2015; Volume 2015.
36. Lee, U.; Balu, E.; Chung, J.N. An experimental evaluation of an integrated biomass gasification and power generation system for distributed power applications. *Appl. Energy* **2013**, *101*, 699–708. [CrossRef]
37. Curaqueo, G.; Meier, S.; Khan, N.; Cea, M.C.; Navia, R. Use of biochar on two volcanic soils: effects on soil properties and barley yield. *J. Soil Sci. Plant Nutr.* **2014**, *14*, 911–924. [CrossRef]
38. Kimetu, J.M.; Lehmann, J.; Ngoze, S.O.; Mugendi, D.N.; Kinyangi, J.M.; Riha, S.; Verchot, L.; Recha, J.W.; Pell, A.N. Reversibility of Soil Productivity Decline with Organic Matter of Differing Quality Along a Degradation Gradient. *Ecosystems* **2008**, *11*, 726–739. [CrossRef]
39. Zhang, Q.; Dijkstra, F.A.; Liu, X.; Wang, Y.; Huang, J.; Lu, N. Effects of biochar on soil microbial biomass after four years of consecutive application in the north China plain. *PLoS ONE* **2014**, *9*, e102062. [CrossRef]
40. Li, B.; Fan, C.H.; Zhang, H.; Chen, Z.Z.; Sun, L.Y.; Xiong, Z.Q. Combined effects of nitrogen fertilization and biochar on the net global warming potential, greenhouse gas intensity and net ecosystem economic budget in intensive vegetable agriculture in southeastern China. *Atmos. Environ.* **2015**, *100*, 10–19. [CrossRef]
41. SAC. *Scottish Agricultural College Farm Management Handbook*; SAC: Penicuik, Scotland, 2016.
42. Jolliet, O.; Margni, M.; Charles, R.; Humbert, S.; Payet, J.; Rebitzer, G.; Rosenbaum, R. IMPACT 2002+: A new life cycle impact assessment methodology. *Int. J. Life Cycle Assess.* **2003**, *8*, 324–330. [CrossRef]
43. Hong, J.; Li, X.; Zhaojie, C. Life cycle assessment of four municipal solid waste management scenarios in China. *Waste Manag.* **2010**, *30*, 2362–2369. [CrossRef]
44. Buratti, C.; Barbanera, M.; Testarmata, F.; Fantozzi, F. Life Cycle Assessment of organic waste management strategies: an Italian case study. *J. Clean. Prod.* **2014**, 1–12. [CrossRef]
45. Marques, R.C.; da Cruz, N.F.; Ferreira, S.F.; Simões, P.; Pereira, M.C. *EIMPack—Economic Impact of the Packaging and Packaging Waste Directive-Environmental Valuation (Literature Review)*; Instituto Superior Técnico: Lisbon, Portugal, 2013.
46. Thi, T.L.N.; Laratte, B.; Guillaume, B.; Hua, A. Quantifying environmental externalities with a view to internalizing them in the price of products, using different monetization models. *Resour. Conserv. Recycl.* **2016**, *109*, 13–23.
47. Finnveden, G.; Eldh, P.; Johansson, J. Weighting in LCA based on ecotaxes: Development of a mid-point method and experiences from case studies. *Int. J. Life Cycle Assess.* **2006**, *11*, 81–88.

48. Ahlorth, S.; Finnveden, G. Ecovalue08—A new valuation set for environmental systems analysis tools. *J. Clean. Prod.* **2011**, *19*, 1994–2003. [CrossRef]
49. Weidema, B.P. Using the budget constraint to monetarise impact assessment results. *Ecol. Econ.* **2009**, *68*, 1591–1598. [CrossRef]
50. Eurostat 2016 Inflation Rate. Annual Average Rate of Change 2016. Available online: https://ec.europa.eu/eurostat/tgm/table.do?tab=table&init=1&language=en&pcode=tec00118&plugin=1 (accessed on 5 June 2019).
51. Pizzol, M.; Weidema, B.; Brandão, M.; Osset, P. Monetary valuation in Life Cycle Assessment: A review. *J. Clean. Prod.* **2015**, *86*, 170–179. [CrossRef]
52. ISO. *14044 Environmental Management-Life Cycle Assessment-Requirements and Guidelines*; ISO: Geneva, Switzerland, 2006.
53. Igos, E.; Benetto, E.; Meyer, R.; Baustert, P.; Othoniel, B. How to treat uncertainties in life cycle assessment studies? *Int. J. Life Cycle Assess.* **2018**, *24*, 794–807. [CrossRef]
54. Batouli, M.; Bienvenu, M.; Mostafavi, A. Putting sustainability theory into roadway design practice: Implementation of LCA and LCCA analysis for pavement type selection in real world decision making. *Transp. Res. Part D Transp. Environ.* **2017**, *52*, 289–302. [CrossRef]
55. Guo, M.; Murphy, R.J. LCA data quality: Sensitivity and uncertainty analysis. *Sci. Total Environ.* **2012**, *435–436*, 230–243. [CrossRef]

Article

Environmental and Economic Analysis of an Anaerobic Co-Digestion Power Plant Integrated with a Compost Plant

Sara Rajabi Hamedani [1], Mauro Villarini [1], Andrea Colantoni [1,*], Maurizio Carlini [2], Massimo Cecchini [1], Francesco Santoro [3] and Antonio Pantaleo [3]

1 Department of Agricultural and Forestry Sciences (DAFNE), Tuscia University of Viterbo, Via San Camillo de Lellis, snc-01100 Viterbo, Italy; sara.rajabi@unitus.it (S.R.H.); mauro.villarini@unitus.it (M.V.); cecchini@unitus.it (M.C.)

2 Department of Economy, Engineering, Society And Business (DEIM), Tuscia University of Viterbo, Via Del Paradiso, 47-01100 Viterbo, Italy; maurizio.carlini@unitus.it

3 Department DISAAT, University of Bari, Via Amendola, 165 70125 Bari, Italy; francesco.santoro@uniba.it (F.S.); antonio.pantaleo@uniba.it (A.P.)

* Correspondence: colantoni@unitus.it; Tel.: +39-3665899883

Received: 22 March 2020; Accepted: 22 May 2020; Published: 28 May 2020

Abstract: Italian power generation through anaerobic digestion (AD) has grown significantly between 2009 and 2016, becoming an important renewable energy resource for the country, also thanks to the generous incentives for produced electricity available in the last years. This work focuses on the economic and environmental issues of AD technology and proposes a techno-economic analysis of investment profitability without government support. In particular, the analysis focuses on an AD power plant fed by zootechnical wastewater and agro-industrial residues coupled to a cogeneration (CHP) system and a digestate-composting plant that produces soil fertilizers. We aim to determine the economic profitability of such AD power plants fed by inner-farm biomass wastes, exploiting digestate as fertilizer, using the cogenerated heat and taking into account the externalities (environmental benefits). Environmental analysis was carried out via a life cycle analysis (LCA), and encompassing the production of biogas, heat/electricity and compost in the downstream process. The un-released environmental emissions were converted into economic benefits by means of a stepwise approach. The results indicate that integrating a compost plant with a biogas plant can significantly increase the carbon credits of the process. The results were evaluated by means of a sensitivity analysis, and they report an IRR in the range of 6%–9% according to the Italian legislative support mechanisms, and possibilities to increase revenues with the use of digestate as fertilizer. The results significantly improve when externalities are included.

Keywords: anaerobic digestion; life cycle assessment; global warming potential; externalities; compost

1. Introduction

Depletion of natural energy resources is compelling our planet to face crucial challenges. Hence, energy production from biowaste plays a critical role in this energy transition [1–8]. In waste management, anaerobic digestion (AD) is a widely implemented technology that has recently drawn attention due to its capability to produce sustainable energy [9–15]. Biogas from AD is a renewable energy-carrier that can substitute conventional fuels in terms of heat and power generation, in the transport sector as biomethane or even for production of biochemicals [16,17]. Despite recent progress in the exploitation of biomethane in the transport and heating sector, Italy's greatest use of biogas has been in the generation of power. Biogas power installed in Italy increased from 2009 to 2016 from 359 to 1352 MW, while electricity generated increased from 1665 to 8259 GWh [18]. In these years, power generation from biogas placed the third position in renewable energy after photovoltaic and wind power excluding hydroelectric power which is a conventional energy resource in Italy.

Biogas brings an added value in terms of circular economy in agriculture. The Italian Biomethane Decree introduces specific subsidies for the use of such fuel in gas networks and in transport [19,20]. In the last twelve years, feed-in-tariff incentive mechanisms to bioelectricity from biogas have been ruled by the Ministerial Decree 18/12/2008, the Ministerial Decree 06/07/2012, the Ministerial Decree 23/06/2016. Furthermore, one more option to incentivize renewable sources systems, until 2012, was the so called mechanism "Certificati Verdi (CV)" established by the Legislative Decree n. 79,16/03/99 20., which adopted the European Directive 96/92/CE.

A key advantage of AD plants is their flexibility for a broad range of output products, as well as their capability to provide programmable renewable electricity to the power system. Hence, they contribute to minimizing the challenges of high penetration of variable intermittent generation into the grid. The potential integration of programmable AD power generation and intermittent solar energy has been investigated in in the Argentinian rural sector [21], as well as in Southern Africa energy systems, with concentrating solar power integration [22]. The thermo-economic optimization and optimal sizing of other hybrid systems composed by biomass and natural gas [23,24] or biomass and concentrating solar [25,26] have been recently proposed in literature. Feedstock availability is another advantage of AD power plants, since biogas can be produced from a wide range of feedstocks. Traditionally, biogas is produced via dedicated herbaceous crops (maize or triticale silage). However, the use of dedicated crops raises concerns regarding food security and overall energetic and environmental balances. Therefore, the recovery of agro-industrial byproducts and zootechnical wastewater is undoubtedly a more sustainable and rational solution [27,28]. Many AD power plants are fed by different kinds of biowastes, such as wastewater [29–31], agricultural residues and food wastes [24,32–40]. On the other hand, development of AD power plants entails a large amount of digestate production as a byproduct. Although digestate—due to its macro and micronutrient content—can be utilized as an organic fertilizer for arable land in place of mineral fertilizer [41–43], its large volume and low dry matter content impose considerable costs for management, storage and spreading onto the soil [44]. Moreover, the storage, transport and application of a huge amount of digestate results in CH_4 and NH_3 emissions, contributing to global warming potential and soil acidification, respectively [45,46]. Therefore, the application of digestate as fertilizer without further treatment raises environmental concerns [47]. Hence, the integration of AD processes with a technology handling digestate is attractive. Among various technologies for digestate management, composting is one of the most reliable technologies, thanks to the enhanced quality of the end-product (compost) through reduction of moisture content, as well as reduction of volatile-compound concentration and phytotoxicity potential [48]. Integrating composting units with AD power plant presents more advantages, such as the improvement of energetic balances of the plant (the energy demand of compost production can be met by AD power plant), leading to the possibility to increase plant revenues and reduce environmental emissions.

However, beyond all above-mentioned benefits, the development of AD power plants requires a comprehensive assessment of environmental and economic benefits in order to indicate to what extent these systems improve sustainability. To date, many studies have addressed techno-economic [49–51] and environmental evaluations [52] of AD power plants. Moreover, technologies of digestate management were analyzed from an economic and environmental point of view [53]. To the best of our knowledge, no study assesses the overall environmental and economic performance of AD power plants, together with the downstream technologies required for their digestate management. This work also estimates external costs associated with production of electricity and compost. External costs or externalities are unaccounted costs arising from production or consumption of a business good or service. The monetization of externalities is based on the conversion of social and biophysical impacts into monetary values by weights mirroring social, ethical and political values. The energy sector and clean energy generation have utilized this economic concept [54–56]. The quantification of externalities into monetary values can complete this economic analysis. Therefore, this work aims to perform a comprehensive economic evaluation with the internalization of the monetized environmental benefits from a co-digestion plant, coupled with a downstream composting system.

The article is organized as follows: Section 2 describes the materials and methods including the LCA methodology, the simulation model, the main components and cost–benefit approach; Section 3 presents and discusses the main results of the work and Section 4 draws the conclusions.

2. Materials and Methods

2.1. LCA Methodology and Global Warming Potential

Life cycle assessment as a standard and comprehensive approach is used for environmental analysis of aa studied plant throughout its life cycle. The goal of this LCA study is to quantify the energy requirements and environmental impacts (in terms of global warming potential (GWP)) of a biogas production system—together with compost-production—starting from co-digestion of mixed solid and liquid biomass, followed by electricity and heat production from biogas in the CHP system, and finally, to production of compost known as organic fertilizer in a downstream process. In-line with LCA guidelines [57,58], this study quantifies all emissions relevant to greenhouse gases (GHG) derived from energy and material use in all above-mentioned phases, including carbon dioxide (CO_2), methane (CH_4) and nitrous oxide (N_2O). The functional unit considered for this study is the electricity produced (1 MWh_e) from biogas combustion, in a combined heat-and-power unit. This process is modeled in SimaPro 9.

Description of the Plant and Data Inventory

The case study refers to a wet-anaerobic fermentation plant coupled to an internal combustion engine and a digestate dehydration (or composting) plant located in the province of Bari. The flowchart of the conversion process is shown in Figure 1.

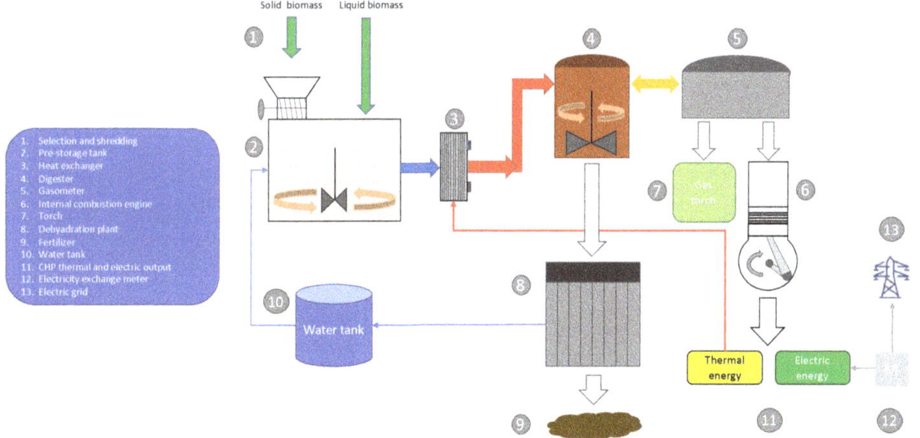

Figure 1. Overview of the production cycle of the plant.

The plant is fed by manure, cheese whey and green crop residues—biomass types widely available in the Bari district. In addition, wastewater from the composting process enters the biomass storage tank to produce biogas from residuals of composting.

The plant is characterized by the following phases:

1. Delivery, pre-treatment, storage: in this step, the raw materials enter into the system, namely semi-solid biomass (manure and green crop residues such as fruit and vegetable waste, foliage,

vegetable mowing, pruning residues, gardening waste) and liquid waste (cheese whey and composting plant wastewater). Liquid wastes are subject to screening and poured into pre-accumulation tanks, while the solid waste is sent to a storage tank where all materials are fluidized (with up to a maximum of 8% solid concentration) and sent to the digesters;

2. Flotation and anaerobic digestion of the serum: The serum from the pre-accumulation tanks is sent to the anaerobic digesters. The suspended solids and possible residues of fat are removed through flotation. Then, the serum remains in the digesters for an optimal period of 20 days. Meanwhile, continuous agitation of the sludge and the anaerobic fermentation produces biogas together with sludge, stabilized with a 95% water-percentage. Recirculation by centrifugal pumps ensures both suspension of the bacterial flora located in the lower part of the reactor and thickness restriction of the biologic layer that forms on the synthetic support of the upper part. The reactor heating is ensured by a heat exchanger.
3. Conventional anaerobic digestion of fluidized greenery: After fluidization of the semi-solid material, the cattle sewage and composting plant wastewater in the pre-accumulation tanks is directed to two digesters, where they are completely mixed with a high retention time. Each reactor is heated by a system with two spiral heat exchangers particularly suitable for sewage with high solid-content.
4. Common gas line, with gasometer and emergency thermal power plant: The treated product then passes from the digester to a third final storage tank, where the biogas is conveyed into two gasometers and subjected to a process of dehumidification and desulfurization in order to obtain a clean and functional chemical composition for the engines.
5. Production of electrical energy and heat with internal combustion engines powered by the biogas: The overall electrical and thermal efficiency is assumed to be 40% and 44%, respectively. The thermal energy (hot water at 80–90 °C) needed to heat up the biomass inside the digester is recycled from engine exhaust gas at 450 °C. The cogenerated heat largely exceeds the digestion process demand.
6. Dehydration, stripping and composting of digestate: For the sludge coming out of the digester reactor, the digested solid is dehydrated in a special centrifuge plant, stripping the ammonia in the dehydration with attached treatments like flocculation and coagulation, to recover the water in the storage tank in order to reuse it in future production cycles. The dehydrated sludge in this phase is deposited in a storage warehouse until it is subject to further stabilization by means of a composting process to obtain pure fertilizer. The refined material may be sent to bulk storage or used for bagging or pelleting, which are not considered in the study.
7. Aerobic biologic process (composting) and serum filtration: According to the stringent regulations in the region of Puglia, the biochemical parameters of BOD (biochemical oxygen demand) and COD (chemical oxygen demand) related to the sludge coming out of the serum digestate are still higher than values permissible for disposal. For this reason, it must undergo a series of purification operations, such as aerobic biologic processes and sand filtration of the various liquid flows for further reduction in the values of BOD and COD. The last processes are secondary flocculation and final disinfection of wastewater with UV rays.

The electricity consumed in the feeding operations accounts for 8% of the total electricity production. As regards the electricity consumption for the composting plant, it is assumed that 1 kWh_e is required per ton of wet organic waste as from literature data [59], and this electricity is withdrawn from the grid. Compost is a composition of N, P and K elements in different concentrations, also present in mineral fertilizers. Hence, the compost can substitute mineral fertilizers (e.g., ammonium nitrate, triple superphosphate and potassium sulfate) in terms of active ingredient contents [60,61]. Therefore, production of theses fertilizers is avoided as in other fields such as biochar application in soil as organic fertilizer [62]. Airborne and waterborne emissions (ammonia, nitrous oxide, nitrogen, nitrate and phosphate) arising from digestate application are excluded from this study as they are neutral in global warming potential. In the case of composting plant wastewater, a supply of about 100 days year^{-1} was considered. In addition, a storage with 20,000 m^3 capacity was assumed for green crop residues due to their seasonality. Cattle farms and dairy factories are within a 20-km radius of the AD power plant, while crop residues are transferred to the plant from a 30-km distance. Carbon dioxide emissions from biogas combustion in the CHP are also excluded from estimation owing to biogenic exemption [27,63,64]. Furthermore, greenhouse gas emissions deriving from the construction, operation and disposal of the plant was not taken into account. Excess heat from cogeneration on top of internal energy requirements was used to match local heat demand and substitute natural gas fuel. The mass and energy flow of the proposed system during 1 year of operation are illustrated in Figure 2. The overall list of energy and material used for 1 MWh electricity production from co-digestion plant is also presented in Table 1. The exploitable heat was not computed between the outputs of Table 1 because it is included among the avoided products as natural gas.

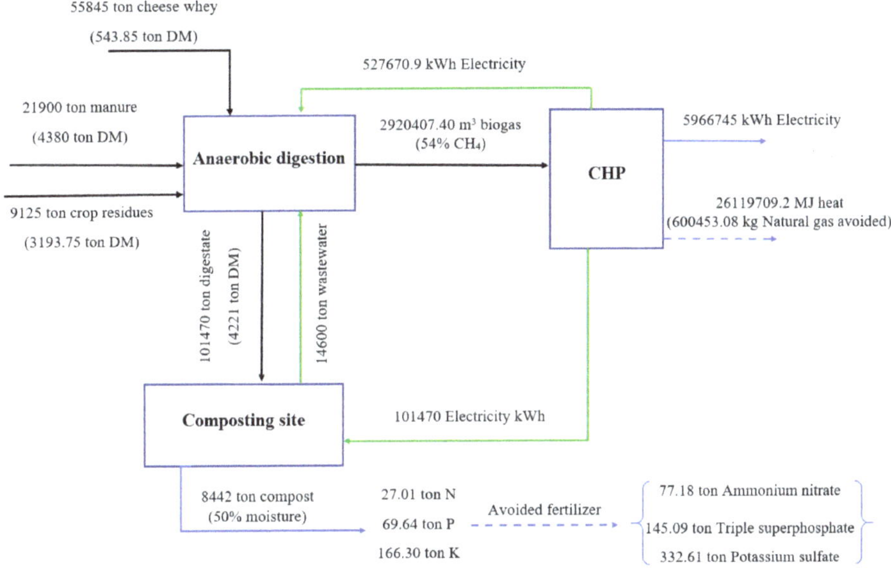

Figure 2. System layout and mass and energy balances referred to 1 year of operation.

Table 1. Global inventory data per 1 MWh$_e$.

Inputs		Outputs	
Manure	3.67 ton	Electricity	1 MWh
Whey from cheese factories	9.36 ton	Compost	1.41 ton
Green crop residues	1.53 ton		
Composting plant wastewater	2.44 ton	Avoided products	
Transport	306.5 tkm *	Natural gas (heat)	50.31 kg
		Ammonium nitrate	12.93 kg
		Triple superphosphate	24.31 kg
		Potassium sulfate	55.74 kg

* tkm = ton-kilometer (unit of transportation measurement).

2.2. Economic Analysis

A cost–benefit approach was applied to assess the investment profitability. The investment cost of the plant and its associated operating expenditures with raw materials and energy consumption were considered. Revenues generated by sales of electricity, heat and compost as well as biophysical sources including externalities were taken into account.

This analysis ignored potentially available incentives (in the form of capital grants, or incentives for avoided primary energy consumption, which could be available in the Italian framework) in order to understand if, and to what extent, the investment was profitable without specific subsidies.

The economic evaluation converted environmental impact to external costs [62,65]. Among different approaches for monetary valuation, budget constraint approach has been recommended by [66] for LCA applications due to its simplicity and its capacity to minimize uncertainty of the monetary value of a human life–year. The used unit is QALY that is a Quality-Adjusted-Life_Year. It represents the monetary value of a life year with high quality. The average annual income is the maximum that a person can purchase an additional life–year and a quality-adjusted life–year (QALY) defines a life–year lived at full wellbeing, then an upper limit for the monetary value of a QALY is provided [67]. The Stepwise 2006 method developed on budget constraint approaches—and specifically designed for life cycle impact assessment—is adapted in this study [68].

Therefore, environmental impact estimated into GWP (kg CO_2) was converted to monetary values in order to internalize social, ethical and political cost of this bioenergy system within economic analysis. Global warming potential GWP was converted to a monetary value by weighting factor of 0.08 Euro/kg CO_2 [62]. Since this factor refers to Eur 2003, it was necessary to use inflation rate to estimate cost in current year.

The internal rate of return (IRR), net present value (NPV) and payback period (PBP) were calculated for a 15-year timeframe.

The investment costs of the plant were estimated through market analysis of plants with similar configurations and also communications with suppliers of technologies similar to those proposed in the study. The costs are summarized in Table 2. The costs assumed for dehydration, stripping and composting treatments of the digestate produced by green residues and manure to obtain fertilizer, as well as those for biologic finishing, ultrafiltration and clariflocculation of serum and plant wastewater to obtain water for fertigation were indicative, achieved from market research and confirmed in literature. The total investment cost of the 1-MW_e system is 4 k€/kW_e, in agreement with previous results [69]. The annual electricity production is 6595 MWh_e, assuming average operating hours in AD powerplants in Italy (GSE statistics, 2020). Costs and revenues of the investment are estimated based on these operating hours.

Biomass costs were determined by the cost of a minority part of biomass consumed respect to the total amount because the project is based on preponderant use of on site available bio-wastes at no cost. The global service costs represent the service cost including maintenance of the system and were determined on the basis of the specific cost of 0.032 €/kWh_e [70]. Staff costs are based on the involvement of 4 employees and overhead expenses are considered on top of the other operating costs. The discount rate—or weighted average cost of capital (WACC)—was set to 8% according to the relevant literature [14,71–74].

Revenues were generated by physical sources, namely sale of electricity, heat and compost as well as externalities. Electricity price was fixed according to a power purchase agreement. External benefit regarding carbon offsets was estimated 13.22 €/MWh_e, assuming a weighting factor 0.08 eur/kg CO_2. The monetary valuation can be applied to LCA results in order to weigh environmental benefits against environmental costs through different approaches. In the present study, the Stepwise 2006 approach was used. Nevertheless, in Section 3, sensitivity analysis of economic parameters respect to variation of the approach was shown. To that end, in addition to the Stepwise 2006 method, ecotax and ecovalue approaches were considered.

Table 2. Operating and investment costs for the plant under study.

Investment Costs	
Cost item	Value (kEur)
Cost of civil works	700
Cost of digesters, tanks and biogas treatment	1150
Cost of electrical system group and cogeneration plant	600
Dewatering, stripping and composting plant cost	750
Cost of filtration and clariflocculation	650
Engineering and development costs	150
Total amount	4000
Operating Costs kEur/year	
Biomass	47
Global service	211
Staff	140
Overhead expenses	60
Total amount	458
Additional Parameters	
Plant useful life	15 years
Discount rate	8%
Heat exploitation	50%
External benefit (Stepwise 2006 method)	13.22 €M Wh_e^{-1}
Electricity selling price	120 €M Wh_e^{-1}
Price of natural gas	75 €M Wh_{th}^{-1}
Price of compost	10 €t^{-1}

3. Result and Discussion

3.1. Environmental Analysis

Global warming potential (GWP) was quantified by IPCC 2013 method converting GHG emissions to kilograms of CO_2. For the studied system, the global warming potential was found to be −167.52 kg CO_2, representing an outstanding carbon offset (Figure 3). Negative values reflect environmental benefits achieved by avoidance of product uses [62]. These benefits were primarily associated with avoided mineral fertilizers consumption.

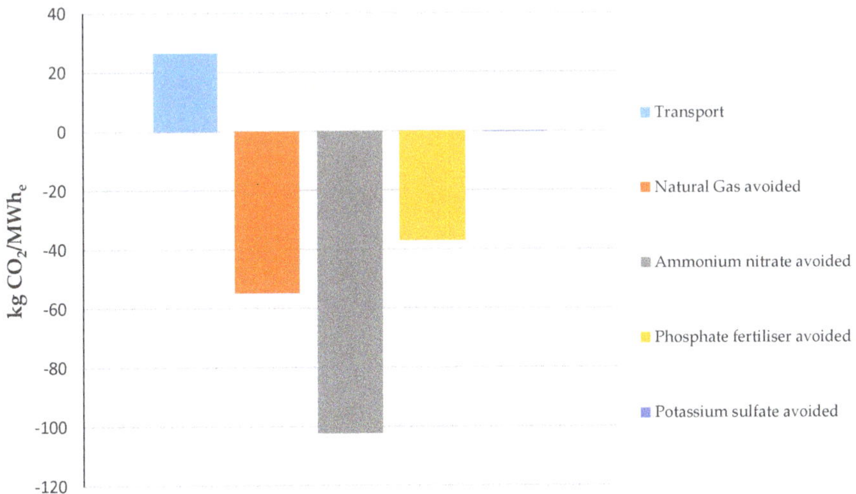

Figure 3. Global warming potential of processes in the studied system.

3.2. Economic Analysis

The main economic results are shown in Table 3. In addition, Figures 4–6 represent results of sensitivity analysis. The investment profitability is obviously lower than in the previous years, when generous incentives for electricity generation were available. This reduction was however partially mitigated by the reduction of the investment costs for the learning curves of well-established technologies and the possibility to purchase biomass at very low cost.

Table 3. Results of the cost-effectiveness analysis.

Economic Index	With Externalities	Without Externalities	Unit of Measurement
Payback time (PBT)	8	10	years
Net present value (NPV)	312	−323	kEur
Internal rate of return (IRR)	9.36	6.54	%

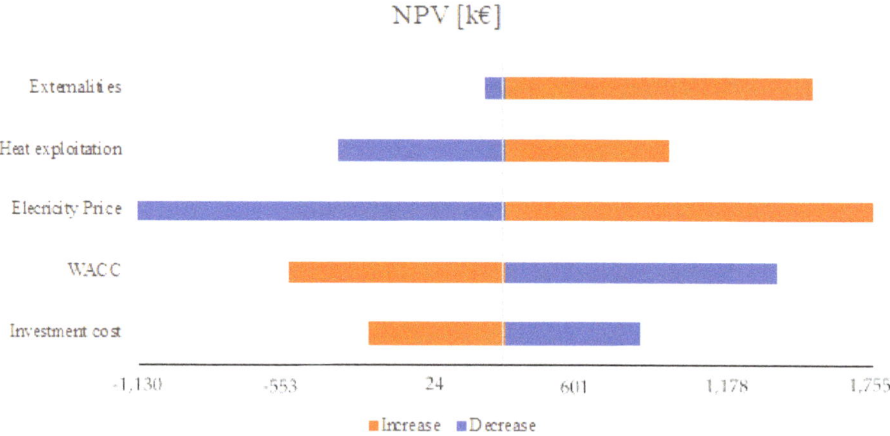

Figure 4. Tornado diagram for representation of sensitivity analysis of net present value (NPV).

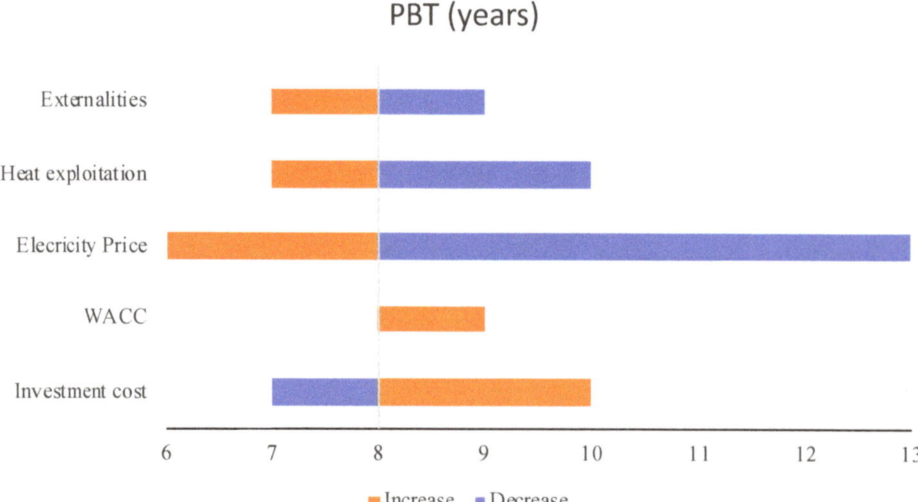

Figure 5. Tornado diagram for representation of sensitivity analysis of payback time (PBT).

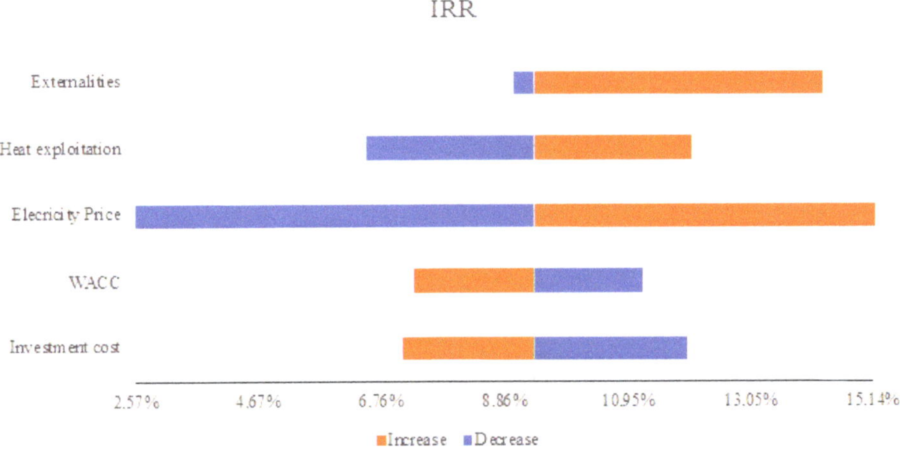

Figure 6. Tornado diagram for representation of sensitivity analysis of internal rate of return (IRR).

Based on Table 3, it was possible to appreciate the different results with and without the economic benefits from environmental evaluation.

Furthermore, a sensitivity analysis aimed to appreciate the response of the economic model to variation of the most impactful parameters was developed and represented by means of a tornado diagram in Figures 4–6 for NPV, PBP and IRR, respectively. Parameters were varied within the following realistic and interesting ranges. Percentages of variation of each considerable parameter were: ± 10% for the investment cost, ± 25% for the WACC, ± 25% for the electricity price, ± 30% for the thermal energy exploitation, while the externalities were varied between the minimum value attainable by the ecotax method corresponding with 11.58 €/MWh$_e$ and the maximum value attainable by the ecovalue method corresponding with 38.37 €/MWh. The tornado diagrams are centered on the values shown in Table 2.

Considering the three tornado diagrams, the range of variation of the most influencing parameter—electricity price—was extended, both in decrease and increase. This parameter had a remarkable impact on all three considered parameters. Its increase allowed reaching the best economic configuration of the project represented by 15.14% IRR, 1.75 M€ NPV and 6 years PBT. The sensitivity of the externalities method was especially relevant to the ecovalue approach, which enables to achieve 14.25% IRR, 1.52 M€ NPV and 7 years PBT. In general, PBT varied between 6 and 10 with the exception of the case of electricity price whose decrease considerably affected PBT: it grew until 13 years. As in several CHP projects, the exploitation of heat produced by the internal combustion engine was important to the good outcome. The sensitivity analysis shows that disadvantageous scenarios are not so far from the base configuration.

4. Conclusions

After a remarkable growth between 2009 and 2016, biogas-sourced electricity generation slowed down significantly, due to the lack of subsidies available. This article describes an economic and environmental analysis of electricity generation from an anaerobic co-digestion plant coupled to a downstream process producing compost from digestate. The aim was to mark out some of the key aspects which could increase the sustainability of this technological application such as the use of low cost biomasses on site available and exploitation of digestate as fertilizer to the soil. A life-cycle assessment was applied to count the global warming potential of the system. Furthermore, the economic concept of externalities expressing environmental and socioeconomic impacts in monetary values was included in this study. The novelty of this work was to consider externalities and to internalize them in the economic assessment. Therefore, economic analysis encompasses not only physical and private costs, e.g., the operating and investment costs of a plant (digestion plant, cogeneration system and compost plant), the incentives available in the Italian legislative scenario, the raw material costs and the sale price of compost, but also biophysical costs as externalities. Results demonstrate economic and environmental profitability of this plant which mainly arises from bioelectricity production. In particular, sustainable economic performance were demonstrated independently of the presence of incentives regarding the electric production. Incidentally, the work was aimed to the evaluation of the system without incentives in order to understand if the system can face the market without any external support. Outstanding environmental benefits were represented by means of the −167.52 kg CO_2 global warming potential. Acceptable economic results were attained in terms of NPV, PBT and IRR, respectively 0.31 M€, eight years and 9.36% for base configuration and a propitious variation of parameters can be crucial for the improvement of economic performance as shown by the sensitivity analysis. These results were much more important if the lack of incentives recognized to the electricity produced by the system was considered. For Externalities contribute propitiously to the project evaluation and this contribution was much more important in case of ecovalue approach. From the sensitivity analysis exigency to choose the controllable expedient conditions ensues. Consequently, these findings make the investment on this type of plants encouraging on condition that parameters are duly selected.

Author Contributions: Conceptualization, A.P. M.V. and S.R.H.; data curation, M.V., S.R.H., A.C. and F.S.; formal analysis, A.C., M.C. (Maurizio Carlini), M.C. (Massimo Cecchini); investigation, A.C. M.V. and F.S.; methodology, S.R.H, A.C., F.S., M.C. (Maurizio Carlini) and M.V.; resources, A.P., M.V.; supervision, A.P., M.C. (Maurizio Carlini) and M.C. (Massimo Cecchini); validation, A.C., M.V., and S.R.H.; visualization, S.R.H., and M.V.; writing—original draft, M.V., S.R.H and A.P.; writing—review and editing, S.R.H., M.V., M.C. (Maurizio Carlini), M.C. (Massimo Cecchini), F.S. All authors have read and agreed to the published version of the manuscript.

Funding: This research was chiefly funded by MIUR (Italian Ministry for Education, University and Research), Law 232/2016, "Department of excellence".

Conflicts of Interest: The authors declare no conflict of interest.

References

1. Achinas, S.; Achinas, V.; Euverink, G.J.W. A Technological Overview of Biogas Production from Biowaste. *Engineering* **2017**, *3*, 299–307. [CrossRef]
2. Carlini, M.; Castellucci, S.; Mennuni, A. Thermal and fluid dynamic analysis within a batch micro-reactor for biodiesel production fromwaste vegetable oil. *Sustainability* **2017**, *9*, 2308. [CrossRef]
3. Malinauskaite, J.; Jouhara, H. The trilemma of waste-to-energy: A multi-purpose solution. *Energy Policy* **2019**, *129*, 636–645. [CrossRef]
4. Pallozzi, V.; Di Carlo, A.; Bocci, E.; Villarini, M.; Foscolo, P.U.; Carlini, M. Performance evaluation at different process parameters of an innovative prototype of biomass gasification system aimed to hydrogen production. *Energy Convers. Manag.* **2016**, *130*, 34–43. [CrossRef]
5. Carlini, M.; Castellucci, S.; Cocchi, S.; Manzo, A. Waste wood biomass arising from pruning of urban green in viterbo town: Energy characterization and potential uses. In *International Conference on Computational Science and Its Applications*; Springer: Berlin/Heidelberg, Germany, 2013; pp. 242–255.
6. Pantaleo, A.; Villarini, M.; Colantoni, A.; Carlini, M.; Santoro, F.; Hamedani, S.R. Techno-economic modeling of biomass pellet routes: Feasibility in Italy. *Energies* **2020**, *13*, 1636. [CrossRef]
7. Villarini, M.; Bocci, E.; Di Carlo, A.; Savuto, E.; Pallozzi, V. The case study of an innovative small scale biomass waste gasification heat and power plant contextualized in a farm. *Energy Procedia* **2015**, *82*, 335–342. [CrossRef]
8. Rajabi Hamedani, S.; Bocc, E.; Villarini, M.; Di Carlo, A.; Naso, V. Techno-economic Analysis of Hydrogen Production Using Biomass Gasification -A Small Scale Power Plant Study. *Energy Procedia* **2016**, *101*, 806–813. [CrossRef]

9. Oreggioni, G.D.; Gowreesunker, B.L.; Tassou, S.A.; Bianchi, G.; Reilly, M.; Kirby, M.E.; Toop, T.A.; Theodorou, M.K. Potential for energy production from farm wastes using anaerobic digestion in the UK: An economic comparison of different size plants. *Energies* **2017**, *10*, 1396. [CrossRef]
10. Browne, J.; Allen, E.; Murphy, J. Assessing the variability in biomethane production from the organic fraction of municipal solid waste in batch and continuous operation. *Appl. Energy* **2014**, *128*, 307–314. [CrossRef]
11. Carlini, M.; Castellucci, S.; Mennuni, A.; Ferrelli, S.; Felicioni, M.A. Application of a circular & green economy model to a ceramic industrial district: An Italian case study. *AIP Conf. Proc.* **2019**, *2123*, 020087. [CrossRef]
12. Carlini, M.; Castellucci, S.; Moneti, M. Anaerobic co-digestion of olive-mill solid waste with cattle manure and cattle slurry: Analysis of bio-methane potential. *Energy Procedia* **2015**, *81*, 354–367. [CrossRef]
13. Carlini, M.; Castellucci, S.; Moneti, M. Biogas production from poultry manure and cheese whey wastewater under mesophilic conditions in batch reactor. *Energy Procedia* **2015**, *82*, 811–818. [CrossRef]
14. Carlini, M.; Mosconi, E.M.; Castellucci, S.; Villarini, M.; Colantoni, A. An Economical Evaluation of Anaerobic Digestion Plants Fed with Organic Agro-Industrial Waste. *Energies* **2017**, *10*, 1165. [CrossRef]
15. Santoro, F.; Anifantis, A.S.; Ruggiero, G.; Zavadskiy, V.; Pascuzzi, S. Lightning protection systems suitable for stables: A case study. *Agriculture* **2019**, *9*, 72. [CrossRef]
16. Scarlat, N.; Dallemand, J.F.; Fahl, F. Biogas: Developments and perspectives in Europe. *Renew Energy* **2018**, *129*, 457–472. [CrossRef]
17. Fagerström, A.; Al Seadi, T.; Rasi, S.; Briseid, T. *The Role of Anaerobic Digestion and Biogas in the Circular Economy*; Murphy, J.D., Ed.; Cork, IE, IEA Bioenergy Task 37: Paris, France, 2018.
18. GSE. Quarta Relazione Dell'Italia in Merito Ai Progressi Ai Sensi Della Direttiva 2009/28/CE. 2017. Available online: https://www.gse.it/documenti_site/Documenti%20GSE/Studi%20e%20scenari/Progress%20Report%20Rinnovabili%20Italia%202017.pdf (accessed on 9 April 2020).
19. Italian Decree Dlgs 2 March 2018, n.d. Available online: https://www.gse.it/documenti_site/DocumentiGSE/Serviziperte/BIOMETANO/NORMATIVA/D.M.MiSE2marzo2018.pdf (accessed on 9 April 2020).
20. Decreto Legislativo 16 Marzo 1999 n. 79. Available online: https://www.gazzettaufficiale.it/eli/id/1999/03/31/099G0136/sg (accessed on 9 April 2020).
21. Blanchet, C.A.C.; Pantaleo, A.M.; van Dam, K.H. A process systems engineering approach to designing a solar/biomass hybrid energy system for dairy farms in Argentina. Comput. *Aided Chem. Eng.* **2019**, *46*, 1609–1614. [CrossRef]
22. Liu, M.; van Dam, K.H.; Pantaleo, A.M.; Guo, M. Optimisation of Integrated Bioenergy and Concentrated Solar Power Supply Chains in South Africa. *Comput. Aided Chem. Eng.* **2018**, *43*, 1463–1468. [CrossRef]
23. Pantaleo, A.M.; Camporeale, S.; Shah, N. Natural gas-biomass dual fuelled microturbines: Comparison of operating strategies in the Italian residential sector. *Appl. Therm. Eng.* **2014**, *71*, 686–696. [CrossRef]
24. Pantaleo, A.M.; Camporeale, S.M.; Shah, N. Thermo-economic assessment of externally fired micro-gas turbine fired by natural gas and biomass: Applications in Italy. *Energy Convers. Manag.* **2013**, *75*, 202–213. [CrossRef]
25. Pantaleo, A.M.; Camporeale, S.M.; Sorrentino, A.; Miliozzi, A.; Shah, N.; Markides, C.N. Hybrid solar-biomass combined Brayton/organic Rankine-cycle plants integrated with thermal storage: Techno-economic feasibility in selected Mediterranean areas. *Renew Energy* **2020**, *147*, 2913–2931. [CrossRef]
26. Pantaleo, A.M.; Camporeale, S.M.; Miliozzi, A.; Russo, V.; Shah, N.; Markides, C.N. Novel hybrid CSP-biomass CHP for flexible generation: Thermo-economic analysis and profitability assessment. *Appl. Energy* **2017**, *204*, 994–1006. [CrossRef]
27. Rajabi Hamedani, S.; Del Zotto, L.; Bocci, E.; Colantoni, A.; Villarini, M. Eco-efficiency assessment of bioelectricity production from Iranian vineyard biomass gasification. *Biomass Bioenergy* **2019**, *127*, 105271. [CrossRef]
28. Rajabi Hamedani, S.; Colantoni, A.; Gallucci, F.; Salerno, M.; Silvestri, C.; Villarini, M. Comparative energy and environmental analysis of agro-pellet production from orchard woody biomass. *Biomass Bioenergy* **2019**, *129*, 105334. [CrossRef]
29. Shen, Y.; Linville, J.L.; Urgun-Demirtas, M.; Mintz, M.M.; Snyder, S.W. An overview of biogas production and utilization at full-scale wastewater treatment plants (WWTPs) in the United States: Challenges and opportunities towards energy-neutral WWTPs. *Renew Sustain. Energy Rev.* **2015**, *50*, 346–362. [CrossRef]

30. Jürgensen, L.; Ehimen, E.A.; Born, J.; Holm-Nielsen, J.B. A combination anaerobic digestion scheme for biogas production from dairy effluent—CSTR and ABR, and biogas upgrading. *Biomass Bioenergy* **2018**, *111*, 241–247. [CrossRef]
31. Al-Addous, M.; Saidan, M.N.; Bdour, M.; Alnaief, M. Evaluation of biogas production from the co-digestion of municipal food waste and wastewater sludge at refugee camps using an automated methane potential test system. *Energies* **2019**, *12*, 32. [CrossRef]
32. Yong, Z.; Dong, Y.; Zhang, X.; Tan, T. Anaerobic co-digestion of food waste and straw for biogas production. *Renew Energy* **2015**, *78*, 527–530. [CrossRef]
33. De Menna, F.; Malagnino, R.A.; Vittuari, M.; Molari, G.; Seddaiu, G.; Deligios, P.A.; Solinas, S.; Ledda, L. Potential biogas production from artichoke byproducts in Sardinia, Italy. *Energies* **2016**, *9*, 92. [CrossRef]
34. Hassaan, M.A.; Pantaleo, A.; Tedone, L.; Elkatory, M.R.; Ali, R.M.; Nemr, A.E.; Mastro, D.G. Enhancement of biogas production via green ZnO nanoparticles: Experimental results of selected herbaceous crops. *Chem. Eng. Commun.* **2019**. [CrossRef]
35. Amirante, R.; Demastro, G.; Distaso, E.; Hassaan, M.A.; Mormando, A.; Pantaleo, A.M.; Tamburrano, P.; Tedone, L.; Clodoveo, M.L. Effects of Ultrasound and Green Synthesis ZnO Nanoparticles on Biogas Production from Olive Pomace. *Energy Procedia* **2018**, *148*, 940–947. [CrossRef]
36. Pantaleo, A.D.E.; Gennaro, B.; Shah, N. Assessment of optimal size of anaerobic co-digestion plants: An application to cattle farms in the province of Bari (Italy). *Renew Sustain. Energy Rev.* **2013**, *20*, 57–70. [CrossRef]
37. Bulgakov, V.; Pascuzzi, S.; Santoro, F.; Anifantis, A.S. Mathematical model of the plane-parallel movement of the self-propelled root-harvesting machine. *Sustainability* **2018**, *10*, 3614. [CrossRef]
38. Bulgakov, V.; Pascuzzi, S.; Anifantis, A.S.; Santoro, F. Oscillations analysis of front-mounted beet topper machine for biomass harvesting. *Energies* **2019**, *12*, 2774. [CrossRef]
39. Guerrieri, A.S.; Anifantis, A.S.; Santoro, F.; Pascuzzi, S. Study of a large square baler with innovative technological systems that optimize the baling effectiveness. *Agric* **2019**, *9*, 86. [CrossRef]
40. Carlini, M.; Castellucci, S.; Mennuni, A.; Selli, S. Poultry Manure Biomass: Energetic Characterization and ADM1-based Simulation. *J. Phys. Conf. Ser.* **2019**, *1172*, 012063. [CrossRef]
41. De Vries, J.W.; Vinken, T.M.W.J.; Hamelin, L.; De Boer, I.J.M. Comparing environmental consequences of anaerobic mono-and co-digestion of pig manure to produce bioenergy – a life cycle perspective. *Bioresour. Technol.* **2012**, *125*, 239–248. [CrossRef] [PubMed]
42. Comparetti, A.; Greco, C.; Navickas, K.; Venslauskas, K. Evaluation of potential biogas production in sicily. *Eng. Rural. Dev.* **2012**, *24*, 555–559.
43. Risberg, K.; Cederlund, H.; Pell, M.; Arthurson, V.; Schnürer, A. Comparative characterization of digestate versus pig slurry and cow manure—Chemical composition and effects on soil microbial activity. *Waste Manag.* **2017**, *61*, 529–538. [CrossRef]
44. Al saedi, T.; Drosg, B.; Fuchs, W.; Rutz, D.; Janssen, R. Biogas digestate quality and utilization. In *The Biogas Handbook*; Woodhead Publishing: Cambridge, UK, 2013; pp. 267–301.
45. Longhurst, P.J.; Tompkins, D.; Pollard, S.J.T.; Hough, R.L.; Chambers, B.; Gale, P.; Tyrrel, S.; Villa, R.; Taylor, M.; Wu, S.; et al. Risk assessments for quality-assured, source-segregated composts and anaerobic digestates for a circular bioeconomy in the UK. *Environ. Int.* **2019**, *127*, 253–266. [CrossRef]
46. Pivato, A.; Vanin, S.; Raga, R.; Lavagnolo, M.C.; Barausse, A.; Rieple, A.; Laurent, A.; Cossu, R. Use of digestate from a decentralized on-farm biogas plant as fertilizer in soils: An ecotoxicological study for future indicators in risk and life cycle assessment. *Waste Manag.* **2016**, *49*, 378–389. [CrossRef]
47. Al Seadi, T.; Lukehurst, C. Quality management of digestate from biogas plants used as fertiliser. *IEA Bioenergy Task* **2012**, *40*.
48. Bustamante, M.A.; Alburquerque, J.A.; Restrepo, A.P.; De Fuente, C.; Paredes, C.; Moral, R.; Bernal, M.P. Co-composting of the solid fraction of anaerobic digestates, to obtain added-value materials for use in agriculture. *Biomass Bioenergy* **2012**, *43*, 26–35. [CrossRef]
49. Martínez-ruano, J.A.; Restrepo-serna, D.L.; Carmona-garcia, E.; Alejandro, J.; Giraldo, P.; Aroca, G.; Cardona, C.A. Effect of co-digestion of milk-whey and potato stem on heat and power generation using biogas as an energy vector: Techno-economic assessment. *Appl. Energy* **2019**, *241*, 504–518. [CrossRef]

50. Martínez-ruano, J.A.; Caballero-galván, A.S.; Restrepo-serna, D.L.; Cardona, C.A. Techno-economic and environmental assessment of biogas production from banana peel (Musa paradisiaca) in a biorefinery concept. *Environ. Sci. Pollut. Res.* **2018**, *25*, 35971–35980. [CrossRef] [PubMed]
51. Hamzehkolaei, F.T.; Amjady, N. A techno-economic assessment for replacement of conventional fossil fuel based technologies in animal farms with biogas fueled CHP units. *Renew Energy* **2018**, *118*, 602–614. [CrossRef]
52. Hijazi, O.; Munro, S.; Zerhusen, B.; Effenberger, M. Review of life cycle assessment for biogas production in Europe. *Renew Sustain. Energy Rev.* **2016**, *54*, 1291–1300. [CrossRef]
53. Vochozka, M.; Marous, A. Obsolete Laws: Economic and Moral Aspects, Case Study—Composting Standards. *Sci. Eng. Ethics* **2016**, 1–6. [CrossRef]
54. Bickel, P.; Friedrich, R.; Droste-Franke, B.; Bachmann, T.; Greßmann, A.; Rabl, A.; Hunt, A.; Markandya, A.; Tol, R.; Hurley, F.; et al. *ExternE—Externalities of Energy, Methodology 2005 Update*; Bickel, P., Friedrich, R., Eds.; Office for Official Publications of the European Communities: Luxembourg, 2005; Volume EUR, ISBN 92-79-00423-9.
55. Bielecki, A.; Ernst, S.; Skrodzka, W.; Wojnicki, I. The externalities of energy production in the context of development of clean energy generation. *Environ. Sci. Pollut. Res.* **2020**, *27*, 11506–11530. [CrossRef]
56. Patrizio, P.; Leduc, S.; Chinese, D.; Kraxner, F. Internalizing the external costs of biogas supply chains in the Italian energy sector. *Energy* **2017**, *125*, 85–96. [CrossRef]
57. *ISO 14040:2016 Environmental Management—Life Cycle Assessment—Principles and Framework*; ISO: Geneva, Switzerland, 2016.
58. *ISO 14044:2006 Environmental Management—Life Cycle Assessment—Requirements and Guidelines*; ISO: Geneva, Switzerland, 2006.
59. Del, A.; Di, F. DCO 2/08 2008. Available online: https://www.arera.it/it/docs/dc/08/080220_2.htm (accessed on 10 April 2020).
60. Poeschl, M.; Ward, S.; Owende, P. Environmental impacts of biogas deployment—Part I: Life cycle inventory for evaluation of production process emissions to air. *J. Clean. Prod.* **2012**, *24*, 168–183. [CrossRef]
61. Lijó, L.; González-garcía, S.; Bacenetti, J.; Fiala, M.; Feijoo, G.; Lema, J.M.; Moreira, M.T. Life Cycle Assessment of electricity production in Italy from anaerobic co-digestion of pig slurry and energy crops. *Renew Energy* **2014**, *68*, 625–635. [CrossRef]
62. Rajabi Hamedani, S.; Kuppens, T.; Malina, R.; Bocci, E.; Colantoni, A.; Villarini, M. Life cycle assessment and environmental valuation of biochar production: Two case studies in Belgium. *Energies* **2019**, *12*, 2166. [CrossRef]
63. Villarini, M.; Rajabi Hamedani, S.; Marcantonio, V.; Colantoni, A.; Cecchini, M.; Monarca, D. Comparison of Environmental Impact of Two Different Bioelectricity Conversion Technologies by Means of LCA. *Innov. Biosyst. Eng. Sustain. Agric. For. Food Prod.* **2020**. [CrossRef]
64. Rajabi Hamedani, S.; Villarini, M.; Colantoni, A.; Moretti, M.; Bocci, E. Life Cycle Performance of Hydrogen Production via Agro-Industrial Residue Gasification—A Small Scale Power Plant Study. *Energies* **2018**, *11*, 675. [CrossRef]
65. Lim, S.R.; Kim, Y.R.; Woo, S.H.; Park, D.; Park, J.M. System optimization for eco-design by using monetization of environmental impacts: A strategy to convert bi-objective to single-objective problems. *J. Clean. Prod.* **2013**, *39*, 303–311. [CrossRef]
66. Weidema, B.P.; Brandão, M.; Pizzol, M. The use of monetary valuation of environmental impacts in life cycle assessment: State of the art, strengths and weaknesses. *SCORE-LCA Rep. Nb 2012-03. Aalborg University.* **2013**.
67. Pizzol, M.; Weidema, B.; Brandão, M.; Osset, P. Monetary valuation in Life Cycle Assessment: A review. *J. Clean. Prod.* **2015**, *86*, 170–179. [CrossRef]
68. Weidema, B. Using the budget constraint to monetise impact assessment results. *Ecol. Econ.* **2009**, *68*, 1591–1598. [CrossRef]
69. Grebe EU Project, n.d. Available online: http://grebeproject.eu/ (accessed on 9 April 2020).
70. Salerno, M.; Gallucci, F.; Pari, L.; Zambon, I.; Sarri, D.; Colantoni, A. Costs-benefits analysis of a small-scale biogas plant and electric energy production. *Bulg. J. Agric. Sci.* **2017**, *23*, 357–362.
71. Lovarelli, D.; Falcone, G.; Orsi, L.; Bacenetti, J. Agricultural small anaerobic digestion plants: Combining economic and environmental assessment. *Biomass Bioenergy* **2019**, *128*, 105302. [CrossRef]

72. Wang, Y.; Zhang, J.; Li, Y.; Jia, S.; Song, Y.; Sun, Y.; Zheng, Z.; Yu, J.; Cui, Z.; Han, Y.; et al. Methane production from the co-digestion of pig manure and corn stover with the addition of cucumber residue: Role of the total solids content and feedstock-to-inoculum ratio. *Bioresour. Technol.* **2020**, *306*, 123172. [CrossRef] [PubMed]
73. Monarca, D.; Cecchini, M.; Guerrieri, M.; Colantoni, A. Conventional and alternative use of biomasses derived by hazelnut cultivation and processing. *Acta Hortic.* **2009**, *845*, 627–634. [CrossRef]
74. Zambon, I.; Colantoni, A.; Carlucci, M.; Morrow, N.; Sateriano, A.; Salvati, L. Land quality, sustainable development and environmental degradation in agricultural districts: A computational approach based on entropy indexes. *Environ. Impact Assess. Rev.* **2017**, *64*, 37–46. [CrossRef]

Article

Techno-Economic Modeling of Biomass Pellet Routes: Feasibility in Italy

Antonio Pantaleo [1], Mauro Villarini [2,*], Andrea Colantoni [2], Maurizio Carlini [3], Francesco Santoro [1] and Sara Rajabi Hamedani [2]

1 Department of Agro-environmental sciences (DISAAT), University of Bari, 70125 Bari, Italy; antonio.pantaleo@uniba.it (A.P.); francesco.santoro@uniba.it (F.S.)

2 Department of Agricultural and Forestry Sciences (DAFNE), Tuscia University of Viterbo, 01100 Viterbo, Italy; colantoni@unitus.it (A.C.); sara.rajabi@unitus.it (S.R.H.)

3 Department of Economy, Engineering, Society and Business (DEIM), Tuscia University of Viterbo, 01100 Viterbo, Italy; maurizio.carlini@unitus.it

* Correspondence: mauro.villarini@unitus.it; Tel.: +39-340-2266196

Received: 28 February 2020; Accepted: 22 March 2020; Published: 2 April 2020

Abstract: Wood and agricultural biomass pellets boost the potential as bio-fuels toward power production in tertiary and residential sectors. The production of pellets, however, is a multi-stage process where the supply-processing phases and the overall energy input strongly depend on the characteristics of the input biomass. In this paper, we describe the key features of the market for pellets in Italy, including national production and consumption data, production costs and prices, the available energy conversion systems, and the current regulatory issues. Moreover, we outline the main technical, economic, and end-user barriers that should be addressed in order to foster the growth of Italian pellet production. Additionally, we propose a methodology to evaluate the profitability of the pellet production chain, by assessing the investment and operation costs as a function of the quality of the raw biomass. The approach is applied to a real case study of a small firm producing wooden frames along with dry wood chips as the main by-product, which can be utilized subsequently for pellet production. Moreover, in order to optimize the size of the pellet production plant, further biomass was purchased from the market, including wood pruning and agricultural residues, wood chips from forestry, and uncontaminated residues of wood processing firms. A sensitivity analysis of the main technical and economic parameters (including the cost and quality of raw material, pellet market value, investment and operational costs, and plant lifetime) indicated that the biomass market price considerably affects the profitability of pellet production plants, particularly where the biomass has a high moisture content. Therefore, a 20% increase in the price of biomass with a high moisture content leads to a 60% fall in profitability index, turning it into negative one. This is due in particular to the costs of pre-treatment and drying of biomass, as well as to the lower energy content of wet biomass. As a result, the use of forestry residues with high moisture and high ash content, high costs of collection/transport, and high costs of pre-treatment and drying is not financially competitive.

Keywords: pellet; agricultural residues; wood chips; market

---

1. Introduction

The global warming impact, the increasing prices of fossil fuel, and the need to produce thermal and electrical energy stimulated the creation of an industry leaning toward energy production by renewable sources. Biomass is a widespread renewable source to provide energy demand in terms of electricity and heat [1–4]. The Renewable Energy Directive 2018/2001/EU (RED II) obligates the European Union (EU) to raise renewable energy consumption to 32% by 2030 [5]. According to

IRENA [6], biomass currently accounts for 14% of the global renewable energy demand, of which almost 70% is utilized for residential applications (e.g., cooking and heating purposes) [7].

Biomass harvesting in some cases faces mechanical difficulties that could be solved by means of accurate studies of a mathematical model that describes the whole machine or parts of it [8,9], as well as the application of new technologies in harvesting machines themselves [10].

The major hurdle to generating renewable energy from biomass is the low bulk density, low energy density, and high moisture content [11–14], which increase biomass transport and handling costs.

The biomass densification process transforming biomass into pellet and briquette can diminish the logistics cost [15].

In recent years, pellets became an important fuel in the production of heat and power in Europe [16]. The consumption of wood pellets grew rapidly during the last decade. The swift increase in the consumption of pellets is mainly due to legislation in several European countries that supports renewable energy [5]. In addition, between 2000 and 2017, global production jumped significantly, particularly in South America, Asia, and Oceania. The contribution of five main areas, i.e., EU 28, North America, Asia and Oceania, and South America, to the global wood pellet production in 2017 accounted for 48%, 32%, 8%, and 2%, respectively [17]. However, in the consumer market, EU 28 is a massive pellet consumer with a 77% of the world's wood pellet consumption [17]. In 2017, European pellet demand experienced a growth of 2.5 million tons, while production raised to 1.4 million tons [17].

Among the top 10 pellet-consuming countries in 2017, Italy took the second position with the consumption of 3.5 million tons in order to meet commercial and residential heat demands [18].

The Italian production of pellets settled, in 2015, about 300 thousand tons, a decrease by 16% from the previous year. The main production sites are Lombardia accounting for 45% of the national supply, followed by Veneto (18%), Friuli Venezia Giulia (16%), and Trentino Alto Adige (8%) [19]. On the other hand, the Italian national demand of pellets was estimated, in 2015, as about 2.25 million tons, experiencing an increase of 35% from the previous year [20], which was met by importing from foreign countries, particularly Austria [21]. The growth of the pellet production industry depends on the economic and energetic efficiency of the pellet plant, which is a function of many variables, namely, woody biomass availability, location, cost of investment, operation and maintenance, plant capacity, logistics, energy costs, and the possibility to locate pellet plants close to a source of low-cost heat for drying purposes (i.e., industrial cogeneration plants), environmental benefits, and financial incentives. Therefore, techno-economic examination of the pellet production systems is essential to evaluate the sustainability of the pelletization schemes and to select key factors affecting its development [22].

Pellet market evaluation indicates that, although the size and efficiency of the pellet production plant affect investment and operational costs of pellet, the production heavily depends on the physical characteristics of the raw biomass, particularly moisture content, and the need for mechanical pre-treatments [23,24]. Among diverse types of biomass (woody, herbaceous, fruity, or mixtures), raw materials for the pelletizing process in Europe are dominated by secondary feedstocks encompassing any by-products from the wood industry and pruning residues [17,18]. Agricultural residues also massively contribute to pellet production, considering that 102,000 kt of these by-products are annually produced in Europe [25]. Hence, some studies focused on different feedstocks to produce pellet. Carone et al. [26] carried out an assessment of technical factors influencing the quality of pellet produced from olive pruning residues and other agricultural waste, by means of an experimental set-up. Sánchez et al. carried out a cost evaluation of the pellet production chain from agri-food and wood industries in Spain. In this study, the cost of pelleting was affected by business fee, biomass transport, profit margin (15%), and pellet transport [27]. Hoefnagels and Junginger investigated the economic potential of wood pellet production from secondary forestry residues to find an optimal size of pellet plant [28]. They showed that optimal size depends on the location and feedstock supply assumptions.

Sultana and Kumar developed a multi-criteria assessment model for large biomass heat and power generation plants. They also revealed the importance of environmental, economic, and technical factors in decision-making regarding five pellets, each produced from a different sustainable biomass feedstock, i.e., wood, straw, switchgrass, alfalfa, and poultry litter [29]. In other research [30], Sultana et al. investigated minimum production cost and optimum plant size for pellet plants fed by agricultural biomass residue from wheat, barley, and oats. Three scenarios involving minimum, average, and maximum yields of straw were considered for developing a techno-economic model. Results showed that the total cost of pellet production is highly affected by field cost and transportation cost. To the best of our knowledge, there is a lack of studies on the effects of raw material characteristics on the investment costs of pellet production. Therefore, this research focuses on the assessment of the influence on pellet production plants in terms of the investment and operational costs of different biomass typologies and supply chains. To this end, operational indices, net present value (NPV), payback time (PBT), and profitability index (PI) are applied. In addition, the uncertainty impact of the key physical, chemical, and economic characteristics of the raw biomass on each indicator is quantified by sensitivity analysis. The target audiences of this study are potential investors interested in the pellet production sector, as well as policymakers evaluating the optimal scale of pellet plants to foster growth of the pellet production sector and biomass supply companies investigating the relationship between the quality and price of the biomass.

2. Legislative Framework for Biomass Pellets

Pellets are classified according to their physical and chemical properties. These properties affect the possibilities of pellet use in energy conversion technologies. For instance, a comparatively low amount of dust and ash in the pellet is an important factor for small heating systems, while larger power systems can cope with higher amounts of dust and ash. Other important parameters include durability, surface smoothness, and resistance to swelling. In Table 1, the main physical–chemical characteristics of pellets are shown, according to European Committee for Standardization (CEN) standards.

Table 1. Pellet physico-chemical characteristics [31].

Parameters	Effects
	Chemical Characteristics
Elements	
CI	Emission of dioxynes and furanoids, corrosion issues
	Emission of NOx, HCN, N_2O
N	Emission of SOx
S	Corrosion issues, low melting point of ashes
K	High melting point of ashes, pollutants in exhaust fumes
	High melting point of ashes, pollutants in exhaust fumes
Mg, Ka, P	
Heavy Metals	
Composition of Ashes	Polluting emissions, ash disposal issues
	Physical Characteristics
Moisture	Storage issues, Low Heating Value (LHV), auto-combustion
Density	Transport and storage issues, combustion properties
Pellet Size	Fluidity, transport safety, production of dust
Mechanical Durability	Changes in pellet quality, leakage

In order to foster the development of the pellet market, the European Committee for Standardization (CEN) issued a set of procedures for the characterization of solid bio-fuels (EN 14961, TC335, EN 17225) and for the quality certification of the bio-fuels (EN 15234), including pellets. Table 2 outlines the regulatory quality standards set within different nations.

Table 2. Quality standards for pellets in different EU Countries and in the USA.

Parameter	Austria (1)	Sweden (2)	Germany (3)	USA (4)
Size (mm)	3 < D < 4L ≤ 100	L = 5D	-	-
Density (kg/m^3)	-	≤500	-	≤639
Durability (%, <3 mm)	-	≤1.5	-	≤0.5
Energy Density (MJ/kg)	≤18.0	≤16.9	17.5–19.5	-
Moisture (% mass)	≤12	≤10	≤12	-
Ashes Content (% mass)	≤0.5	≤1.5	≤1.5	≤1
Sulfur (% mass)	≤0.04	≤0.08	≤0.08	-
Nitrogen (% mass)	≤0.3	-	≤0.3	-
Chlorine (% mass)	≤0.02	≤0.03	≤0.03	-
Additives (glues)	Not allowed	To be declared	-	-

Notes: (1) ONORM M 7135 [32], 2 categories for wood (pellet) and bark (briquettes). (2) SS 187120[33], three groups, having L = 4D, L = 5D, and L = 6D. (3) DIN 51731[34], five categories having L between 5 and 30 cm. (4) Pellet Fuel Institute [35], two categories (standard and premium), having ash content between 1% and 3%.

In Italy, in accordance with legislative decree 152/2006, the only kind of raw material allowed toward the production of pellets is biomass derived from mechanical processes applied within agriculture and forestry production, pruning residues, and lumber-mill by-products from raw wood.

Following requirements for the development of a voluntarily certified pellet quality certification in Italy, the Pellet Gold system, including a brand statement and quality assurance, was recently developed by AIEL (Associazione Italiana Energia dal Legno) [36]. The procedure Pellet Gold involves a series of tests, performed according to stringent quality parameters. The process to obtain and maintain a certificate of quality involves audits in companies, with sampling, testing, and process control. The fundamental assumption is that the pellet product is composed of virgin wood not contaminated with paints, additives, or other chemical adhesives. The requirements to Pellet Gold are similar to those indicated by the more stringent regulations CEN/TS 14961 [37], DIN plus, and ONORM M 7135, and they are aligned to the limits set by the Pellet Fuel Institute (PFI) [35].

On 21 July 2011, Italy adopted the European standard (EN 14961-2) to define the quality characteristics of pellets for non-industrial use. This standard was updated in 2014 (UNI EN ISO 17225:2014), which includes a series for pellets from woody biomass and another one for pellets from non-woody biomass. The standard introduces three quality classes:

- Class A1, which corresponds to a higher quality, and maximum ash content of 0.7%;
- Class A2, characterized by an ash content of 1.5%;
- Class B, characterized by a maximum ash content of 3.5%, which can be produced either from sawdust by the cortex, destined to centralized plants of greater dimensions, for commercial or pseudo-industrial application.

Since March 2012, the certificate Pellet Gold ensures compliance with UNI EN 17225-2. Therefore, companies certified must deal with the ash content of the pellets from their product. In addition to conforming to the European Pellet Gold certification, the determination of formaldehyde content and radioactivity was supplemented as criteria for the manufacturer. Technical specifications and the classification of wood-based pellets from woody biomass, as well as those for non-industrial applications, are indicated in Table 3.

Table 3. Pellet quality standard according to UNI EN ISO 17225-2.

Features	Category					
	A1 Tree Trunks and/or Untreated Wood without Bark (No Additives)	A2 Tree Trunks and/or Untreated Wood without Bark (No Additives)		B Forestry Wood, Wood Processing By-Products, Used Wood		
Diameter D (mm) and Length L (mm)	D = 6–8 L = 3.15–40	D = 6–8 L = 3.15–40	8 ± 0.5	6 ± 0.5	8 ± 0.5	From D > 10 ± 1.5 To D < 25 ± 1.0
Moisture (%)	10	10		10		18
Ashes (%)	0.7	1.5		3		To be declared
Durability (%)	1	1		1		To be declared
Additives (%)	Not allowed	To be declared (1)		To be declared (1)		To be declared (1)
Sulfur (%)	0.05	0.05		0.05		To be declared
Nitrogen (%)	0.3	0.3		0.3		To be declared
Chlorine (%)	0.03	0.03		To be declared		To be declared
Density (kg/m^3)	620–720	620–720		620–720		550
LHV (MJ/kg)	16.9	16.9		16.2		To be declared

(1) Permissible additives (glues) are maize starch, raw vegetable oil extracted from purely mechanical pressing, molasses, and natural paraffin. No artificial substances are allowed. The nature and quantities of additive must be declared.

3. Biomass Pellet Routes and Agro-Pellet Main Issues

Figure 1 outlines the phases of the pellet production process. In this section, these phases are shortly discussed, in order to highlight the main technical barriers of the pellet chain and to define the costs assumed in the successive economic assessment.

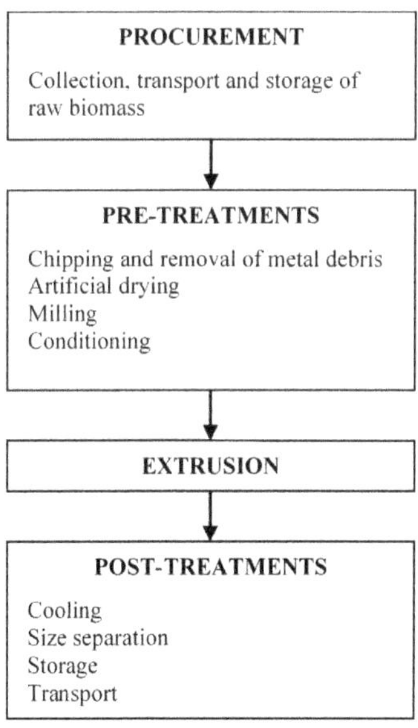

Figure 1. Phases of the pellet production process.

3.1. Biomass Supply

This phase includes the collection, transport, and storage of biomass to the collection point; green wood is mechanically removed, and pruning residues are air-dried.

3.2. *Biomass Pre-Treatments*

This phase includes the mechanical processes such as converting wood into wood chips (except when starting from sawdust), the removal of ferrous material, drying, milling, and conditioning. The wood chips are typically dried in heaters fed with conventional fuel, although sometimes the heaters use wood chips as fuel. The most common driers use rotating drums with flow of air, in which the wet biomass does not stick to the drum surfaces and over-heating is minimized. If the biomass can withstand contact with the combustion products, the simplest and cheapest system is a direct-heating drier, in which the wet biomass is in contact with hot combustion gases. Alternatively, the wet biomass can be dried using hot air. The dried biomass is further milled to obtain wood chips with average length of 3 mm and to homogenize the end product. The most common mills use rotating hammers, and the material is shifted through the machine using compressed air. Sometimes, the secondary milling is bypassed by a simple tilted-plane mechanical size selector. The pre-conditioning involves exposure of the biomass to an appropriate mix of environmental conditions (temperature, moisture, length of exposure time) to optimize its behavior in the subsequent extrusion. A common pre-conditioning process exposes the biomass to rapid heating using hot water vapor, with the effect of softening the wood chips and obtaining a partial decomposition of starch and cellulose in simpler sugars, which allow easier compacting. The short time of exposure to hot vapor minimizes significant increases in the moisture content of the biomass. Additives (such as molasses, starch, fats, oils, glues, etc.) aimed at improving biomass quality and extrusion behavior can be added to the raw feed during this phase [38,39].

3.3. *Extrusion*

This phase involves the physical production of the pellet by applying mechanical pressure on the biomass through a suitable holed plate, to obtain pellets with diameters in the 2–12 mm range and heights in the 12–18 mm range. The main technical parameters of the pelletizer are as follows: canal geometry, number and speed of pressurizing drums, ratio between diameter and length of canals, and distance between drum and holed plate. The devices may use a vertical cylindrical holed drum or a plane plate.

3.4. *Post-Treatments*

These phases include cooling, selection of pellet size, collection, and storage in silos or sacks for subsequent sale. The cooling phase is critical for the stabilization of the product, since, during the extrusion, the pellets reach comparatively high temperatures (90–95 °C) and are typically obtained via forced exposure to air at room temperature. Pellets with a non-standard size are mechanically removed to minimize development of dust in the storage areas.

4. Methodology for the Economic Evaluation

The economic evaluation is carried out by calculating the net present value (NPV) and profitability index (PI) of the investment. Subsequently, a sensitivity analysis is performed considering following parameters: biomass moisture content, pellet market value, cost of raw biomass, and average biomass transport distance.

The cost of investment ($C_{Investment}$) is calculated as follows:

$$C_{Investment} = C_{pell} + C_{dry} + C_{chip} + C_{store} + C_{inst} + C_{eng}, \tag{1}$$

where C_{pell} is the cost of pelletizing the plant, C_{dry} is the cost of drying the plant, C_{chip} is the cost of pre-treatment processes, C_{store} is the cost of storage, C_{inst} is the cost of plant installation, and C_{eng} is the plant engineering cost.

The annual operation and maintenance costs ($C_{Operation}$) are calculated as follows:

$$C_{Operation} = C_{biomass} + C_{transport} + C_{drying} + C_{electricity} + C_{personnel} + C_{maintenance}\ (€/\text{year}), \quad (2)$$

where $C_{biomass}$ is the cost of raw biomass, $C_{transport}$ is the cost of transport, C_{drying} is the biomass drying cost, $C_{electricity}$ is the cost of electricity, $C_{personnel}$ is the personnel cost, and $C_{maintenance}$ is the maintenance cost. Details on each parameter are outlined below.

$$C_{biomass} = \sum_{j=1}^{m} P_{A,j} \times Q_{B,j} (€/\text{year}), \quad (3)$$

where m is the number of times of biomass feeding in the plant, $P_{A,j}$ is the purchase price of the j-th biomass feed (€/t), and $Q_{B,j}$ is the amount of the j-th biomass feed (t/year).

The pellet production ($Q_{Pellets}$) is equal to

$$Q_{Pellets} = Q_{max} \times i_U \times H.\ (\text{t/year}), \quad (4)$$

where Q_{max} is the maximum production capacity of the plant (t/hour), i_U is the production load factor, H is the annual production time (hours/year).

The number of production hours per year is calculated by

$$\mathrm{H} = \mathrm{n_{shifts}} \times 8 \times 12 \times \mathrm{d_{working}}, \quad (5)$$

where $\mathrm{n_{shifts}}$ is the number of daily shifts (each lasting 8 h), and $\mathrm{d_{working}}$ is the number of working days per month.

The amount of biomass requisite to produce $Q_{Pellets}$ also depends on the moisture content of both the raw biomass and the final pellet, according to

$$Q_{Pellets} = \sum_{j=1}^{m} \left(\frac{1 + m_{pellet}}{1 + m_{j-biomass}} \right) \cdot Q_{B,j}\ (\text{t/year}), \quad (6)$$

where m_{pellet} is the pellet moisture content, and $m_{j\text{-}biomass}$ is the j-th biomass moisture content.

If a portion of raw biomass is utilized to dry the remaining biomass, the biomass drying cost (C_{drying}) is equal to

$$C_{drying} = \sum_{j=1}^{m} \left(m_{j-biomass} - m_{pellet} \right) \cdot k_{drying} \cdot P_{A,j} \cdot Q_{B,j}\ (€/\text{year}), \quad (7)$$

where k_{drying} represents a dimensionless coefficient, equivalent to the additional mass of raw biomass (kg) needed to decrease the moisture content of 1 kg of biomass by one percentage point. The value of the coefficient also depends on the efficiency of the drying plant.

The cost of electricity is equal to

$$C_{electricity} = P_{electricity} \cdot \left(E_{pellet} \cdot Q_{Pellets} + \sum_{j=1}^{m} E_{chip,j} \cdot \overline{Q_{B,j}} \right) (€/\text{year}), \quad (8)$$

where: $P_{electricity}$ is the price of electricity (€/MWh), E_{pellet} is the electricity needed during the pre-loading, conditioning, extrusion, cooling, and size selection phases of the pellet production chain (MWh/t), $E_{chip,j}$ is the electricity needed during the mechanical pre-treatment of the j-th raw biomass feed, and

$\overline{Q_{B,j}}$ is the mass amount of j-th raw biomass feed undergoing mechanical treatments, with the net of biomass amounts used for drying purposes.

$$\overline{Q_{B,j}} = Q_{B,j} \cdot \left[1 + k_{drying} \cdot \left(m_{j-biomass} - m_{pellet}\right)\right] \text{ (t/year)}. \tag{9}$$

The cost of transport is equal to

$$C_{transport} = c_{transport} \left(\sum_{j=1}^{m} d_j \overline{Q_{B,j}} \right) \text{ (€/year)}, \tag{10}$$

where $c_{transport}$ is the cost of transport (€/t·km), and d_j is the average transport distance of j-th raw biomass feed (km).

The personnel cost is equal to

$$C_{personnel} = n_{unit/shift} \cdot n_{shifts} \cdot c_{unit} \text{ (€/year)}, \tag{11}$$

where $n_{unit/shift}$ is the number of personns employed for each shift, n_{shifts} is the number of daily shifts (each lasting 8 h), and c_{unit} is the annual per-unit cost of personnel.

The maintenance costs are equal to

$$C_{maintenance} = (C_{pell} + C_{dry} + C_{chip}) \cdot k_M, \tag{12}$$

where k_M is a coefficient reflecting the ordinary and extraordinary maintenance costs for plants and machinery.

The total revenues are equal to

$$\text{Revenues} = P_{pellet} \times Q_{pellets} \text{ (€/year)}, \tag{13}$$

where P_{pellet} (€/t) is the market value of pellets.

The total costs (costs of goods sold + overheads and interests) are equal to

$$Costs_{total} = \frac{C_{I-year} + C_{Operation}}{Q_{pellets}} \text{ (€/t·year)}, \tag{14}$$

where $C_{I\text{-year}}$ is the annual financial charge, equal to

$$C_{I-year} = \frac{C_{Investment} \cdot r}{1 - (1/(1+r))^n} \text{ (€/year)}, \tag{15}$$

where r is the annual real discount rate, and n is the lifetime of the plant (year).

The *NPV* (net present value) of the investment is

$$NPV = \sum_{i=1}^{n} \frac{CF_i}{(1+r)^i} - C_{Investment} \text{ (€)}, \tag{16}$$

where CF_i is the cash flow generated at the i-th year, and it is equal to

$$CF_i = (Revenues - C_{Operation}) \text{ (€/year)}. \tag{17}$$

The profitability index (*PI*) is calculated according to

$$PI = \frac{NPV}{C_{Investment}}. \tag{18}$$

5. Application to the Case Study

The economic evaluation was carried out for a firm producing door and window frames in laminated wood. This firm generates 3000 t/year of waste virgin wood residues, characterized by a small size and 13–15% moisture content. A small portion of this biomass (500 t/year) is currently used as bio-fuel to meet heat demand within the production site, while the remaining portion is considered as waste and disposed. The evaluation was aimed at assessing the economic viability of employing the waste biomass for pellets production. The analysis was carried out within four scenarios:

- In the base case, pellets are produced using by-products of the wood industry, similar to those available in-house by the firm under exam; however, in this scenario, they are purchased externally.
- In this scenario, all the waste virgin wood residues available in-house by the firm (and conveniently mixed with other similar biomass sourced locally, where appropriate) are used for the production of pellets.
- In this scenario, the waste virgin wood residues available in-house by the firm are mixed with other wood residues having a higher moisture content.
- In this scenario, the waste virgin wood residues available in-house by the firm are mixed with lumber mill residues and pruning residues, which all require suitable mechanical pre-treatments, before the drying and extrusion phases.

Tables 4 and 5 outline the technical and economic input parameters, as well as investment and operating costs considered within the four scenarios. In particular, Table 4 reports the technical and economical parameters which are constant within the four scenarios, while Table 5 reports the parameters varying across the different case studies. The pellet market value is intended as the pellet selling price to the retailer, excluding transport costs, assuming a final selling price for the end user of 220 €/t and considering the income for the distributor of the pellet.

Table 4. Technical and economic parameters held constant within the four scenarios.

Parameters	Unit	Value
Maximum production capacity Q_{max}	t/hour	1.25
Production load factor i_U	%	80
Hourly production capacity Q	t/hour	1
Number of daily shifts n_{shifts}	-	2
Number of annual production hours H	hours/year	3840
Annual pellet production Q_{pellet}	t/year	3840
Pellet moisture content m_{pellet}	%	12
Cost of transport $c_{transport}$ (1)	€/t·km	0.15
Drying coefficient k_{drying} (2)	-	0.015
Price of electricity $P_{electricity}$	€/MWh	150
Electricity needed per t of pellet E_{pellet} (3)	MWh/t	0.15
Annual cost of personnel c_{unit}	€/year·person	20,000
Maintenance coefficient k_M (4)	%	10
Lifetime of the plant n	years	8
Real discount rate r	%	5
Pellet market value P_{pellet}	€/t	135

(1) The quoted value is an average of the fares charged by the Italian Road Transport Operators' Association, relating to a distance of 100 km and a load of 20 t (see: www.confartigianatotrasp.com); (2) The value is obtained by considering an average consumption of 0.3 kg of biomass with 32% moisture content in order to dry 1 kg of biomass with 12% moisture content; (3) According to Reference [40]; (4) Average value for maintenance costs of the drying plant and machinery for chipping and extrusion [41].

Table 5. Technical and economic parameters varying across the four scenarios.

Parameters	Unit	Values			
	Scenario	A	B	C	D
Biomass amount (1)	t/year	4070	2420	2420	2420
Biomass price (1)	€/t	35	0	0	0
Moisture content (1)	%	14	14	14	14
Average transport distance (1)	Km	60	0	0	0
Electricity needed for chipping (1) (*)	MWh/t	0.02	0.02	0.02	0.02
Biomass amount (2)	t/year	0	1800	3300	3300
Biomass price (2)	€/t	0	35	30	25
Moisture content (2)	%	0	18	50	50
Average transport distance (2)	km	0	60	60	60
Electricity needed for chipping (2) (*)	MWh/t	0	0.02	0.02	0.065
Personnel units per shift	units	2	2	2	3

(1) Biomass feed available in-house by the firm; (2) Biomass feed sourced/purchased externally by the firm; (*) Authors' elaboration, based on average values of energy required for chipping and in relation to the quality of raw biomass, size of chipping machinery, and quality of the extruded material [42].

In Tables 6 and 7, the investment and operational costs for each scenario are reported. As seen, operation and maintenance (O&M) costs in case A are high because of the cost of biomass, while, in cases C and D, they are higher than in case B, due to the need to reduce the initial moisture content of the raw biomass from 50% to 12%. Furthermore, in case D, the raw biomass needs to undergo a pre-chipping phase, prior to being dried, which leads to an increase in investment costs and requirement for one additional personnel unit.

Table 6. Investment costs for the chosen scenarios.

Investment Costs (1000 €)								
Scenario	A		B		C		D	
C_{pell}	420	57%	420	57%	420	52%	420	45%
C_{dry}	30	4%	30	4%	100	12%	100	11%
C_{chip}	40	5%	40	5%	40	5%	150	16%
C_{store} (1)	200	27%	200	27%	200	25%	200	22%
C_{inst} (2)	30	4%	30	4%	30	4%	30	3%
C_{eng} (3)	20.7	3%	20.7	3%	22.8	5%	26.1	3%
Total	740.7	100%	740.7	100%	812.8	100%	926.1	100%

Notes: (1) Including cost of land, building, and storage facilities; the salvage value of the storage facilities accounts for 60% of the investment cost, and this value was accounted for as lump sum income generated in the last year of the expected operating life of the plant; (2) Authors' elaboration, based on two technicians employed for 30 working days and charging 500 €/person·day; (3) Authors' elaboration, based on design costs as 5% of the investment costs, net of installation costs.

Table 7. Operation and maintenance (O&M) costs for the chosen scenarios.

Operation and Maintenance Costs (1000 €/year)								
Scenario	A		B		C		D	
$C_{biomass}$	138.3	40%	55.8	23%	63	20%	52.5	14%
$C_{transport}$	36.6	11%	16.2	7%	29.7	10%	29.7	8%
C_{drying}	4.1	1%	5.2	2%	35.9	12%	29.9	8%
$C_{electricity}$	38.1	11%	38.6	16%	43.1	14%	65.3	18%
$C_{personnel}$	80	23%	80	32%	80	26%	120	33%
$C_{maintenance}$	49	14%	49	20%	56	18%	67	18%
Total	346.1	100%	246.7	100%	307.7	100%	364.4	100%

6. Results

The key results are shown in Table 8. The most convenient situation is indeed that outlined in scenario B, where the availability of abundant and good-quality biomass at zero cost results in a pellet production cost of 94 €/t, a PI of 1.53, and a PBT (payback time) lower than three years.

Table 8. Results of the financial analysis. PBT—payback time; NPV—net present value; PI—profitability index. Total production cost excludes row biomass supply cost

Scenario		A	B	C	D
Production cost	€/t	44.6	45.0	56.0	73.5
Total production cost	€/t	120.0	94.1	112.9	132.2
Cash flow	k€/year	172.2	271.7	210.7	154.0
PBT	Year	4.3	2.7	3.9	6.0
NPV	k€	492.7	1135	669.3	189.1
PI	-	0.67	1.53	0.82	0.20

According to scenario A, all raw biomass (having the same characteristics of the by-product available as firm by-product of the normal production cycle) is sourced externally at a cost of 35 €/t (excluding transport costs). Consequently, the cost of production rises to 120 €/t, while the PI drops to 0.67, and the PBT is higher than four years.

In case C, where externally sourced wet biomass is used as integrating feed for the biomass internally available, the economic values are worse than those of case B, because of the need to dry the wet biomass.

Scenario D is the least convenient of all, with a cost of production of 132 €/t, PI of 0.2, and PBT of six years; this is due to the extra costs related to the pre-treatments required for the raw biomass. Figures 2–5 outline the results of the sensitivity analysis associated with moisture content, average transport distance of raw biomass, pellet market value, and annual maintenance coefficient.

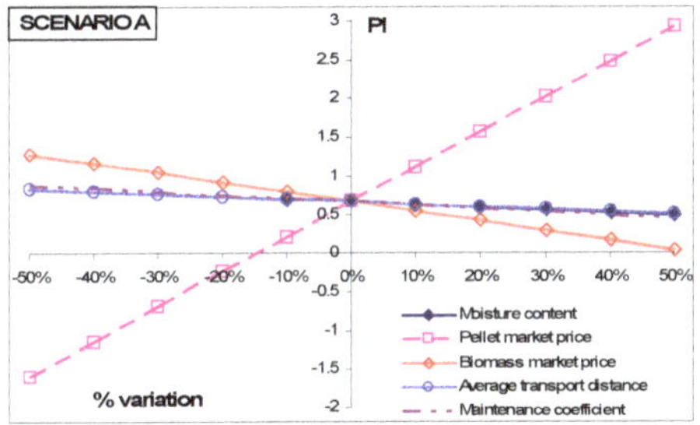

Figure 2. Sensitivity analysis on PI values for scenario A.

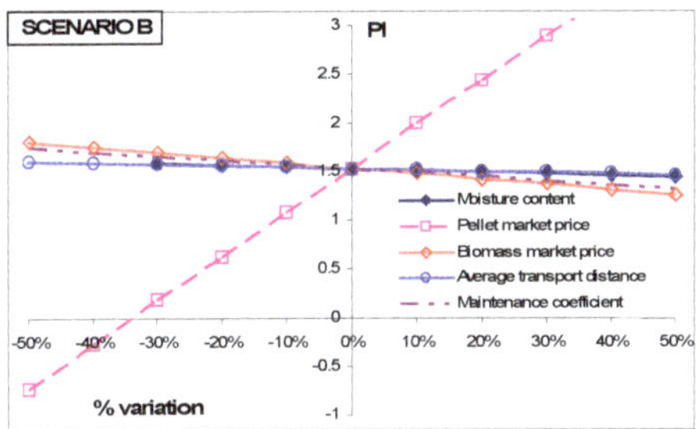

Figure 3. Sensitivity analysis on PI values for scenario B.

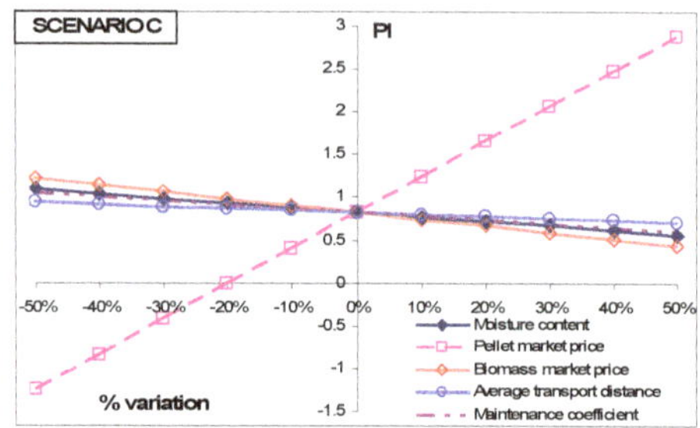

Figure 4. Sensitivity analysis on PI values for scenario C.

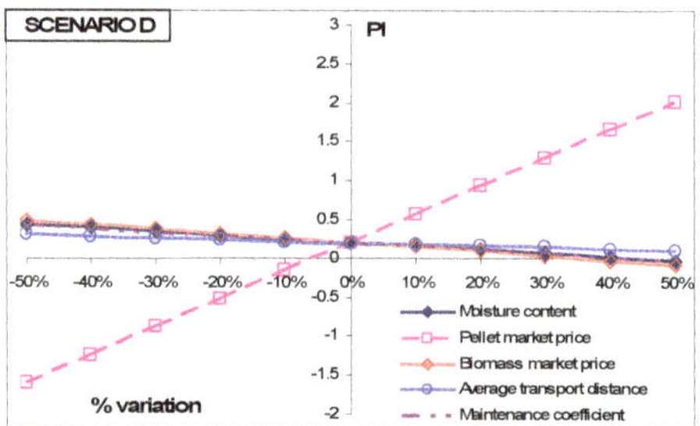

Figure 5. Sensitivity analysis on PI values for scenario D.

The sensitivity analysis allows drawing the following considerations:

- The market value of pellets is the parameter with the highest impact on the PI. The PI becomes negative for pellet prices of 115, 90, 108 and 127 €/t within scenarios A, B, C, and D, respectively.
- The biomass market price significantly affects the PI, particularly in cases A (where all biomass is sourced externally) and D (where the biomass has a high moisture content). Variations of ±20% in the price of biomass result in variations of the PI of ±36%, ±7%, ±20%, and ±60% within scenarios A, B, C, and D, respectively. In scenario A, the PI becomes negative if the price of biomass is higher than 54 €/t.
- An increase in the moisture content of the raw biomass results in a decrease of the PI, since more biomass is needed, and the costs of drying and transport all correspondingly increase. A variation of ±20% in the moisture content results in variations of the PI of ±10%, ±2%, ±12%, and ±50% within scenarios A, B, C, and D, respectively.
- A variation of ±20% in the average transport distance of the raw biomass results in variations of the PI of ±18%, ±4%, ±10%, and ±45% within scenarios A, B, C, and D, respectively. Notably, a reduction of the transport costs by zero results in an increase in PI by about 46%.

7. Conclusions

This paper describes the state of the art and the Italian regulation related to the Italian market for wood pellets. The phases of the pellet production chain are outlined, and a detailed financial appraisal model is put forward with the aim of assessing the financial viability of undertaking the production of pellets. The financial model is applied to an existing firm whose main products are wood door and window frames, which has a sizeable by-product of good-quality biomass (small wood residues) that could be used to manufacture wood pellets. The analysis is carried out within four scenarios, which reflect the main biomass supply options currently available to the firm managers; for each scenario, the NPV, PI, and PBT indices are calculated, and then a sensitivity analysis is carried out, assessing the impact of variations in the main parameters over the PI of the investment.

Based on the hypotheses of this study, it appears that the use of logging residues and bark is not financially competitive, due to the comparatively high costs of pre-treatment and drying. The most promising business opportunities for pellet production lie where an existing high-quality biomass by-product is added to a limited amount of low-moisture and low-cost biomass. It should be noted that, to the author's knowledge, this is one of the first researches comparing the pellet production costs with different biomass supply chains; thus, it is difficult to compare economic profitability and cost figures with previous studies. Moreover, further analyses should be carried out to optimize in an integrated manner the biomass collection area, the biomass processing/pelletization location and sizing, and the technologies for the final energy conversion to match the end users' demand, as already proposed in previous researches focused on sustainable energy systems in urban and peri-urban areas [43,44]. Such approaches could also be linked to the assessment of biomass energy potentials, in order to explore how to best use the resources of the territory in distributed vs. centralized processing and conversion plants, using intermediate bio-fuels (such as pellet) to improve the energy balances, the conversion efficiency, and the logistics of the routes; an example of this approach was proposed in Reference [45], for a case study of the Puglia region.

Finally, further researches should be devoted to the assessment of the potential market segments in the industrial, residential, commercial, and rural sectors, where different typologies of pellet could be used, considering the trade-offs between high-quality/high-cost pellet (from selected woody biomass) and low-quality/low-cost biomass (from agricultural or forestry residues), which could be better used in industrial applications and large-scale combustion/gasification plants able to manage the lower quality of the biofuel, in comparison to domestic stoves or heating plants for the commercial sector.

Author Contributions: Conceptualization, A.P. and F.S.; data curation, A.P., S.R.H., and F.S.; formal analysis, A.C.; investigation, A.P. and F.S.; methodology, A.P., M.C. and S.R.H.; resources, A.C.; supervision, A.P., M.V. and M.C.; validation, A.P., A.C., M.V., and S.R.H.; visualization, S.R.H., and M.V.; writing—original draft, A.P.; writing—review and editing, S.R.H., M.V., A.P., and A.C. All authors have read and agreed to the published version of the manuscript.

Funding: This research was partially supported by MIUR (Italian Ministry of Education, University and Research), Law 232/2016, "Department of Excellence".

Conflicts of Interest: The authors declare no conflicts of interest.

References

1. Nunes, L.J.R.; Matias, J.C.O.; Catalaõ, J.P.S. Mixed biomass pellets for thermal energy production: A review of combustion models. *Appl. Energy* **2014**, *127*, 135–140. [CrossRef]
2. Carlini, M.; Mosconi, E.M.; Castellucci, S.; Villarini, M.; Colantoni, A. An economical evaluation of anaerobic digestion plants fed with organic agro-industrial waste. *Energies* **2017**, *10*, 1165. [CrossRef]
3. Carlini, M.; Castellucci, S.; Moneti, M. Anaerobic co-digestion of olive-mill solid waste with cattle manure and cattle slurry: Analysis of bio-methane potential. *Energy Procedia* **2015**, *81*, 354–367. [CrossRef]
4. Pallozzi, V.; Di Carlo, A.; Bocci, E.; Villarini, M.; Foscolo, P.U.; Carlini, M. Performance evaluation at different process parameters of an innovative prototype of biomass gasification system aimed to hydrogen production. *Energy Convers. Manag.* **2016**, *130*, 34–43. [CrossRef]
5. Council of the European Union. *Directive of the European Parliament and of the Council on the Promotion of the Use of Energy from Renewable Sources*; Council of the European Union: Brussels, Belgium, 2018.
6. IRENA. International Renewable Energy Agency n.d. Available online: https://www.irena.org/ (accessed on 26 March 2020).
7. IRENA. *REmap 2030—Doubling the Global Share of Renewable Energy: A Roadmap to 2030*; IRENA: Abu Dhabi, UAE, 2020; Volume 60.
8. Bulgakov, V.; Pascuzzi, S.; Santoro, F.; Anifantis, A.S. Mathematical model of the plane-parallel movement of the self-propelled root-harvesting machine. *Sustainability* **2018**, *10*, 3614. [CrossRef]
9. Bulgakov, V.; Pascuzzi, S.; Anifantis, A.S.; Santoro, F. Oscillations analysis of front-mounted beet topper machine for biomass harvesting. *Energies* **2019**, *12*, 2774. [CrossRef]

10. Guerrieri, A.S.; Anifantis, A.S.; Santoro, F.; Pascuzzi, S. Study of a large square baler with innovative technological systems that optimize the baling effectiveness. *Agriculture* **2019**, *9*, 86. [CrossRef]
11. Sahoo, K.; Mani, S. Techno-economic assessment of biomass bales storage systems for a large-scale biorefinery. *Biofuels Bioprod. Biorefin.* **2017**, *11*, 417–429. [CrossRef]
12. Parkhurst, K.; Saffron, C.; Miller, R. An energy analysis comparing biomass torrefaction in depots to wind with natural gas combustion for electricity generation. *Appl. Energy* **2016**, *179*, 171–181. [CrossRef]
13. Carlini, M.; Castellucci, S.; Moneti, M. Biogas production from poultry manure and cheese whey wastewater under mesophilic conditions in batch reactor. *Energy Procedia* **2015**, *82*, 811–818. [CrossRef]
14. Carlini, M.; Castellucci, S.; Mennuni, A.; Selli, S. Poultry manure biomass: Energetic characterization and ADM1-based simulation. *J. Phys. Conf. Ser.* **2019**, *1172*, 012063. [CrossRef]
15. Tumuluru, J.; Wright, C.; Hess, J.; Kenney, K. A review of biomass densification systems to develop uniform feedstock commodities for bioenergy application. *Biofuels Bioprod. Biorefin.* **2011**, *5*, 683–707. [CrossRef]
16. European Commission. *Brief on Biomass for Energy in the European Union*; European Commission: Brussels, Belgium, 2019; pp. 1–8.
17. AEBIOM. *European Bioenergy Outlook*; AEBIOM: Bruxelles, Belgium, 2018.
18. Hawkins Wright. *European Pellet Council (EPC) Survey*; Hawkins Wright: Richmond, UK, 2018.
19. Paniz, A. *Development of the Italian Pellet Market*; Associazione Italiana Energie Agroforestali: Vancouver, BC, Canada, 2014.
20. IEA. *Global Wood Pellet Industry and Trade Study*; IEA: Paris, France, 2017.
21. GSE. *Energy from Renewable Sources in Italy*; 2018 Statistical Report; GSE: Kingstree, SC, USA, 2017.
22. Thek, G.; Obernberger, I. *The Pellet Handbook. The Production and Thermal Utilisation of Biomass Pellets*; Routledge: London, UK, 2012.
23. Junginger, M.; Sikkema, R. *The global wood pellet trade—Markets*; Barriers and Opportunities Workshop Summary: Utrecht, The Netherlands, 2008.
24. Castellano, J.M.; Gómez, M.; Fernández, M.; Esteban, L.S.; Carrasco, J.E. Study on the effects of raw materials composition and pelletization conditions on the quality and properties of pellets obtained from different woody and non woody biomasses. *Fuel* **2015**, *139*, 629–636. [CrossRef]
25. Monforti, F.; Lugato, E.; Motola, V.; Bodis, K.; Scarlat, N.; Dallemand, J.F. Optimal energy use of agricultural crop residues preserving soil organic carbon stocks in Europe. *Renew. Sustain. Energy Rev.* **2015**, *44*, 519–529. [CrossRef]
26. Carone, M.T.; Pantaleo, A.; Pellerano, A. Influence of process parameters and biomass characteristics on the durability of pellets from the pruning residues of Olea europaea L. *Biomass Bioenergy* **2011**, *35*, 402–410. [CrossRef]
27. Sánchez, J.; Dolores Curt, M.; Sanz, M.; Fernández, J. A proposal for pellet production from residual woody biomass in the island of Majorca (Spain). *AIMS Energy* **2015**, *3*, 480–504. [CrossRef]
28. Hoefnagels, R.; Junginger, M. The economic potential of wood pellet production from alternative, low-value wood sources in the southeast of the U.S. *Biomass Bioenergy* **2014**, *71*, 443–454. [CrossRef]
29. Sultana, A.; Kumar, A. Ranking of biomass pellets by integration of economic, environmental and technical factors. *Biomass Bioenergy* **2012**, *39*, 344–355. [CrossRef]
30. Sultana, A.; Kumar, A.; Harfield, D. Development of agri-pellet production cost and optimum size. *Bioresour. Technol.* **2010**, *101*, 5609–5621. [CrossRef]
31. Alakangas, E. *New European Pellets Standards*; EUBIONET: Jyväskylä Finland, 2011; Volume 4, pp. 1–17.
32. Home-Austrian Standards. Available online: https://www.austrian-standards.at/home/ (accessed on 26 March 2020).
33. Standard-Biofuels and Peat-Fuel Pellets SS 187120. Available online: https://www.sis.se/en/produkter/petroleum-and-related-technologies/fuels/solid-fuels/ss187120/ (accessed on 26 March 2020).
34. Bockris JOM. The hydrogen economy: Its history. *Int. J. Hydrogen Energy* **2013**, *38*, 2579–2588. [CrossRef]
35. Pellet Fuels Institute. Available online: https://www.pelletheat.org/ (accessed on 26 March 2020).
36. Aiel-Associazione Italiana Energie Agroforestali. Available online: https://www.aielenergia.it/chi_siamo.php (accessed on 26 March 2020).
37. International Organization for Standardization. EN 14961-2. Solid Biofuels-Fuel Specification and Classes, Part 1-Wod Pellets for Non-Industrial Use 2010. ISO: Geneva, Switzerland, 2010.

38. Proskurina, S.; Alakangas, E.; Heinimo, J.; Mikkila, M.; Vakkilainen, E. A survey analysis of the wood pellet industry in Finland: Future perspectives. *Energy* **2017**, *118*, 692–704. [CrossRef]
39. Organization for the Promotion of Energy Technologies (OPET). *Wood Pellets in Finland—Technology, Economy, and Market*; OPET: Istanbul, Turkey, 2002.
40. EUBIA. Economics, Applications and Standards. Available online: https://www.eubia.org/cms/wiki-biomass/biomass-pelleting/economics-applications-and-standards/ (accessed on 26 March 2020).
41. Nolan, A.; Mc Donnell, K.; Devlin, G.J.; Carroll, J.P.; Finnan, J. Economic analysis of manufacturing costs of pellet production in the republic of ireland using non-woody biomass. *Open Renew. Energy J.* **2010**, *3*, 1–11. [CrossRef]
42. Di Fulvio, F.; Eriksson, G.; Bergström, D. Effects of wood properties and chipping length on the operational efficiency of a 30 kW electric disc chipper. *Croat. J. For. Eng.* **2017**, *36*, 85–100.
43. Pantaleo, A.M.; Giarola, S.; Bauen, A.; Shah, N. Integration of biomass into urban energy systems for heat and power. Part I: An MILP based spatial optimization methodology. *Energy Convers. Manag.* **2014**, *83*, 347–361. [CrossRef]
44. Pantaleo, A.M.; Giarola, S.; Bauen, A.; Shah, N. Integration of biomass into urban energy systems for heat and power. Part II: Sensitivity assessment of main techno-economic factors. *Energy Convers. Manag.* **2014**, *83*, 362–376. [CrossRef]
45. Pantaleo, A.; Pellerano, A.; Carone, M.T. Potentials and feasibility assessment of small scale CHP plants fired by energy crops in Puglia region (Italy). *Biosyst. Eng.* **2009**, *102*, 345–359. [CrossRef]

Article

Evaluation of Compressor Heat Pump for Root Zone Heating as an Alternative Heating Source for Leafy Vegetable Cultivation

Chiara Terrosi [1], Sonia Cacini [1], Gianluca Burchi [1], Maurizio Cutini [2,*], Massimo Brambilla [2], Carlo Bisaglia [2], Daniele Massa [1] and Marco Fedrizzi [3]

1 CREA Research Centre for Vegetable and Ornamental Crops, Consiglio per la ricerca in agricoltura e l'analisi dell'economia agraria (Council for Agricultural Research and Economics), Via dei Fiori 8, 51017 Pescia (PT), Italy; chiara.terrosi@crea.gov.it (C.T.); sonia.cacini@crea.gov.it (S.C.); gianluca.burchi@crea.gov.it (G.B.); daniele.massa@crea.gov.it (D.M.)

2 CREA Research Centre for Engineering and Agro-Food Processing, Consiglio per la ricerca in agricoltura e l'analisi dell'economia agraria (Council for Agricultural Research and Economics), Via Milano, 43 24047 Treviglio (BG), Italy; massimo.brambilla@crea.gov.it (M.B.); carlo.bisaglia@crea.gov.it (C.B.)

3 CREA Research Centre for Engineering and Agro-Food Processing, Consiglio per la ricerca in agricoltura e l'analisi dell'economia agraria (Council for Agricultural Research and Economics), Via della Pascolare, 16 00015 Monterotondo (Rome), Italy; marco.fedrizzi@crea.gov.it

* Correspondence: maurizio.cutini@crea.gov.it; Fax: +39-069-062-5591

Received: 18 December 2019; Accepted: 6 February 2020; Published: 8 February 2020

Abstract: Protected horticulture is a high energy-consuming sector in which the optimization of energy use and cost for heating facilities is strategic in achieving high environmental and economic sustainability of production. The main aim of the project was to evaluate the use of a heat pump for basal heating as an alternative technology to grow crops with reduced canopies, such as basil. During the test, an area of the greenhouse contained two systems of coaxial pipes circulating warm water from a heat pump and a condensing boiler. These pipes were placed above the growing media. At the same time, a separate area of the same greenhouse contained a traditional heating system consisting of an air heater, the solution commonly used to heat greenhouses. Microclimatic conditions and energy consumption were analyzed for the three heating technologies. The energy analysis of the three experimental heating options showed that all of them could ensure suitable thermal conditions for cultivation in the winter period. Overall, the results confirmed the energy saving resulting from the adoption of the heat pump, underlining the importance of this device in terms of the support that the energy-saving goal receives.

Keywords: energy saving; efficiency; greenhouse; controlled environment

---

1. Introduction

Worldwide, protected agriculture (glass and plastic greenhouses, tunnels) covers an area of at least 900,000 hectares. 70% of it (mostly located in Asian countries such as Japan, China and Korea) uses greenhouses made of flexible plastic films. Within the Mediterranean basin, protected agriculture reaches up to 400,000 hectares, concentrated above all in Spain, Italy, Egypt, France, Greece and Turkey [1]. Heating the greenhouse is also a feasible option: it is a well-established practice in central-northern regions and an increasingly widespread one in southern areas. According to Campiotti [1], throughout the Mediterranean basin the energy consumption of greenhouse systems ranges between 5 and 7 kg of oil equivalents/m^2/y (1 kg$_{oe}$ = 11.63 kWh), or 60–80 kWh/m^2/y, while in central and northern Europe it may reach 40–80 kg$_{oe}$/m^2/y (460–930 kWh/m^2/y).

Energy costs to heat a greenhouse represent a significant expense: in cold regions they represent the second-largest cost item after the workforce [2]. According to Campiotti [1], the greenhouse equipment (light-supplementation, dehumidification, heating, cooling and actuators) needed to achieve maximum yields may result in energy costs in the order of 30%–40% of total production costs. As matter of fact, focusing on energy consumption partitioning, Runkle and Both [3] assessed that heating requires approximately 65%–85% of the total energy consumed in a greenhouse, with the remaining portion used for electricity and transportation. Reducing energy consumption for heating has therefore become an important challenge. Many efforts have allowed the optimization of the daily average temperature and humidity as well as that of solar radiation and CO_2 concentration [4–6]. Furthermore, studies have proposed innovative strategies such as structural design, the use of energy-efficient covers, improved heating and ventilation systems, management of indoor micro-climates and use of renewable energy sources based on the location of greenhouses. These techniques are mostly focused on reducing the energy requirement of farming, increasing the efficient use of energy and reducing the costs of the required power. There are a variety of heaters commonly used in greenhouses: air heaters (either wall-mounted or free-standing) to warm the inside volume of greenhouses, water heaters and boilers (gas or electric powered) for basal (or root-zone) heating of crops both in soil or soilless cultivation; furthermore, electric convectors, wood or pellet stoves and heat pumps (HPs) are also used [7–11]. However, focusing only on heating devices technologies may be misleading: the balance between the agronomic needs of plants and the energy-saving potential of each heating technique requires attention as well [12]. Growers, researchers, and manufacturers require that the information on the energy-efficient strategies and their effect on plants refers to the economic feasibility of the existing heating energy-saving technologies for conventional greenhouses. To this extent, Sethi and Sharma [13] and Ahamed et al. [14] reviewed and evaluated passive heating technologies available worldwide for protected cultivation, with the main aims being to increase the heat gain and reduce heat loss from the greenhouse. Researchers examined the energy-saving potential that renewable and sustainable solutions (e.g., photovoltaic modules, solar thermal collectors, hybrid photovoltaic/solar thermal collectors and systems, energy-efficient HPs, innovative ventilation technologies and efficient lighting systems) may have for greenhouse systems [15,16]. However, concerning HPs, their use in agriculture still mainly only refers to fruit drying [17,18]. Concerning greenhouse heating, studies have focused on HPs only in the case of geothermal source HPs, on integrated systems [11,16,19–22] or they have investigated their financial and environmental viability using simulation tools [10]. There are many reasons why HPs have not garnered great interest in protected agriculture applications, mostly due to the traditional design in building experiences. Indeed, conventional heat pumps have the best efficiency with a relatively low temperature of supply water that cannot exceed 45–50 °C, so they are only thought to be efficient in cases of well insulated structures, mainly associated with low-temperature heating systems, such as floor heating.

As renewable power sources (wind, solar, hydro, and geothermal) do not consume fuel, the energy sources they rely on cannot be accounted for in the same manner as for fossil fuel sources. Choosing the right methodology for these power technologies is, therefore, essential to achieve an unbiased estimation of the source-based building energy use and, at the same time, to provide overall energy metrics (e.g., energy productivity) [23].

This study aims to investigate the possibility of using conventional heat pumps as an energy source and to compare their efficiency compared to traditional heating systems. Furthermore, this study aims to verify in an existing ordinary greenhouse i) the applicability of a commercial air to water heat pump and ii) test the energy efficiency of the heat pump without geothermal or photovoltaic solutions. Commonly, moderate temperature heating systems such as HPs engender a limitation on the heating output of the system resulting in a partial coverage of the heating requirements of large volume facilities during cold periods. Following this, the HP technology was used to provide basal heating to a small leafy vegetable (sweet basil) by means of root zone heating.

With these premises, this study presents a comparison between three heating technologies available at retailers (two with a basal heating system: an air-to-water heat pump and a condensing boiler; and the third with an oil-fired air heater for heating the total volume of the greenhouse). This work focuses entirely on the energy consumption of the tested systems with specific reference to energy efficiency meant as the ratio between direct energy consumption per unit of product or per unit of crop cycles, where the energy consumption refers to the primary consumption of fuel and/or electricity [1]. Moreover, insights into the use of the captured energy and the fossil fuel equivalence approach are also presented to discuss the *impact* each heating technology has when varying the energy source.

2. Materials and Methods

2.1. Design of the Experiment

The experimental plan involved the use of three heating technologies: a condensing boiler (CB), a heat pump (HP) and an air-to-air heater (AH). CB and HP provided basal heating to basil seedlings while the AH represented the standard reference heating system (control) for the entire volume of the greenhouse. Each treatment affected two benches and had four replicates (two per growing bench with 70 seedlings per replicate for a whole of 280 seedlings/treatment). All the plants underwent the same agronomic treatments (fertigation and pest control), according to basil requirements. Growing media composition was peat: perlite 50:50 $v \cdot v^{-1}$.

Inside the three differently heated environments, an experiment on a small leafy crop was carried out following a completely randomized design. 840 sweet basil seedlings (*Ocimum basilicum* L.), one of the aromatic plants with greatest consumption worldwide [24], underwent simultaneous hand transplanting in 6 growing benches on 12 February 2018. Each bench contained 5 rows of seedlings with 28 plants per row. February is the month of the year that historically has the coldest temperatures: Figure 1 reports in detail the mean temperature trend of 2018 and the period of the experimental test.

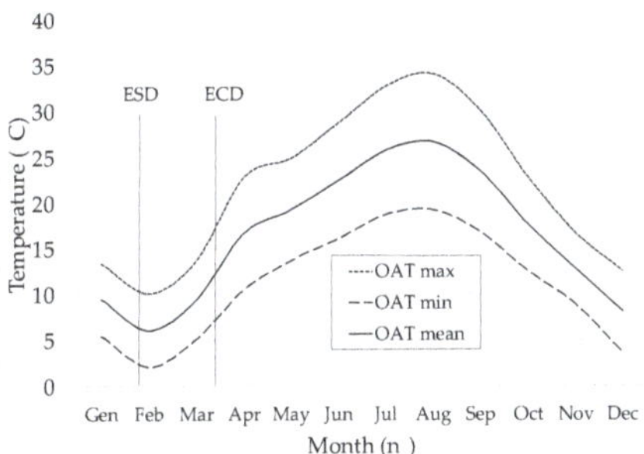

Figure 1. Monthly means of the external temperature at the experimental site and period of testing (OAT: outer external temperature; ESD: Experiment starting date; ECD: Experiment completion date).

During the growing period, the plants underwent two harvests: the first one took place 36 days after transplant (DAT) and the second 13 days after the first (49 DAT), assessing plant biomass production (i.e., yield) in both cuts, measuring shoot fresh weight and shoot dry weight (shoots dried in oven at 75 °C, until constant weight reached). The amounts of direct energy consumption (kcal) was calculated keeping into account the two different heated surfaces (10.2 m^2 for HP and CB; 100 m^2 for AH) and referring to unit of product (fresh $g_{f.w.}$ and dried $g_{d.w.}$). Energy-use efficiency (EUE, $kcal \cdot g^{-1}$) was calculated as described by Equation (1):

$$\text{EUE} = \frac{\text{kcal}}{\text{g}} \tag{1}$$

The EUE was calculated for both the fresh (EUE_f_Yeld) and the dry (EUE_d_Yeld) weight of harvested plant biomass.

2.2. The Greenhouse and the Growing Benches

The trial took place in a greenhouse of the CREA Research Center for Vegetable and Ornamental Crops (Pescia, Italy, 43°53′13″ N, 10°41′18″ E; degree day: 1.877). The greenhouse had a total surface area of 200 m^2, a ridge height of 3.50 m and it is North/East to South/West longitudinally oriented. It consisted of a supporting structure made of galvanized iron, roofing in polycarbonate slabs and walls of polyethylene sheets equipped with a fully automatized opening system. It bordered on open spaces to the North, South and West and with another greenhouse eastward. Inside the greenhouse there were regular prismatic concrete benches (0.70 m width × 7.25 m length × 0.30 m height, with 3 cm thickness polystyrene panels coating on the walls and on the bottom) placed on cement blocks that raised them 0.35 m above the soil level. A vertical polyethene sheet divided the inner space of the greenhouse into two parts, one with the air-heating system (AS) and the other with a basal-heating system (BS), to test the three heating technologies running simultaneously (Figure 2). Figure 3 shows the schematic layout of the experimental greenhouse.

Figure 2. Global layout of the experimental greenhouse (basal side on the left and air side on the right).

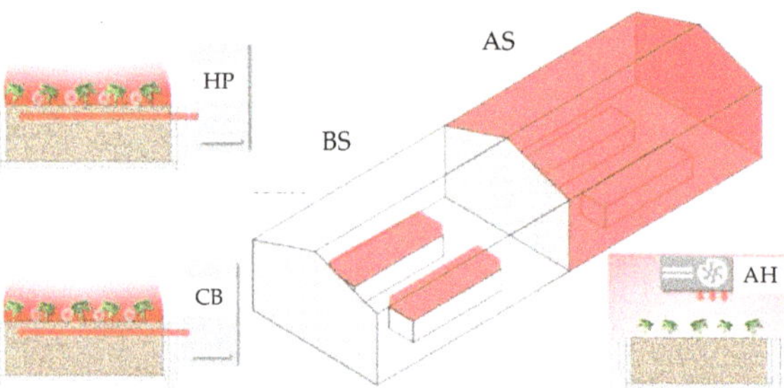

Figure 3. Schematic layout of the experimental test.

Coaxial pipes [25] placed on the surface of the growing media among the plants' rows provided the basal heating (Figure 4). This position was chosen to check if this kind of heating system could also heat the plant canopy as well as the growing media and the root zone.

Figure 4. Detail of the coaxial pipes and basil plant layout.

2.3. The Heating System

The test aimed to compare three heating technologies (Table 1) to point out the extent of both energy-saving and consumption and to assess their suitability to heat a leafy vegetable crop, in an existing greenhouse with standard insulation.

Table 1. The main characteristics of the three heating technologies.

Label	Type	Energy Source	Heating Distribution System
CB	Condenser boiler	LPG	Canopy and root zone
HP	Heat pump	Electricity	Canopy and root zone
AH	Oil fired air heater	Diesel	Air (i.e., entire greenhouse volume, or rather canopy)

The air to air AH represented the reference system of the experiment: it consisted of a floor-mounted fan-cooled diesel generator cabinet Gentili Junior 85 SP(Gentili Generatori S.r.l., Pescia, Italy) (Table 2). A thermostat, placed at about 1 m from the surface of the growing media, switched on/off the AH in accordance with the established temperature setpoint of 15 °C. The CB basal system relied on an LPG (Liquified petroleum gas) fueled condensation boiler Ferroli EcoConcept 15A(Ferroli S.p.A.,Verona, Italy), while the HP basal heating system was an Aermec HSI 140CT (Aermec S.p.A., Verona, Italy) (Table 2). The boiler exhaust fumes were not released into the greenhouse. The thermostat for managing CB and HP systems, set at 15° C, had the temperature probe placed inside the canopy, just above the centre of the hydraulic line.

Table 2. Technical declared data of the three heating systems.

Heat Pump	Airmec HSI 140CT	
Maximum thermal power	kW	12.8
Maximum cooling capacity	kW	10.00
Maximum absorbed electrical power (external unit)	kW	5.2
Coefficient of performance (COP, radiators, external air t = 7 °C, water in/out t = 40/45 °C)		3.6
Condenser boiler	Ferroli EcoConcept 15A	
Maximum thermal power	kW	15.3–3.6
LPG gas flow	Nm^3	1.19–0.28
Maximum absorbed electrical power	W	140
Efficiency (80–60 °C)	%	98.1–97.5
Efficiency (50–30 °C)	%	104.9–106.7
Efficiency 30% P_{max}	%	109.3
Air heater	Gentili Junior 85 SP	
Maximum heating power	kW	118
Declared efficiency	%	91

Figure 5 reports a detailed layout of the experimental greenhouse, while Figure 6 reports the layout of CB and HP systems.

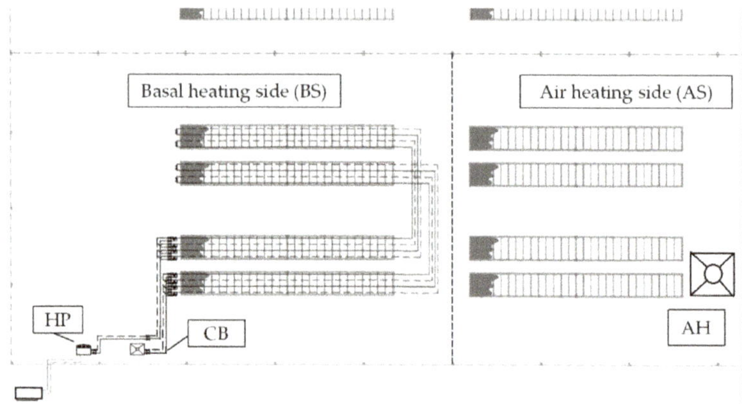

Figure 5. Detailed layout of the experimental trial: in the left part the dashed and full lines represent the pipes used to heat the growth substrate.

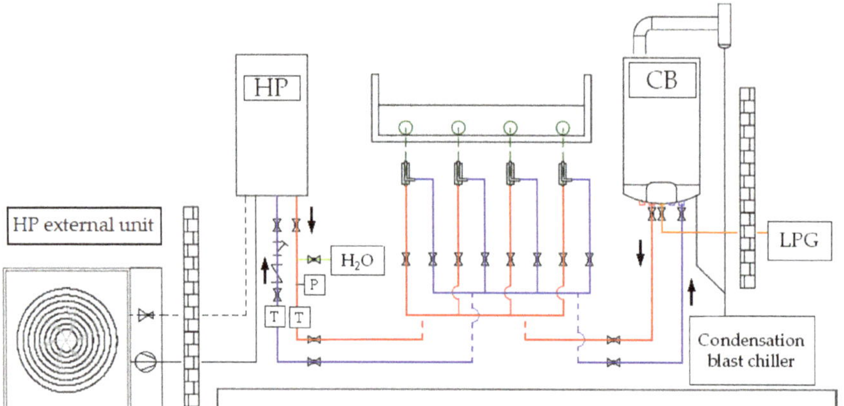

Figure 6. Detail of the scheme adopted at the basal side (heat pump (HP) and condensing boiler (CB) were connected to different benches).

2.4. Instruments

To investigate the effect of the different heating treatments, probes recorded greenhouse internal air, canopy and growing media temperatures, respectively. Canopy temperature monitoring was performed contemporarily in each area of the greenhouse using 24 Testo (Testo SE & Co. KGaA, Titisee-Neustadt, Germany) mod. 175 data loggers: 8 for each heating treatment. They were positioned 10 cm above the growing media. Growing media temperature was recorded by 4 sensors for each heating treatment, positioned 10 cm below the surface of the growing media. Greenhouse temperature and relative humidity was monitored by installing a datalogger Testo mod. 175H1 at 2 m above ground level. The list of the adopted instruments is reported in Table 3.

Table 3. Instruments. BS: basal-heating greenhouse section; OAT: greenhouse outside air temperature; IAT: greenhouse inside air temperature; AS: air-heating greenhouse section; HP: heat pump; CB: condensing boiler; AH: air heater.

Brand/Model	Measured Values	Position/Purpose	n°
Testo 175 T1	Temperature	Canopy	12
Testo 175 T2	Temperature	Canopy	12
Testo 175 T2	Temperature	Growing media BS	8
Testo 175 T2	Temperature	Irrigation water	1
Testo 175 H1	Temp, Humidity	OAT, IAT	2
VP-3	Temp, Humidity	Greenhouse	2
Testo 175 T3	Temperature	Growing media AS	4
QSO-S PAR	Solar radiation	Greenhouse	2
Orno OR-WE-505	Electricity consumption	HP	1
Elkrogas BK-G4 P	LPG consumption	CB	1
Aqua metro VZO 4	Diesel consumption	AH	1

2.5. Methodology to Relate the Energy Efficiency to Fossil/Renewable Energy Sources

As mentioned above, energy from non-combustible sources of renewable power cannot be accounted for in the same manner as it is for fossil fuel sources. Subsequently, the fossil fuel equivalency approach and the captured energy methodology were used to compare the three heating technologies under testing to consider the energy conversion occurring from both fossil fuel and renewable sources [23]. The fossil fuel equivalency approach (FFE) considers the average heat rate of fossil generators (currently 9510 BTU/kWh or about 35% efficiency) and assigns it as the heat rate for non-combustible renewable electricity (RE) generation. This value represents the source energy value of the fossil generation that RE generation displaces. The captured energy (CE) methodology assumes that the energy source is precisely equal to the electricity produced without losses before its transmission and distribution: in this case, the heat conversion rate is 3412 BTU/kWh and corresponds to a conversion efficiency of 100%. It is noted that these are not the complete set of the possible methodological choices for non-combustible source energy accounting. Another method would assume that non-combustible renewable generation consumes no source energy (e.g., 0 BTU/kWh) [23].

2.6. Data Processing

The experimental data underwent statistical processing with Minitab 17 statistical software [26] to calculate the descriptive statistics for each treatment (arithmetic mean ± standard deviation). To evaluate the statistically significant differences between the treatments, data underwent analysis of variance (ANOVA) followed by post-hoc multiple comparisons (Tukey honest significant difference ((HSD) test, $p < 0.05$).

3. Results

3.1. Thermal and Environmental Conditions

The mean recorded OAT during the entire experiment was 8.9 °C, the daily average minimum temperatures was 4.5 °C with an absolute minimum of −6 °C. The average air temperature inside (IAT) of the BS during the experiment was 15.5 °C, with a mean daily minimum temperature of 10.6 °C and an absolute minimum temperature of 4.1 °C. At the same time, the average IAT of the AS was 17.6 °C, with a daily mean minimum of 12.9 °C and an absolute minimum of 9.1 °C. Figure 7 shows, as example, a scatterplot of the air temperature inside (IAT) and outside (OAT) the greenhouse registered during the three coldest days of the experiment. As expected, the minimum temperatures recorded in the AS section were higher than those recorded in the BS zone (CB and HP).

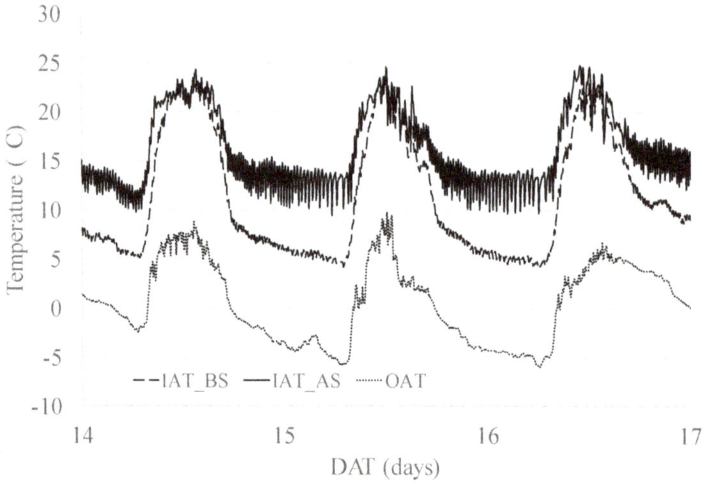

Figure 7. Plot of the greenhouse inner and outer environmental temperatures during the three coldest days of the experimental period.

Significant differences were observed among the different heating technologies (Table 4), temperature density distribution are reported in Figure 8 to highlight the differences that occurred between the two environments.

Table 4. Main measured temperature outside (OAT) and inside (IAT) the greenhouse. Data are expressed as average ± standard deviation. Means that do not share a letter are significantly different.

Environment	Temperature (°C)		
	Mean	**Mean Daily Minimum**	**Absolute Minimum**
OAT	8.9 ± 4.2 c	4.5 ± 3.9 c	−6
BS-IAT	15.5 ± 5.3 b	10.6 ± 2.6 b	4.1
AS-IAT	17.6 ± 4.3 a	12.9 ± 1.6 a	9.1

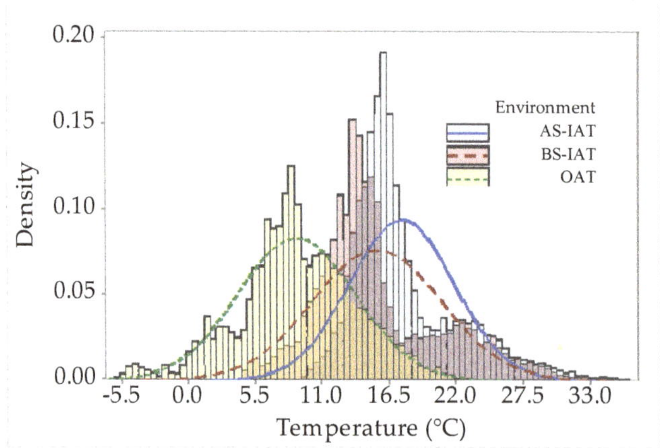

Figure 8. Density of temperature distribution of the OAT (outside air temperature) and the IAT (inside air temperature) in the AS (air-heating section) and BS (basal-heating section) of the greenhouse.

Inside canopy mean and mean daily minimum temperatures were not significantly different for the three heating technologies (Table 5). Inside growing media temperatures in the BS system were significantly higher than in the AS system, both for CB and HP treatments, with mean registered temperature of 20–21 °C, compared to 16 °C registered in AS system (Table 5). Measured temperatures inside canopy and inside growing media are also reported as a boxplot in Figure 9.

Table 5. Main measured temperatures inside canopy and inside growing media for the three heating systems. Data are expressed as average ± standard deviation. Means that do not share a letter are significantly different. CB: condensing boiler; AH: air heater; HP: heat pump.

Heating Technology	Temperature (°C)					
	Canopy			Substrate		
	Mean	Mean daily minimum	Absolute minimum	Mean	Mean daily minimum	Absolute minimum
CB	16.6 ± 2.2 a	12.4 ± 2.8 a	5.1	21.0 ± 1.2 a	20.0 ± 1.3 b	16.8
AH	17.1 ± 2.0 a	13.1 ± 1.5 a	9.3	15.9 ± 2.0 b	14.5 ± 1.8 c	10.7
HP	16.4 ± 2.2 a	12.2 ± 2.7 a	4.9	21.6 ± 1.2 a	20.8 ± 1.2 a	18.3

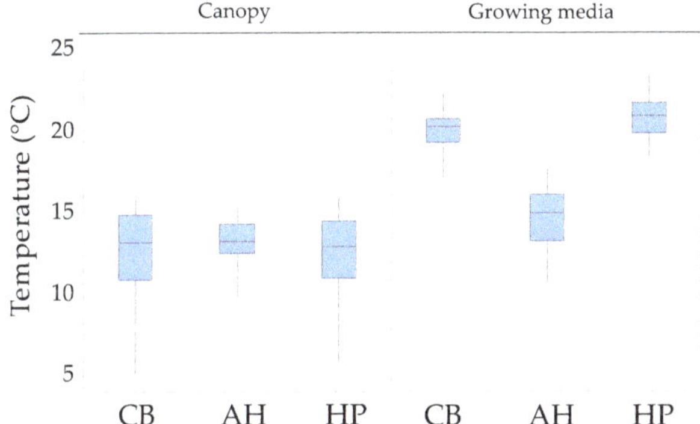

Figure 9. Boxplots of the daily minimum temperatures measured inside canopy and inside the growing media for the different heating systems. CB: condensing boiler; AH: air heater; HP: heat pump.

Relative humidity mean values registered inside and outside the greenhouse during the trial are shown in Table 6.

Table 6. Mean values of humidity inside and outside the greenhouse during the experimental test. BS: basal heating system; AS: air heating system.

	Outside Humidity %	BS Humidity %	AS Humidity %
Min	48.9	35.6	31.6
Mean	72.4	64.0	53.5
Max	90.6	79.6	69.0

3.2. Basil Biomass Production

Figure 10 shows the basil biomass harvested at the first and second cut (36 and 49 DAT). Significant differences of biomass production (g $_{f.w.}$·m^{-2}) were found between the AS and the BS, while no differences were found between the CB and HP system both in each cut and in the total harvested biomass. Moreover, further processing pointed out that the total harvested biomass (g·m^{-2}) positively and significantly correlates with the minimum temperatures recorded in the growing media.

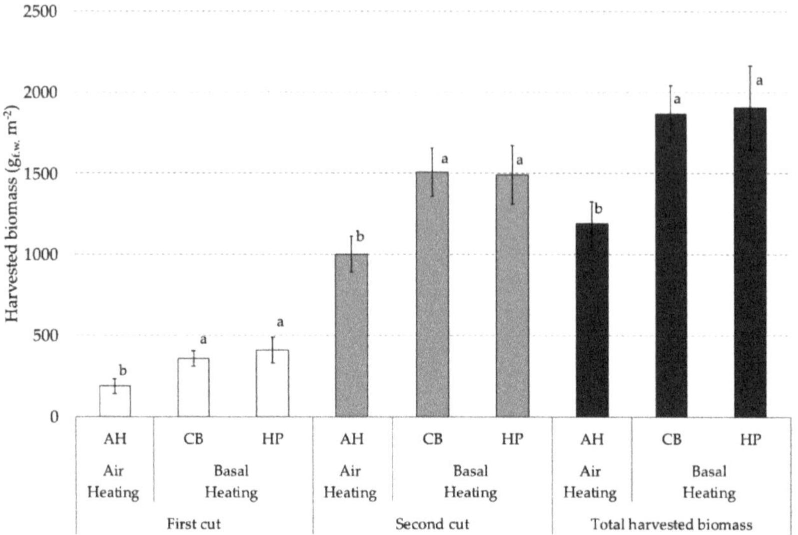

Figure 10. Average values of harvested biomass ($g_{f.w.}$.m^{-2}) of the first, and second cut, and of the total harvested biomass, reported for each tested heating systems: AH (air-heater–air section); CB (condensing boiler–basal section); HP (heat pump–basal section). Data are expressed as average ± standard deviation. Means that do not share a letter are significantly different.

The daily growth rate of the basil seedlings in the first and in the second growing period for each heating system is presented in Figure 11.

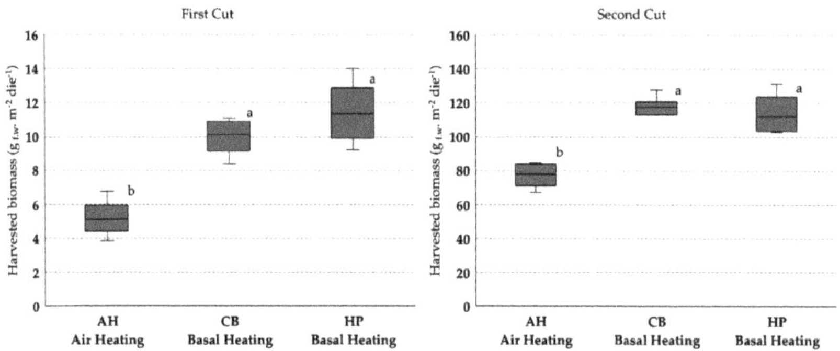

Figure 11. Boxplots representing the daily growth rate of the basil seedlings in the first and in the second growing period for each heating technologies. Means that do not share a letter are significantly different.

3.3. Energy Consumption

Values were measured in liters for the AH, cubic meters for the LPG and kilowatt-hours for the HP. With the aim of making the three different systems uniform, each measuring unit was converted to kWh. Table 7 reports the energy consumption measured during the period of the first and the second growing cycles and the total value. The value is reported as global and specific heat consumption, where the latter is referred to as the heated square meter. The heated square meter resulted the two different heated surfaces (10.2 m^2 of the benches for HP and CB; 100 m^2 of the greenhouse section dedicated to AH)

Table 7. Global and specific heat consumption for the studied heating technologies. CB: condensing boiler; AH: air heater; HP: heat pump.

Harvests	Heating Technology	Global Heat Consumption (kWh)	Specific Heat Consumption (kWh/m^2)
First cut	AH	13.99×10^3	139.99
	CB	2.45×10^3	240.07
	HP	13.3×10^6	130.59
Second cut	AH	3.59×10^3	35.92
	CB	0.61×10^3	60.18
	HP	0.29×10^3	28.82
Total consumption	AH	17.59×10^3	175.91
	CB	3.06×10^3	300.25
	HP	1.63×10^3	159.41

The energy consumption per unit of product is reported in Table 8. Results show that the CB–BS system per gram of product resulted similar compared with the AH–AS. The HP–BS system per gram of product resulted in 44% energy saving than the AH–AS and CB–BS (0.08 instead of 15 kWh/$g_{f.w}$).

Table 8. Kilowatt-hours expended for gram of harvested biomass. CB: condensing boiler; AH: air heater; HP: heat pump; EUE_{f_Yeld}: Energy-use efficiency expressed for fresh weight of harvested plant biomass; EUE_{d_Yeld}: Energy-use efficiency expressed for dry weight of harvested plant biomass; $g_{f.w}$. grams of fresh weight; $g_{d.w.}$: grams of dry weight.

Harvest	Heating Technology	EUE_{f_Yeld} (kWh/ $g_{f.w}$)	EUE_{d_Yeld} (kWh/ $g_{d.w.}$)
First cut	AH	0.74	7.37
	CB	0.66	7.50
	HP	0.32	3.44
Second cut	AH	0.04	0.40
	CB	0.04	0.45
	HP	0.02	0.98
Total consumption	AH	0.15	1.63
	CB	0.15	1.80
	HP	0.08	0.98

Table 9 reports the source-related energy-use efficiencies (kWh/kg $_{f.w.}$) resulting from the correction of the global energies consumption (Table 8) with FFE (9510 BTU/kWh) and CE rates (3412 BTU/kWh). In this way the importance of the energy source and of the adopted approach arises in objective manner: the HP–FFE shows the related fossil fuel required but also the fossil fuel displaced in case of RE.

Table 9. Kilowatt-hours expended for kilogram of fresh harvested biomass of the three tested heating technologies. CB: condensing boiler; AH: air heater; HP: heat pump; CE: captured energy; FFE: fossil fuel equivalency.

Heating Technology	$SREUE_{f\_Yeld}$ (kWh/$kg_{f.w}$)
AH	148.07
CB	154.77
HP–CE	83.68
HP–FFE	233.09

4. Discussion

Reducing the energy costs in protected environments is possible in order to: i) reduce energy needs, ii) use energy more efficiently and, iii) use less expensive energy sources. The balance between the agronomic needs of plants and the energy-saving potential of different techniques should to be considered to achieve the target of energy saving. For this reason, this work was focused on the evaluation of thermal energy saving for leafy vegetables production in wintertime. The experiment consisted in a comparison between an air heater, which heated the whole greenhouse volume; a basal heating system powered by a condensing boiler and a basal heating system powered by an air-to-water heat pump. The temperature trends show that the minimum air temperatures recorded inside the greenhouse are 6.1 (BS) and 8.4 °C (AS) higher than those occurring outside. Such increments are of the same order of magnitude of those (5.3–7.3 °C) reported in a similar experience carried out in a greenhouse equipped with a solar combined air source heat pump system for strawberry production [27]. At both the harvest times, the AS production ($g_{f.w.}$ m^{-2}) was significantly lower than that of the BS, heated both with CB and HP. The biomass produced in the AS section was lower, on average, from 51% (first cut) to 33.3% (second cut) of that harvested in the BS. Biomass productions resulting from HP and CB in the BS do not differ significantly. As a matter of fact, increasing soil/growing media temperature results in heat accumulation at root zone level and this can be associated with an increase in yields [28,29]. The dry matter content was of 8.70 ± 0.30%, on average, and it is in line with other results reported for sweet basil by Walters and Currey [30]. The increase in biomass production of basil between the first and the second cut was significantly higher for the AH system as compared to the BS system. In particular, the HP resulted in the lowest percentage increase of biomass, even if the first cut achieved the highest production. As matter of fact, basal heating seems to allow accelerated plant growth [28]. Further support for these results can be found in the positive significant correlation that fresh basil biomass showed compared to minimum temperatures recorded in the growing media, which in the BS were higher than in the AS treatment. The energy consumption of the condensing boiler and of the air heater resulted similar. The energy consumption of the heat pump resulted in 45% energy saving than the air heater and than the condensing boiler system confirming that significant thermal energy saving is possible.

These results confirm the possibility of using conventional heat pumps in agriculture even as a simple installation in an existing greenhouse and their potential efficiency with respect to traditional heating systems [18]. Other authors confirmed that this efficiency can be higher if the heat pump is associated with geothermal and photovoltaic systems [19]. The fact that the heat pump was able to heat the basil plant during the entire period can be attributed to the reported climatic conditions and to the use of a basal system with a small-canopy leafy vegetable. In other layouts of cultivation or species, it should be considered that underfloor or basal heating systems with a moderate temperature (approximatively 40 °C) engender a limitation on the heating output of the radiant system. Therefore, they may not cover all the heating requirements in cold periods, meaning that they should be coupled with another heating system [31]. It is important to consider that even if heat pumps could be used for greenhouse systems, and were more efficient than other conventional heating sources, this would not automatically be translated into financial and environmental benefits [19,22,32]. In fact, the financial cost depends on the unit cost of the energy of electricity or fuel as LPG, methane or diesel, on taxes, on excise duties and on the fixed rate of the supplier. Moreover, the environmental assessment of the layout of the heat pump plant (conventional, geothermal or with photo-voltaic), using electricity from the electricity grid, does not allow an unequivocal environmental assessment [19].

The FFE and CE rates point out that the higher efficiency of HP shall be also evaluated in light of the used energy source: when fossil fuels are the primary energy source, the $SREUE_{f\_Yeld}$ points out that such technology is not the most efficient. However, in case the primary energy source is a renewable one, the results enhance the fossil fuel saving potential (Table 9). This opens issues on the fact that energy evaluation shall refer also to social, ecological and strategic values [33]; this shifts attention towards the infrastructures required for energy demand, production, capture and conversion. A thorough analysis of these aspects would enable the setup of a concrete energy-efficient machining system, tailored to meet the enterprise's needs [34]. For example, in Italy, during 2017, powerplants had 41.5% of efficiency [35], which corresponds to 8222 BTU/kWh, that would have resulted in a $SREUE_{f\_Yeld}$ for HP–FFE equal to 201.64 kWh/ $kg_{f.w}$. that is 13.5% lower than the value of Table 9.

However, given the 35.1% average contribution of renewable source [0 BTU/kWh] to the National electric power production, relating the efficiency to the sole fossil fuel contribution would have resulted in a global heat rate of 5355 BTU/kWh and, subsequently, in a $SREUE_{f\_Yeld}$ of 130.986 kWh/$kg_{f.w}$.

In any case, it is also interesting in relation to the approach of the European Council for an Energy Efficient Economy, which recently issued a document about the "energy sufficiency" concept, based on a recognition that energy-efficiency policies alone are not enough to turn around the rising demand for environmentally-costly energy services [36]. The simplest definition of "sufficiency" is: "an amount of something that is enough for a particular purpose". When this refers to energy (defined as the ability to do work, or to bring about change) and to energy efficiency (meant as the measure of the ratio between energy outputs and inputs) it is clear that *energy sufficiency* becomes a higher-order idea according to which "sufficient" production facilities or processes are those leading, by definition, to the lowest energy requirements in absolute terms. The present research and the achieved results comply with this recent concept.

5. Conclusions

The energy analysis of the three experimental heating options showed that all of them provide a greenhouse with suitable root zone heating in the winter period. In particular, the results show that the adoption of a conventional heat pump for greenhouse heating leads to remarkable energy savings (–45% energy consumption compared to the condensing boiler and the air heater). On the one hand, these savings underline the key role that HP technology plays in supporting the energetic sustainability of horticultural and floricultural greenhouse farming; on the other, they open issues of the extension of HP applicability in agriculture even without its coupling with geothermal or photovoltaic solutions. The source of the electrical power shall, however, be considered as it affects the overall efficiency of the system.

Author Contributions: Conceptualization, M.F., G.B., M.C., D.M., S.C.; methodology, M.F., C.T., S.C., G.B., M.C., D.M.; formal analysis, M.F., C.T., S.C., M.C., M.B., D.M.; investigation, C.T.; resources, M.F., G.B., C.B.; data curation, M.F., C.T., S.C., M.B., C.B.; writing—original draft preparation, writing, C.T., M.F., S.C., M.C., M.B., D.M., G.B.; supervision, M.F., C.B.; project administration, M.F., G.B., C.B. All authors have read and agreed to the published version of the manuscript.

Funding: This work was supported by the Italian Ministry of Agricultural, Food, Forestry and Tourism Policies, MIPAAFT) under the AGROENER project (D.D. n. 26329, 1 April 2016)—http://agroener.crea.gov.it/.

Conflicts of Interest: The authors declare no conflict of interest. The funders had no role in the design of the study; in the collection, analyses, or interpretation of data; in the writing of the manuscript; or in the decision to publish the results.

Nomenclature

LPG	Liquified petroleum gas
CB	Condensing boiler
HP	Compressor heat pump
COP	Coefficient of performance of heat pump
AH	Oil fired air heater
OAT	Outside air temperature
IAT	Inside air temperature
ESD	Experimental starting date
ECD	Experimental completion date
DAT	Days after transplant
BS	Basal heating section
AS	Air heating section
EUE	Energy-use efficiency
SREUE	Source-related energy-use efficiencies

kg_{oe}	kg of oil equivalents
FFE	Fossil fuel equivalency
CE	Captured energy
RE	Renewable energy

References

1. Campiotti, A.C.; Bibbiani, C.; Dondi, F.; Viola, C. Efficienza energetica e fonti rinnovabili per l'agricoltura protetta. *Ambiente Risorse Salute* **2010**, *126*, 6–12. Available online: https://www.researchgate.net/publication/285776483_Efficienza_energetica_e_fonti_rinnovabili_per_l'agricoltura_protetta (accessed on 2 September 2019).
2. Hemming, S.; Balendonck, J.; Dieleman, J.A.; De Gelder, A.; Kempkes, F.L.K.; Swinkels, G.L.A.M.; De Visser, P.H.B.; De Zwart, H.F. Innovations in greenhouse systems—Energy conservation by system design, sensors and decision support systems. *Acta Hortic.* **2015**, *1170*, 1–15. [CrossRef]
3. Runkle, E.; Both, A. Greenhouse Energy Conservation Strategies. *Michigan State University-Extension Buletin.* February 2012. Available online: https://www.canr.msu.edu/resources/greenhouse_energy_conservation_strategies (accessed on 2 September 2019).
4. Shen, Y.; Wei, R.; Xu, L. Energy Consumption Prediction of a Greenhouse and Optimization of Daily Average Temperature. *Energies* **2018**, *11*, 65. [CrossRef]
5. Dieleman, J.A.; Marcelis, L.F.M.; Elings, A.; Dueck, T.A.; Meinen, E. Energy saving in greenhouses: Optimal use of climate conditions and crop management. *Acta Hortic.* **2006**, *718*, 203–210. [CrossRef]
6. Elings, A.; Kempkes, F.L.K.; Kaarsemaker, R.C.; Ruijs, M.N.A.; Van de Braak, N.J.; Dueck, T.A. The energy balance and energy-saving measures in greenhouse tomato cultivation. *Acta Hortic.* **2005**, *691*, 67–74. [CrossRef]
7. Bailey, B.J. Limiting the relative humidity in insulated greenhouses at night. *Acta Hortic. Energy Prot. Cultiv.* **1983**, *148*, 1–11. [CrossRef]
8. Kozai, T. Thermal performance of an oil engine driven heat pump for greenhouse heating. *J. Agric. Eng. Res.* **1986**, *35*, 25–37. [CrossRef]
9. Garcia, J.L.; De la Plaza, S.; Navas, L.M.; Benavente, R.M.; Luna, L. Evaluation of the feasibility of alternative energy sources for greenhouse heating. *J. Agric. Eng. Res.* **1998**, *69*, 107–114. [CrossRef]
10. Aye, L.; Fuller, R.J.; Canal, A. Evaluation of a heat pump system for greenhouse heating. *Int. J. Therm. Sci.* **2010**, *49*, 202–208. [CrossRef]
11. Esen, M.; Yuksel, T. Experimental evaluation of using various renewable energy sources for heating a greenhouse. *Energy Build.* **2013**, *65*, 340–351. [CrossRef]
12. Paradiso, R.; De Pascale, S. Effects of Plant Siza, Temperature, and Light Intensity on Flowering of Phalaenopsis Hybrids in Mediterranean Greenhouses. *Sci. World J.* **2014**, *2014*, 9. [CrossRef] [PubMed]
13. Sethi, V.P.; Sharma, S.K. Survey and evaluation of heating technologies for worldwide agricultural greenhouse applications. *Sol. Energy* **2008**, *82*, 832–859. [CrossRef]
14. Ahamed, M.S.; Guo, H.; Tanino, K. Energy saving techniques for reducing the heating cost of conventional greenhouses. *Biosyst. Eng.* **2019**, *178*, 9–33. [CrossRef]
15. Cuce, E.; Harjunowibowo, D.; Cuce, P.M. Renewable and sustainable energy saving strategies for greenhouse systems: A comprehensive review. *Renew. Sustain. Energy Rev.* **2016**, *64*, 34–59. [CrossRef]
16. Anifantis, A.S.; Colantoni, A.; Pascuzzi, S. Thermal energy assessment of a small scale photovoltaic, hydrogen and geothermal stand-alone system for greenhouse heating. *Renew. Energy* **2017**, *103*, 115–127. [CrossRef]
17. Teeboonma, U.; Tiansuwan, J.; Soponronnarit, S. Optimization of heat pump fruit dryers. *J. Food Eng.* **2003**, *59*, 369–377. [CrossRef]
18. Goh, L.J.; Othman, M.Y.; Mat, S.; Ruslan, H.; Sopian, K. Review of heat pump systems for drying application. *Renew. Sustain. Energy Rev.* **2011**, *15*, 4788–4796. [CrossRef]
19. Hui, F.; Qichang, Y.; Ji, S. Application of ground-source heat pump and floor heating system to greenhouse heating in winter. *Trans. Chin. Soc. Agric. Eng.* **2008**. Abstract no.12.
20. Russo, G.; Anifantis, A.S.; Verdiani, G.; Mugnozza, G.S. Environmental analysis of geothermal heat pump and LPG greenhouse heating systems. *Biosyst. Eng.* **2014**, *2014*, 11–23. [CrossRef]

21. Anifantis, A.S.; Colantoni, A.; Pascuzzi, S.; Santoro, F. Photovoltaic and Hydrogen Plant Integrated with a Gas Heat Pump for Greenhouse Heating: A Mathematical Study. *Sustainability* **2018**, *10*, 378. [CrossRef]
22. D'Arpa, S.; Colangelo, G.; Starace, G.; Petrosillo, I.; Bruno, D.E.; Uricchio, V.; Zurlini, G. Heating requirements in greenhouse farming in southern Italy: Evaluation of ground-source heat pump utilization compared to traditional heating systems. *Energy Effic.* **2016**, *9*, 1065–1085. [CrossRef]
23. Donohoo-Vallett, P. Accounting Methodology for Source Energy of Non-Combustible Renewable Electricity Generation. U.S. Department of Energy, October 2016; DOE/EE 1488. Available online: https://www.energy.gov/sites/prod/files/2016/10/f33/Source%20Energy%20Report%20-%20Final%20-%2010.21.16.pdf (accessed on 7 January 2020).
24. Makri, O.; Kintzios, S. Ocimum sp. (basil): Botany, cultivation, pharmaceutical properties, and biotechnology. *J. Herbs Spices Med. Plants* **2008**. [CrossRef]
25. Fedrizzi, M.; Cacini, S.; Burchi, G. Root zone heating optimization in ornamental plant production. *GreenSys* **2009**, *893*, 389–395. [CrossRef]
26. Minitab. *Minitab 17 Statistical Software*; Minitab: State college, PA, USA, 2010.
27. Tang, Y.; Ma, X.; Li, M.; Wang, Y. The effect of temperature and light on strawberry production in a solar greenhouse. *Sol. Energy* **2020**, *195*, 318–328. [CrossRef]
28. Flores Velasquez, J.; Ibarra, J.L. Cultivation of peppers using plastic mulch with coloured films and nutrient irrigation. *Plasticulture* **1998**, *1998 116*, 16–26.
29. Elad, Y.; Israeli, L.; Fogel, M.; Rav David, D.; Kenigsbuch, D.; Chalupowicz, D.; Maurer, D.; Lichter, A.; Silverman, D.; Biton, S.; et al. Conditions influencing the development of sweet basil grey mould and cultural measures for disease management. *Crop Prot.* **2014**, *64*, 67–77. [CrossRef]
30. Walters, K.J.; Currey, C.J. Hydroponic Greenhouse Basil Production: Comparing Systems and Cultivars. *HortTechnology* **2015**, *25*, 645–650. [CrossRef]
31. Fabrizio, E. Energy reduction measures in agricultural greenhouses heating: Envelope, systems and solar energy collection. *Energy Build.* **2012**, *53*, 57–63. [CrossRef]
32. Esen, H.; Inalli, M.; Esen, M. Technoeconomic appraisal of a ground source heat pump system for a heating season in eastern Turey. *Energy Convers. Manag.* **2006**, *47*, 1281–1297. [CrossRef]
33. Stern, P.C.; Aronson, E. National Research Council. In *Energy Use: The Human Dimension*; The National Academies Press: Washington, DC, USA, 1984.
34. Peng, T.; Xun, X. Energy-efficient machining systems: A critical review. *Int. J. Adv. Manuf. Technol.* **2014**, *72*, 1389–1406. [CrossRef]
35. Caputo, A. Fattori di emission atmosferica di gas a effetto serra nel settore elettrico nazionale e nei principali Paesi Europei. ISPRA, Rapporti 303/2019; ISBN 978-88-448-0945-4. Available online: http://www.isprambiente.gov.it/files2019/pubblicazioni/rapporti/R_303_19_gas_serra_settore_elettrico.pdf (accessed on 14 January 2020).
36. Darby, S.; Fawcett, T. Energy Sufficiency: An Introduction. *ECEEE and the Authors*. 2018. Available online: https://www.energysufficiency.org/libraryresources/library/items/energy-sufficiency-an-introduction/ (accessed on 2 September 2019).

Article

Wood Chip Drying through the Using of a Mobile Rotary Dryer

Angelo Del Giudice [1], Andrea Acampora [1], Enrico Santangelo [1], Luigi Pari [1], Simone Bergonzoli [1], Ettore Guerriero [2], Francesco Petracchini [2], Marco Torre [2], Valerio Paolini [2] and Francesco Gallucci [1,*]

1 Consiglio per la ricerca in agricoltura e l'analisi dell'economia agraria (CREA)—Centro di ricerca Ingegneria e Trasformazioni agroalimentari (CREA-IT)—Via della Pascolare 16, 00015 Monterotondo (Rome), Italy; angelo.delgiudice@crea.gov.it (A.D.G.); andrea.acampora@crea.gov.it (A.A.); enrico.santangelo@crea.gov.it (E.S.); luigi.pari@crea.gov.it (L.P.); simone.bergonzoli@crea.gov.it (S.B.)

2 National Research Council of Italy, Insitute of Atmospheric Pollution Research, via Salaria km 29,300, 00015 Monterotondo (RM), Italy; ettore.guerriero@iia.cnr.it (E.G.); petracchini@iia.cnr.it (F.P.); marco.torre@iia.cnr.it (M.T.); v.paolini@iia.cnr.it (V.P.)

* Correspondence: francesco.gallucci@crea.gov.it; Tel.: +39-0690675238; Fax: +39-0690625591

Received: 19 March 2019; Accepted: 17 April 2019; Published: 26 April 2019

Abstract: Drying is a critical point for the exploitation of biomass for energy production. High moisture content negatively affects the efficiency of power generation in combustion and gasification systems. Different types of dryers are available however; it is known that rotary dryers have low cost of maintenance and consume 15% and 30% less in terms of specific energy. The study analyzed the drying process of woody residues using a new prototype of mobile rotary dryer cocurrent flow. Woodchip of poplar (*Populus* spp.), black locust (*Robinia pseudoacacia* L.), and grapevine (*Vitis vinifera* L.) pruning were dried in a rotary drier. The drying cycle lasted 8 h for poplar, 6 h for black locust, and 6 h for pruning of grapevine. The initial biomass had a moisture content of around 50% for the poplar and around 30% for grapevine and black locust. The study showed that some characteristics of the biomass (e.g., initial moisture content, particle size distribution, bulk density) influence the technical parameters (i.e., airflow temperature, rate, and speed) of the drying process and, hence, the energy demand. At the end of the drying process, 17% of water was removed for poplar wood chips and 31% for grapevine and black locust wood chips. To achieve this, result the three-biomass required 1.61 (poplar), 0.86 (grapevine), and 1.12 MJ $kg_{dry\ solids}^{-1}$ (black locust), with an efficiency of thermal drying (η) respectively of 37%, 12%, and 27%. In the future, the results obtained suggest an increase in the efficiency of the thermal insulation of the mobile dryer, and the application of the mobile dryer in a small farm, for the recovery of exhaust gases from thermal power plants.

Keywords: rotary dryer; drying process; thermal energy; wood chips

---

1. Introduction

The use of biomass for energy purposes is related to its moisture content, availability, and pre-treatments such as the drying process [1].

The moisture content of the biomass used for energy production is a key parameter for the proper management of the power plant or in the densification process [2–5]. Generally, the wet wood biomass has a moisture content, on a wet basis, higher than 50% [6,7] and the natural drying process hardly lowers the moisture contents under 35% in 3–4 months of storage [8,9]. High moisture content of fuels increases the cost of transport, reduces the combustion efficiency [10], and decreases the potential energy input for steam generation. Consequentely, a reduction in the calorific value of the fuel gas produced in gasification is experienced, with a negative effect on the efficiency of power generation in combustion, gasification systems, and pyrolysis processes [5,11,12]. Concerning human health, higher biomass moisture content causes an increase of CO and VOC emission [7] as well as the formation of carcinogenic compounds from wood combustion [13,14]. The fine particles may be responsible for severe diseases, like invasive pulmonary infections or broncho-pulmonary allergies [15], whereas the larger particles one may have a role in air and soil contamination [16]. Forced hot air drying is a process for the conditioning of biomass (firewood and/or wood chips) which allows increasing the efficiency and flexibility of combustion, transportation, and storage process [17]. It may increase the calorific value, lower the emissions [18] and save fuel [19,20]. The principles of biomass drying can also be applied to increase the time to preserve food [21].

However, the choice of a suitable drying system and drying conditions is critical to achieve the required final moisture content [22,23]. Although the forced drying process is a suitable alternative to natural drying [6], it presents higher production costs. The drying process consumes a significant amount of energy, so it would be very important to implement energy saving strategies to reduce energy consumption during the drying process [24]. This one involves the use of hot and dry air as a drying fluid, fed by a fan with a working temperature which can vary between 20 °C and 100 °C [25]. The process depends on several factors such as the particle size of the biomass [10], the temperature and speed of drying air [26,27] and the temperature inside the container [28]. The drying fluid is characterized by a low relative humidity so that the air-water contact causes the evaporation of free water contained in the pores of the biomass particles, the water bounded in the intercapillary spaces and/or the water adsorbed on the surface of the product [29].

In the case of hot air, the fluid can be introduced inside the system directly through a dedicated thermal system, or by using low-cost or even free heat, co-produced and recovered from cogeneration plants by injecting hot air at 80 °C [25]. Of course, the supply of hot air through a dedicated heat plant or the recovery of thermal waste from cogeneration plants leads to different energetic and economic costs. In this way the share of thermal energy recovered by cogeneration plants, almost always dissipated, can be exploited to dry firewood or wood chips [25].

The most common industrial systems for drying biomass are conveyor dryers, rotary dryers of single or multiple passes, fixed and mobile bed dryers, perforated floor bin dryers, direct and indirect fired rotary dryers, cascade dryers, superheated steam dryers, microwave dryers, fluidized bed dryers, screw conveyor dryers, and flash or pneumatic dryers [5,23,27]. However, it is known that rotary dryers have a low cost of maintenance and consume 15% and 30% less in terms of specific energy that the pneumatic and cascade types, respectively [6]. An exhaustive description of the drying systems is present in Mujumdar [30].

The rotary dryer is the most diffuse system for drying small-sized woody biomass [31,32]. Considering the method of heat transfer, rotary dryers, can be classified as direct, indirect-direct, indirect, and special types [33]. The direct rotary dryers consist of a slightly inclined metal hollow cylinder, rotating around its axis. The internal space is designed to ensure direct contact between the biomass and the drying fluid, usually hot air.

Rotating dryers have the advantage of being less sensitive to particle size and can accept the hottest exhaust gases of any type of dryer. They have lower maintenance costs and greater capacity than any type of dryer. The drying process of the wet biomass in a rotary dryer can be challenging owing to the prolonged time for the uniform drying of the biomass [19], this can also increase the fire hazard inside the dryer [5,18].

The heat transfer between the hot air and the biomass is improved by a series of flights on the inner surface of the cylinder which serve to increase the contact of the two flows (air/solid). During the rotation, the action of the flights lifts and drops the biomass regularly from top to bottom through the flow of hot air. In this way, each portion of the biomass is invested by the flow of hot air [32,33]. Depending on how the hot air is introduced, there are two types of dryer, cocurrent and countercurrent. In this type of system, the solid fluid (wood chips), kept in constant movement by the rotation of the cylinder, is mixed with the drying fluid (hot air) favouring its drying [31]. A quick drying process reduces considerably the drying time [23], compared to conventional storage methods (piles), in which the natural drying process lasts from 6 to 12 months [18]. If the piles of woodchips are covered with a special fabric, the moisture content can reach about 35% [34]. The energy required to dry 1 kg of wood chips in a rotary dryer is 3.1 MJ [35,36], while the heat needed to evaporate 1 kg of water from wet biomass fuel can exceed 2.6 MJ kg^{-1} depending on initial and final moisture content and temperature of drying [7]. However, the energy required may vary depending on the type of biomass and the homogeneity of the material [37].

The CREA-IT of Monterotondo (RM) in collaboration with the CNR-IIA tested a new prototype of mobile rotary dryer, for the exploitation of biomass in the field. This preliminary study analyzed the variation of the thermal requirements and the drying profile as a function of the physical characteristics of the biomass. The objective of the study was to evaluate the applicability of the system for the waste heat recovery resulting from combustion plants.

2. Materials and Methods

2.1. Rotary Dryer Prototype

The prototype was a cocurrent rotary dryer with drum designed for wood chips composed of a metallic rotating cylinder of 5 m length and 0.8 m diameter and a volume of 2.5 m^3. The cylinder was provided with four openings (Figure 1a): two for the loading and unloading of the product and two for the inlet and outlet of the drying fluid (hot air). The dryer was placed on a mobile floor and was equipped with a system of ventilation of the MZ aspirator (Italy) and dust recovery with two bag filters connected in parallel with a 50 cm diameter, a height of 1 m, and a total dust collection efficiency of 98% (Figure 1b).

Figure 1. Rotary dryer prototype (**a**); dust recovery system (**b**).

The wet biomass was loaded into a hopper equipped with a 150 mm diameter screw conveyor of Pelltech (Germany) for the transport inside the cylinder. Here, the advancement and the subsequent unloading of the biomass was favored by the rotation at 5 revolutions per minute and by an angle of the cylinder of 2° to the horizontal axis.

The rotation of the cylinder was regulated trough an electrical board and a gear reducer. The adjustment of the height occurred trough the setting of two support legs powered by 2 electrics motors. The internal metal structure was provided with flights (48) differently shaped (Figure 2) which favored the mixing of the mass. The set of wings was fixed on a supporting structure, removable from the cylindrical body, which rotates with the drum.

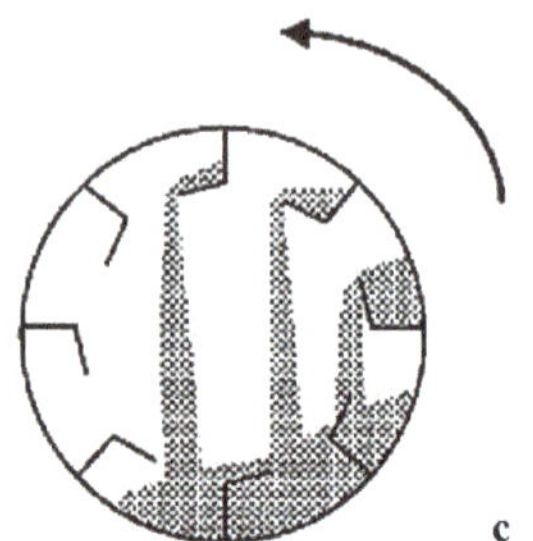

Figure 2. Intertwined flights inside of cylinder (**a**,**b**); schematic representation of the system (**c**).

The prototype was connected to a commercial hot air generator of 80 kW (Company D'Alessandro Termomeccanica mod. GSA) through a 250 mm insulated pipe (Figure 3). The generator was equipped with a centrifugal fan, set at a flow rate of about 1100 m^3 h^{-1} (maximum load 5500 $m^3 \cdot h^{-1}$) for the diffusion inside the cylinder of a drying fluid at 80 °C. The volume of the hopper for the supply of the hot air generator was 0.19 m^3.

Figure 3. Mobile rotary dryer connected to a boiler of the power of 80 kW.

The internal temperature of the cylinder was monitored by two k-type termocouples with a thermal range from −60 to 350 °C and a resolution of 0.2 °C; the first positioned at the entry point of the fresh biomass, the second placed at the exit (Figure 4). Both values were displayed on the electric panel. The cylinder was closed inside an insulating structure to limit thermal dispersion and to reduce the risk of contact with hot surfaces.

2.2. Characterization of the Biomass

The characterization of the biomass was carried out at the LAS-ER-B laboratory of CREA-IT in Monterotondo (RM).

The moisture content was monitored at regular intervals of two hours following the European Standard UNI EN 14774-2:2010 [38]. In all tests, three samples of approximately 300 g each were collected in plastic bags, sealed, labeled and transported to the laboratory where they were dried in an oven with forced ventilation for 24 h at 105 ± 2 °C.

The bulk density of the biomass was determined before and after drying, making five weighing with a normalized cylinder of 0.026 m^3, according to the requirements of the European standard UNI

EN 15103: 2010 [39]. Before the drying tests the analysis of particle size distribution was performed on the incoming biomass (UNI CEN/TS 15149-1:2006 [40]). A representative sample of 12 L was divided into sub-samples of 3 L each, each subsample was analyzed by means of a mechanical sieve shaker of the Fritsch mod. Analysette 18 with normalized sieves according to ISO 3310-2 [41].

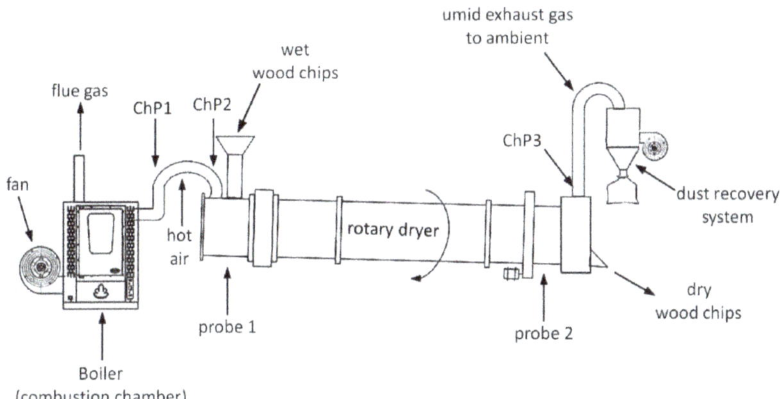

Figure 4. Schematic representation of the system and the points (checkpoint and probe) where the temperature and the airflow were recorded during the test.

The energy content of the biomass referred to a given moisture content (w%), which was calculated with the following Equation (1) [42], referring to the pre and post-drying humidity conditions:

$$H_{u(w)} = \frac{H_{u(wf)} \times (100 - w) - 2.443 \times w}{100} \quad (1)$$

where, $H_{u(wf)}$: calorific value of the wood dry matter in the "anhydrous" 18.5 MJ kg^{-1} [43]. w: moisture content of the wet biomass. 2.443 MJ·kg^{-1}: energy required for preheating and evaporation of the water.

The air permeability of the feedstock was calculated following the method reported by Manzone [43] and Pari [44] by means of Equation (2):

$$A = 19125\ (\text{Mean particle size, mm})^{-0.874} \quad (2)$$

where "A" is a coefficient describing the pressure resistance of the heaped chips to airflow.

As reported by the cited authors, if the mean particle size can be calculated using geometric means, which partly compensate for such skewness towards the lower size classes. The mean particle size is therefore obtained by a weighted average of all particle classes, as represented by their geometric mean calculated with the following Equation (3):

$$\text{Geometric mean} = e\hat{}((\ln b - \ln a)/2) + \ln a) \quad (3)$$

where a and b are respectively the lower and the upper limits of the given size class.

2.3. Experimental Procedure and Characterization of the Airflow

The feedstock used in the study were woodchip of poplar (*Populus* spp.), black locust (*Robinia pseudoacacia* L.), and grapevine (*Vitis vinifera* L.) pruning. The drying cycle had a constant air flow and lasted 8 h for poplar, 6 h for black locust, and 6 h for pruning of grapevine.

Poplar plants were processed with a chipper FARMI mod. CH 260 to obtain two particle sizes (named Poplar 1 and Poplar 2). The black locust was processed with a drum chipper of the Pezzolato mod. PTH 700/660. The wood chip of grapevine pruning was provided by the company ONG s.n.c. of Castel Bolognese (RA) who made the chipping through the prototype mod. PC50 [45]. The quantities of processed biomass were as follows: 250 kg for both the poplar tests, 207 kg for the black locust, and 144 kg for the grapevine. It should be noted that the pruning of vineyard underwent a preliminary drying in open field for ten days before chipping.

During the tests, biomass sampling was carried out every two hours. Along the system (Figure 4), temperature, rate, and speed of the airflow were measured at three control points (checkpoint, ChP). The first (ChP1) was positioned on the duct conveying the hot air from the boiler to the rotary dryer. It was chosen a point far from the boiler five-fold the diameter of the duct, to avoid the influence of turbulence when reading the data. The second (ChP2) was placed in the same duct immediately before the entrance of the drying fluid in the dryer, while the third (ChP3) corresponded to the output of exhausted fluid from the system. The values of temperature, speed, and rate of the airflow were detected at each ChP with a wire thread anemometer (TSI mod. 9535-A). The temperature inside the cylinder was measured by two probes (Probe 1 and 2) and displayed by the control panel. During the drying process, to optimize economic and environmental performance, the dryer generator was fed with the same dried biomass (Poplar) as was obtained from the dryer [29].

2.4. Energy Analysis

Energy analysis was carried out for assessing the efficiency of the process. To this aim, the thermal energy (Q) needed for drying was calculated applying Equation (4) of Zhou [46]:

$$Q = M \times C_u \times (\Delta T) \quad (4)$$

where, Q: thermal energy (kJ·h^{-1}). M: drying air flow (kg·h^{-1}). C_u: specific heat of air (J·kg^{-1}·K). ΔT: temperature difference air at the dryer inlet and outlet.

In Equation (4), M is expressed as mass flow, that is, the volumetric flow rate multiplied by the air density (1.03 kg/m^3) referred to the working temperature. The following formula is used to determine the amount of heat needed to evaporate moisture from moist woody biomass. The efficiency of thermal drying (η) was calculated as the ratio between the heat used to evaporate the moisture from woody biomass (Q) and the primary energy input (E_{fuel}) in the system, by applying the Equation (5) of Meza [6] and Tippayawong [47]:

$$\eta = \frac{Q}{E_{fuel}} \quad (5)$$

2.5. Statistical Analysis

The data of particle size distribution, bulk density of the biomass before and after drying as well as the thermodynamic parameters were analyzed with the software PAST. After checking the data for normality, the data were subjected to the ANOVA, and the differences tested according to Tukey's HSD test.

The effect of the specie at different ChP on the drying process was analyzed using the 50-50 MANOVA [48]. The method is a modified variant of classical MANOVA that integrates the Principal Component Analysis (PCA) in its algorithm. The software was applied for the elaboration of PCA. In a PCA, a set of uncorrelated variables (principal components, PCs) is obtained as a linear combination of the original interrelated variables, in such a way that the first PCs explain the largest fraction of the original data variability.

3. Results and Discussion

3.1. Biomass Characterization

Except for grapevine pruning, the amount of dried biomass was above 200 kg (Table 1). The reduced amount used for the grapevine pruning (143.7 kg) was imposed by the physical characteristics of the biomass, because a higher amount would have increased the risk of clogging at the flight system during the movement of the biomass inside the cylinder. During the drying process the deflectors cyclically moved the solid fluid by lifting and dropping it through the drying fluid. In this way the hot air flow blended directly with the wood chips inside the cylinder, and then released its heat. The heat was mostly conveyed during the drop of the product. However, the system presented some losses caused by the contact of the biomass with the cylinder walls, when the heat was transferred by conduction and irradiation [29].

Table 1. Main characteristics of the wood chips (mean ± SD) before drying.

Biomass	Amount (kg)	Moisture Content (%)	Calorific Value (MJ·kg^{-1})	Bulk Density (kg/m^3)
Poplar 1	250.0	54.4 ± 1.4	7.1 ± 0.3	280.6 ± 8.4
Poplar 2	250.0	52.5 ± 0.6	7.5 ± 0.1	285.2 ± 3.2
Grapevine	143.7	33.3 ± 1.5	11.5 ± 0.3	199.3 ± 8.2
Black locust	207.0	31.2 ± 0.3	12.0 ± 0.1	358.5 ± 10.5

The biomass used in the study showed two levels of moisture content (Table 1): above 50% for poplar and above 30% for grapevine and black locust. This allowed testing of the rotary drier in presence of biomass requiring highly different drying power. The storage period of about 10 days lowered the moisture content of the vineyard chips to 33.3%, while the humidity of the black locust (31.2%) must be considered normal considering the physiological characteristics of the species. In energy plantations, two-year old plants of black locust can show a moisture content close to 45% [49,50] while for older plants the water content ranges between 39% [49] and 32% [3].

It should be noted as the figures confirmed the inverse relationship between moisture content and net calorific value reported by Hellrigl [51]: as the moisture content was much higher the net calorific value was lower. The bulk density of the comminuted biomass decreased with the order black locust > poplar 2 > poplar 1 > grapevine and appeared influenced by the characteristic of the wood morphology rather than the moisture content. Bulk density is a parameter extremely important in handling and storage of biomass because it directly influences the transport costs, the storable amount, the storage conditions, and the final quality of the fuel [52].

The slight difference between the two poplar types probably also reflected the different particle size distribution (Figure 5A). Almost 80% of poplar biomass was concentrated in the classes 3.15–8 mm and 8–16 mm. For these fractions, the difference reached a statistical significance. The fraction 3.15–8 mm accounted for in 37.5% Poplar 2 and 30.1% in Poplar 1, while the fraction 8–16 mm showed an opposite behaviour: higher in Poplar 1 (48.7%) than in Poplar 2 (41.16%).

Such differentiation led to a significant higher mean particle size for Poplar 1 then Poplar 2 and even more to a lower pressure resistance of poplar 1 to air (Table 2).

Table 2. Mean particle size and resistance to air flow (mean ± SD) before drying. For each column, different letters indicate a significant difference at the level of $P \leq 0.01$ after Tuckey's HSD test.

Biomass	Mean Particle Size (mm)	A
Poplar 1	64.0 ± 8.7 B	511.1 ± 68.9 B
Poplar 2	39.2 ± 6.0 C	786.0 ± 100.2 A
Grapevine	145.0 ± 45.4 A	263.9 ± 78.8 C
Black locust	68.3 ± 7.8 B	480.4 ± 46.6 B

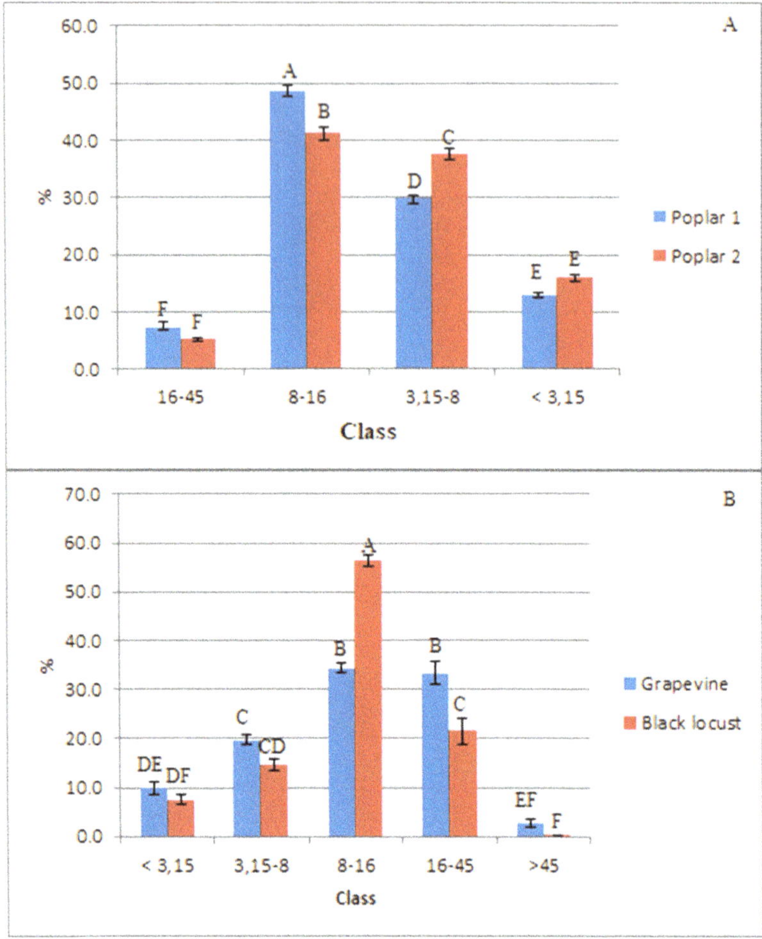

Figure 5. Particle size distribution (mean ± SE) of the wood chips: (**A**) Poplar 1 and Poplar 2; (**B**) grapevine and black locust. Within each size class, different letters indicate a significant difference at the level of $P \leq 0.01$ after Tuckey's HSD test. Before the ANOVA analysis the data were transformed as square root of the arcsine.

Unlike poplar, the distributions of the wood chips of grapevine and black locust were shifted towards the longer fractions (Figure 5B), and, both had the highest percentage within the classes 8–16 mm and 16–45 mm. To one side, this has led to the highest mean particle size for grapevine, coupled to the lowest pressure resistance (Table 2). On the other side, the biomass of black locust showed the same characteristics of Poplar 1. The increase in particle size leads to a reduction of the pressure resistance to air, given by the A value [44]. In general, low values mean a good circulation of air and in this specific case, grapevine pruning appeared to be feedstock more prone to facilitating the air movement inside the biomass. In this way, it must be recognized how important it is to achieve the right particle size distribution for woody biomass, because, beside the storage behavior and the handling properties [52], the chip size largely influences the drying speed. Therefore, the comminution phase upwards from the drying process may address to some extent the final result.

3.2. Drying Process

The type of biomass and the characteristics of the drying fluid influenced each other leading to different conditions inside the cylinder and, in turn, to a different pattern of drying (Figure 6A). In eight hours, the moisture content of the Poplar 1 decreased from 54.4% to 43.3%, while the Poplar 2 passed from 52.5% to 34.1%, with a statistically significant difference starting from the fourth hour. The temperature curves inside the cylinder where rather similar, with higher values for Poplar 2 than Poplar 1.

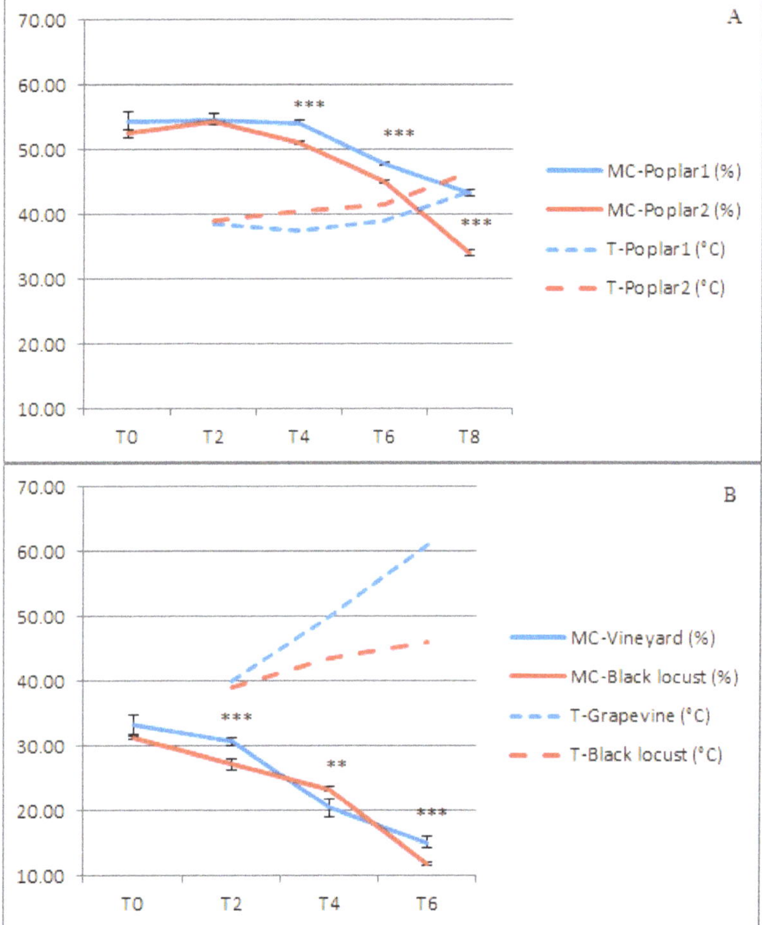

Figure 6. Mean (±SD) of the moisture content (MC, solid line) and the temperature (T, dotted line) inside the cylinder at each time-point of the drying cycle: (**A**) Poplar 1 and Poplar 2; (**B**) grapevine and black locust. For the moisture content, significant differences between treatment within each time were determined by Student's *t*-test. Where reported, ** $P < 0.01$ and *** $P < 0.001$ probability level.

On the other side, after a drying cycle of 6 h the moisture content of the biomass dropped to 15.3% for the grapevine and 11.8% for the black locust (Figure 6B). The two dehydration curves showed a similar trend although in all point of data recording the difference between black locust and the vine pruning residues value was significant. However, these values were obtained with a trend of the temperature inside the cylinder that was completely different with respect to the poplar and between the grapevine and the black locust too. During drying of the grapevine, the temperature inside the cylinder rose from 40 °C to 61 °C while for the black locust it increased from 39 °C to 45 °C.

These results confirm the observations of Kocsis [26] about the influence of the temperature of the drum on the drying rates of biomass. With increasing temperature, the rate and the time required to lower the moisture content decline. The heat provided at the beginning increases the temperature of particles. In our case, this increment was affected by initial moisture content (higher in poplar than in black locust and grapevine) and a higher amount of heat was spent to warm up the particles of poplar. In the following phase, the drying rate of particles was affected only by drying conditions [28] involved in transferring the water layered on the surface of particles to gas flow. In the case of vineyard and black locust a role in increasing the level of the temperature inside the cylinder was played by the different air permeability of the biomass. In fact, as described previously, the higher mean particle size of the grapevine with respect to the black locust lowered the pressure resistance to air leading to a more pronounced increase of the temperature into the drier. Although, with the increase of temperature inside the cylinder, the time required to achieve the same moisture content may decline [28] in the present study the higher temperature during the grapevine drying did not bring an improvement of the drying rate of the grapevine (see below). Other characteristics of the woody particles like the thickness and weight [10,28] may have affected such a result.

3.3. Characteristics of the Airflow

The data shown previously, gave some clues about the influence of the feedstock on the parameters involved in the drying process. The results of the Anova showed as the rate and the speed of the airflow were significantly affected by the species (Table 3), while all the parameters (temperature, rate, and speed) showed significant differences at the ChP. The interaction between species and ChP resulted significant only for the temperature and the speed of the airflow.

Table 3. Main results of the Anova on the characteristics of the airflow.

Source of Variation	Temperature		Rate		Speed	
	F	p	F	p	F	p
Species	1.72	0.184	39.86	0.000	48.66	0.000
ChP	147.30	0.000	145.60	0.000	8.81	0.001
Species X ChP	4.33	0.003	0.68	0.666	4.34	0.003

On average, the airflow temperature, rate, and speed were higher in poplar than in grapevine and black locust. The reader must be aware about the differences of the biomass in terms of initial moisture content (higher in poplar) as well as their particle size distribution and mean particle size (higher for grapevine). As a general behavior, there was an abatement of all the variables going through ChP1 and ChP2 (before the entry) to ChP3 (at the exit), showing a clear interaction between the energy provided during the drying process and the resident biomass.

However, the extent of such a decrease was different (Table 4). The reduction of the airflow rate from ChP1 to ChP3 was comparable between poplar (37%), but different from grapevine (59%) and black locust (51%). Similarly, temperature and speed of the airflow showed a defined trend. The difference in airflow rate and temperature registered at the ChP3 compared with the values at entry (ChP1 and ChP2) reflected the loss of energy employed for drying the biomass. The poplar drying required a remarkable expenditure of temperature that decreased from 74–77 °C to 29–31 °C, with about 60% reduction. The pruning registered a diminution of the temperature at ChP3 limited to 48.3% for the black locust and 30.5% for the grapevine. In fact, for the latter, the difference among the ChPs that resulted was not significant. This behavior upheld the previous observation about the higher heat expenditure required to increase the temperature of the wetter particles of poplar.

Table 4. Features of the airflow monitored during the drying test (mean ± SD). For each variable, different letters indicate a significant difference at the level of $P \leq 0.01$ after Tuckey's HSD test.

Variable	ChP	Poplar 1	Poplar 2	Grapevine	Black Locust
Temperature (°C)	1	73.75 ± 4.22 A	76.85 ± 1.14 A	60.40 ± 9.69 AC	64.30 ± 11.97 AB
	2	71.23 ± 4.23 A	72.15 ± 2.85 A	60.27 ± 10.00 AC	67.47 ± 48.79 AB
	3	28.58 ± 1.15 D	31.10 ± 0.84 D	41.97 ± 10.31 BD	34.87 ± 2.99 CD
Airflow rate ($m^3 \cdot h^{-1}$)	1	1035.99 ± 47.64	1004.40 ± 70.59	809.52 ± 155.50	811.80 ± 49.98
	2	948.24 ± 16.13	978.03 ± 47.06	734.88 ± 139.58	775.80 ± 26.82
	3	659.79 ± 49.67	625.41 ± 36.56	332.88 ± 54.05	396.31 ± 6.64
Speed ($m \cdot s^{-1}$)	1	5.79 ± 0.33 A	5.70 ± 0.37 A	4.63 ± 0.91 AC	4.59 ± 0.28 AC
	2	5.28 ± 0.05 AB	5.49 ± 0.26 A	4.08 ± 0.73 AC	4.52 ± 0.11 AC
	3	5.76 ± 0.41 A	5.53 ± 0.36 A	2.95 ± 0.46 C	3.48 ± 0.02 C

A similar, but opposite trend was observed for the airflow speed. When comparing the speed at ChP3, the difference was significant between the poplar and the grapevine. However, it remained almost constant in each ChP for poplar, while sharply declined at ChP3 for black locust and grapevine, although also in this case, the difference among ChP was never significant.

A clearer glance of the outcomes can be provided by the analysis of the corresponding PCA (Figure 7). PCA can graphically show differences and similarities between the elements, by projecting them in a 2-dimensional plan defined by the two main components: the more the objects are similar, the closer they are. Visually, the effect of the species was expressed by the divergent gap among poplar from one side and grapevine and black locust on the other. The common feature was a sort a gradient were the three variables (temperature, rate and speed of the airflow) tended to decrease from ChP1 to ChP3. The airflow rate was strictly associated with the principal component which explained 74% of the variability, while the airflow temperature and speed showed a greater variance associated with the second component. The analysis showed four distinct groups. In the first two, correspondent to the ChP1 and ChP2, the group of poplar (both particle size distribution) was clearly distinguishable from the group of grapevine and black locust. Both groups appeared more sensible to the effect of the airflow rate, rather than the temperature or the airflow speed. The third and the fourth groups were referred to as ChP3 and separated the poplar from the other species. For poplar, the airflow speed had a more pronounced effect, while for the wood chips of grapevine and black locust, the airflow rate remained probably the most important parameter.

Overall, the data showed as drying required different patterns of heat transfer depending. Each patten was influenced by the characteristic of the biomass which driven the energy demand during the essiccation process as well as the time needed to dry. To this aim, it should be noted that the initial moisture contents of the biomass was greatly different (Table 1), those of Poplar 1 and Poplar 2 being higher than 50% and those of grapevine and black locust slightly above 30%. Moreover, even if the moisture was comparable, grapevine and black locust differed for the bulk density and the particle size distribution. High values of moisture content appeared to influence more the heat content of the airflow rather than the rate or its speed. When using drier biomass, the flow of the air led to a higher loss of both the rate and the speed of the airflow. Viewing the process as a chain (recovery of waste energy at small scale for drying residual biomass), the best results can be obtained through the optimization of each step involved. This means that a trade-off must be sought between the improvement of the residue's properties which influence the drying efficiency (moisture content, bulk density, particle size distribution) and the variables affecting the drying rate which are determined by the characteristics of the device (in this case, temperature and rate of the drying fluid, thermal insulation of the dryer, systems of recirculation of the drying air).

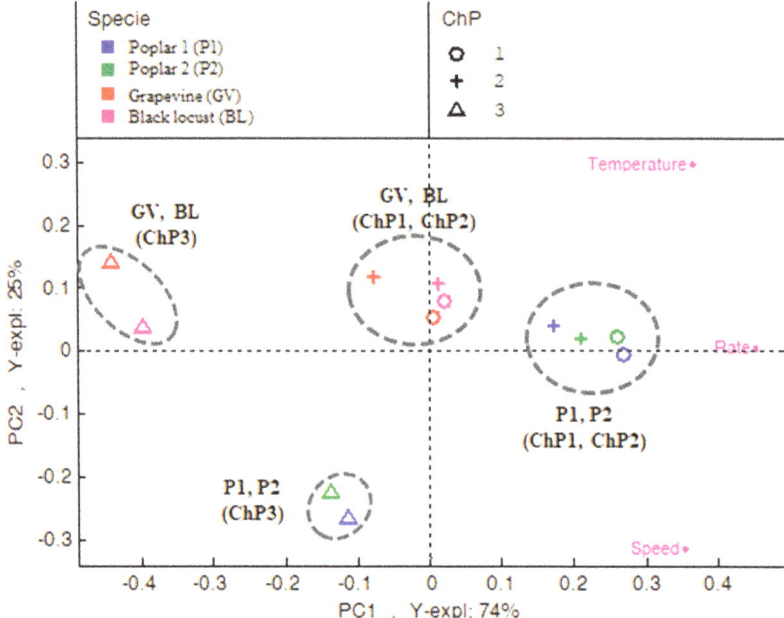

Figure 7. Principal component analysis of the thermodynamic variables monitored during the drying of the wood chips of poplar, grapevine and black locust.

3.4. Energy Balance

The drying made it possible to remove 17% (41.7 kg) of water from the poplar wood chips, while for the grapevine and black locust wood chips drying removed the 31% of water, corresponding to 44.9 and 64.8 kg respectively. (Table 5). Considering the drying cycles of 8 h for the poplar and 6 h for the grapevine and the black locust, the hourly drying performances of the system are equal to 5.2 kg·h^{-1} for the poplar, 7.5 kg·h^{-1} for the vine and 10.8 kg·h^{-1} for black locust. Furthermore, dried biomass increased its calorific value compared to the starting content, with an increase of 51.2% (Poplar 2), 33.1% (Grapevine) and 43.0% (Black locust).

Table 5. Energy parameters (average data for each drying cycle).

Parameter	Poplar 2	Grapevine	Black Locust
Biomass fuel used (kg)	80	60	52
Dried biomass (kg)	208.3	98.8	142.2
Quantity of H_2O evaporated (kg)	41.7	44.9	64.8
Heating value * (MJ·Kg^{-1})			
- wet basis	7.50	11.53	11.96
- dry basis	11.34	15.35	17.10

* Energy content related to the moisture content of the biomass after the drying process.

The energy consumption of conventional dryers is represented by thermal energy and electricity. The thermal energy represents 95% of total energy consumption, the amount of electricity useful for the handling of the cylinder, of the fan and the supply auger, is instead considered equal to 5% of the heat used to dry one kg of wood chips [6].

For the energy balance, we considered only thermal energy (Table 6). For each feedstock, starting from the primary energy input, we calculated the thermal energy (Q) used during drying and the losses. The energy losses during the drying process can be divided into the following parts: transmission losses through the dryer; leakages that mainly arise when the kiln has been opened during the biomass loading [53]. The data indicate that the heat for drying amounted at 1.61 MJ·$kg_{dry\ solids}{}^{-1}$ for the poplar 2, 0.86 MJ·$kg_{dry\ solids}{}^{-1}$ for the grapevine and 1.12 MJ·$kg_{dry\ solids}{}^{-1}$ for black locust. As for the efficiency of thermal drying, the results obtained in this study indicate a drying efficiency (η) of 37% for poplar 2, 12% for grapevine and 27% for black locust.

Table 6. Thermal energy balance (average data for each drying cycle).

Parameter	Poplar 2	Grapevine	Black Locust
Item	Thermal Energy Utilization		
Energy input (MJ·$kg_{dry\ solid}{}^{-1}$)	4.35	6.88	4.14
Heat for drying Q (MJ·$kg_{dry\ solid}{}^{-1}$)	1.61	0.86	1.12
Losses (MJ·$kg_{dry\ solid}{}^{-1}$)	2.74	6.02	3.02
Drying efficiency η (%)	37	12	27

Considering the consumption of electrical energy as 5% of the overall energy, based on the experimental results of this study, the energy input for drying 1 kg of poplar wood chips, grapevine and black locust increases respectively to 4.57, 7.22 and 4.35 MJ·$kg_{dry\ solid}{}^{-1}$.

The loss of energy (Table 6) was 2.74 MJ·$kg_{dry\ solid}{}^{-1}$ (63%), 6.02 MJ·$kg_{dry\ solid}{}^{-1}$ (88%), and 3.02 MJ·$kg_{dry\ solid}{}^{-1}$ (73%) for Poplar 2, Grapevine, and Black locust, respectively. On average, the energy losses of the system were about 75%, a value agreeing with that reported by Johansson and Westerlund [54], which indicate average energy losses of 78%.

The drying process is an energy intensive process and can easily account for up to 15% of industrial energy utilisation [55]. Consequently, in many industrial drying processes, a large fraction of energy is wasted [56]. Drying the biomass applying a low temperature process, by means of secondary heat flows prior to combustion/gasification, is a very reasonable way of increasing the efficiency of heat and power generation. A simple and handy apparatus like the rotary drier tested in the present study can result particularly useful for the recovery of waste energy at small scale. Further research on increasing the energy efficiency of this drier, but also other type of dryers should be directed to specific conditions that provide for the recovery of thermal waste from thermal power plants. The heat request can be provided with greater precision if we consider variables such as the fibrous structure of the biomass, the geometry of the biomass, the part of the plant considered, the free water, and the drying conditions.

From a practical point of view, the use of the rotary dryer for the exploitation of flue gas can be feasible paying attention to some issues. One is linked to the specific rotary dryer analized. Its thermal efficiency requires the improvement of the thermal insulation to increase further the performance. A second issue concern the assessing of the environmental aspects of the drying process for minimizing the emission of volatile organic compounds (VOC) in the air and those condensed into the affluents of liquid waste from the drying process. For instance burning into the boiler very dry biomass (e.g., $<10\%_{w,wb.}$) the CO and the total particulate emissions may increase [57]. The last aspect requires the control of processing parameters with reference to the temperature of the airflow. Values higher than 100 °C are preferable for preventing the condensation of acid and resins.

4. Conclusions

The characteristics of the biomass have shown to influence the technical parameters of the drying process. The moisture content of the biomass as well as the particle size distribution and the bulk density determined a difference in the intensity of airflow temperature, rate, and speed, and this in turn affected the energy demand of the rotary drier. In the present study, the drying process allowed a reduction of the moisture content of 35%, 53%, and 63% respectively for poplar, grapevine, and black locust, with a corresponding increase in the energy content of the biomass of the 52.1%, 33.1%, and 43.0%. On the other hand, at the same operating thermodynamic conditions, the data indicate a thermal efficiency for the grapevine of 12% compared to 37% of poplar and 27% of black locust.

Based on the results, in our opinion the rotary drier presented and assessed in the present study may be viewed as an interesting device for the small farms equipped with energy plants (biogas, gasifiers, and cogeneration). The main strengths of the prototype are the the simplicity of the design, the small size, and its easy handling and transportability. In agricultural contexts where the environmental awareness favours the adoption of energy approaches of self-consumption, the prototype may provide the opportunity to dry residual biomass at low cost through the recovery of waste heat from the energy plant. This choice may also entitle to access at incentive rates for the recovery of residual heat. Being a prototype, the drier is susceptible of further improvements increasing its efficience: these should concern the recirculation of the drying air, the thermal insulation of the dryer, and the increase in the temperature of the drying fluid.

Author Contributions: Conceptualization, A.D.G., A.A., E.S., L.P. and F.G.; Data curation, A.D.G. and E.S.; Formal analysis, E.S.; Funding acquisition, L.P. and F.G.; Investigation, A.D.G., A.A. and E.S.; Resources, F.P., V.P. and F.G.; Supervision, L.P. and F.G.; Writing—original draft, A.D.G. and E.S.; Writing—review & editing, S.B., E.G., F.P., M.T. and V.P.

Acknowledgments: This research was funded by the Ministry of Agriculture, Food, Forestry and Tourism (MiPAAFT) as part of grant D.M. N° 4056 of 24/07/2008 (FAESI project), D.M. N° 2420 of 20/02/2008 (BTT project) and D.D. N° 26329 of 01/04/2016 (AGROENER project).

Conflicts of Interest: The authors declare no conflict of interest.

References

1. Colantoni, A.; Longo, L.; Gallucci, F.; Monarca, D. Pyro-Gasification of hazelnut using a downdraft gasifier for concurrent production of syngas and biochar. *Contemp. Eng. Sci.* **2016**, *9*, 1339–1348. [CrossRef]
2. Nystrom, J.; Dahlquist, E. Methods for determination of moisture continent in wood chip for power plant—A review. *Fuel* **2003**, *83*, 773–779. [CrossRef]
3. Maltoni, A.; Mariotti, B.; Tani, A. *La Robinia in Toscana. La Gestione dei Popolamenti, L'impiego in Impianti Specializzati, il Controllo della Diffusione, Regione Toscana—DEISTAF: Dipartimento di Economia, Ingegneria, Scienze e Tecnologie Agrarie e Forestali Università di Firenze*; Regione Toscana: Firenze, Italy, 2012.
4. Wimmerstedt, R. Recent advances in biofuel drying. *Chem. Eng. Prog.* **1999**, *38*, 441–447. [CrossRef]
5. Kaplan, O.; Celik, C. An experimental research on woodchip drying using a screw conveyor dryer. *Fuel* **2018**, *215*, 468–473. [CrossRef]
6. Meza, J.; Gil, A.; Cortés, C.; González, A. Drying cost of wood biomass in a semi-industrial exsperimental rotary dryer. In Proceedings of the 16th European Conference & Exhibition on Biomass for Energy, Biomass Resources, Valencia, Spain, 2–6 June 2008.
7. Svoboda, K.; Martinec, J.; Pohorely, M.; Baxter, D. Integration of biomass drying with combustion/gasification technologies and minimalization of organic compounds. *Chem. Pap.* **2009**, *63*, 15–25. [CrossRef]
8. Spinelli, R.; Magagnotti, N. Il cippato forestale. In *Foresta, Legno, Energia. Linee guida per lo sviluppo di un modello di utilizzo del cippato forestale a fini energetici*; Press Service Firenze: Florence, Italy, 2007; pp. 69–103.
9. Spinelli, R. Cippato: La raccolta della materia prima. *Mondo Macchina* **2008**, *2*, 56–59.
10. Cuevas, M.; Martínez-Cartas, M.L.; Perez-Villarejo, L.; Hernandez, L.; García-Martín, J.F.; Sanchez, S. Drying kinetics and effective water diffusivities in olive stone and olive-tree pruning. *Renew. Energy* **2019**, *132*, 911–920. [CrossRef]
11. McIlveen-Wright, D.R.; Williams, B.C.; McMullan, J.T. A re-appraisal of wood-fired combustion. *Bioresour. Technol.* **2001**, *76*, 183–190. [CrossRef]
12. Schuster, G.; Löffler, G.; Weigl, K.; Hofbauer, H. Biomass steam gasification—An extensive parametric modelling study. *Bioresour. Technol.* **2001**, *77*, 71–79. [CrossRef]
13. Belis, C.A.; Cancelinha, J.; Duanea, M.; Forcina, V.; Pedronia, V.; Passarella, R.; Tanet, G.; Douglas, K.; Piazzalunga, A.; Bolzacchini, E.; et al. Sources for PM air pollution in the Po Plain, Italy: I. Critical comparison of methods for estimating biomass burning contributions to benzo(a)pyrene. *Atmos. Environ.* **2011**, *45*, 7266–7275. [CrossRef]

14. Petracchini, F.; Romagnoli, P.; Paciucci, L.; Vichi, F.; Imperiali, A.; Paolini, V.; Liotta, F.; Cecinato, A. Influence of transport from urban sources and local biomass combustion on the air quality of a mountain area. *Environ. Sci. Pollut. Res.* **2017**, *24*, 4741–4754. [CrossRef] [PubMed]
15. Barontini, M.; Crognale, S.; Scarfone, A.; Gallo, P.; Gallucci, F.; Petruccioli, M.; Pesciaroli, L.; Pari, L. Airborn fungi in biofuel wood chip storage sites. *Int. Biodeter. Biodegr.* **2014**, *90*, 17–22. [CrossRef]
16. Chapela, S.; Porteiro, J.; Gómez, M.A.; Patiño, D.; Míguez, J.L. Comprehensive CFD modeling of the ash deposition in a biomass packed bed burner. *Fuel* **2018**, *234*, 1099–1122. [CrossRef]
17. Klavinia, K.; Cinis, A.; Zandeckis, A. A study of pressure drops in an experimental low temperature wood chip dryer. *Agron. Res.* **2014**, *12*, 511–518.
18. Li, H.; Chen, Q.; Zhang, X.; Finney, K.N.; Sharifi, V.N.; Swithenbank, J. Evaluation of a biomass drying process using waste heat from process industries: A case study. *Appl. Therm. Eng.* **2012**, *35*, 71–80. [CrossRef]
19. Fredrikson, R.W. Utilisation of Wood Waste as Fuel for Rotary and Flash Tube Wood Dryer Operation. In Proceedings of the Biomass Fuel Drying Conference Proceedings, Superior, WN, USA, 8 August 1984; Office of Special Programs, University of Minnesota: Minneapolis, MN, USA, 1984; pp. 1–16.
20. Hulkkonen, S.; Parvio, E.; Raiko, M. An Advanced Fuel Drying Technology for Fluidized Bed Boilers. In Proceedings of the 13th International Conference on Fluidized Bed Combustion, Orlando, FL, USA, 7–10 May 1995; pp. 399–403.
21. Lamidi, R.O.; Jiang, L.; Pathare, P.B.; Wang, Y.D.; Roskilly, A.P. Recent advances in sustainable drying of agricultural produce: A review. *Appl. Energy* **2019**, *233–234*, 367–385. [CrossRef]
22. Ozollapins, M.; Kakitis, A.; Nulle, I. Stalk biomass drying rate evaluation. In Proceedings of the 12th International Scientific Conference Enginering for Rural Development, Jelgava, Latvia, 23–24 May 2013.
23. Pang, S.; Mujumdar, A.S. Drying of woody biomass for bioenergy: Drying technologies and optimization for an integrated bioenergy plant. *Dry. Technol.* **2010**, *28*, 690–701. [CrossRef]
24. Vigants, E.; Vigants, G.; Veidenbergs, I.; Lauka, D.; Klavina, K.; Blumberga, D. Analysis of Energy Consumption for Biomass Drying Process. In Proceedings of the 10th International Scientific and Practical Conference on Environment, Technology and Resources, Rezekne, Latvia, 18–20 June 2015; Volume II, pp. 317–322.
25. AIEL. Sistemi di essiccazione. In *Legna, Cippato e Pellet, Produzione, Requisiti Qualitativi, Compravendita*; Associazione Italiana Energie Agroforestali (AIEL): Legnaro (PD), Italy, 2012; pp. 93–100.
26. Kocsis, L.; Herdovics, M.; Deákvári, J.; Fenyvesi, L. Corn drying experiments by pilot dryer. *Agron. Res. Biosyst. Eng. Special Iss.* **2011**, *1*, 91–97.
27. Grimm, D.; Elustondo, M.; Mäkelä, M.; Segerström, G.; Kalén, L.; Fraikin, A.; Léonard, S.; Larsson, H. Drying recycled fiber rejects in a bench-scale cyclone: Influence of device geometry and operational parameters on drying mechanisms. *Fuel Process. Technol.* **2017**, *167*, 631–640. [CrossRef]
28. Gu, C.; Yuanb, Z.; Sunb, S.; Guanb, L.; Wub, K. Simulation investigation of drying characteristics of wet filamentous biomass particles in a rotary kiln. *Fuel Process. Tchnol.* **2018**, *178*, 344–352. [CrossRef]
29. Zuccarello, B. La co-densificazione di sanse e residui di potature: Fattibilità tecnico-economica e sviluppo. In Proceedings of the Conference Progetto ECODENS—Ecostabilizzazione delle sanse mediante densificazione, Misura 124-PSR Sicilia, 2007–2013, Università degli Studi di Palermo, Palermo, Italy, 30 January 2014.
30. Mujumdar, A.S. *Handbook of Industrial Drying*; CRC Press: Boca Raton, FL, USA, 2006.
31. Kaplan, O.; Celik, C. Woodchip drying in a screw conveyor dryer. *J. Renew. Sustain. Energy* **2012**, *4*, 063110. [CrossRef]
32. Xu, Q.; Pang, S. Mathematical modeling of rotary drying of woody biomass. *Dry. Technol.* **2008**, *26*, 1344–1350. [CrossRef]
33. Krokida, M.; Marinos-Kouris, D.; Mujumdar, A.S. Rotary drying. In *Handbook of Industrial Drying*; Mujumdar, A.S., Ed.; CRC Press: Boca Raton, FL, USA, 2006; pp. 151–172.
34. Lazdāns, V.; Lazdiņš, A.; Zimelis, A. Technology of biofuel production from slash in clear-cuts of spruce stands. *For. Sci.* **2009**, *19*, 109–121.
35. Williams-Gardner, A. *Industrial Drying*; Leonard Hill Books Ltd.: London, UK, 1971.
36. Brammer, J.; Bridgwater, A. Drying technologies for an integrated gasification bio-energy plant. *Renew. Sustain. Energy Rev.* **1999**, *3*, 243–289. [CrossRef]

37. Maciejewska, A.; Veringa, H.; Sanders, J.; Peteves, C. *Co-Firing of Biomass with Coal: Constraints and Role of Biomass Pre-Treatment*; JRC's Institute for Energy: Petten, The Nederlands, 2006.
38. UNI EN 14774-2. *Solid Biofuels, Determination of Moisture Content—Oven Dry Method. Part 2: Total Moisture—Simplified Method*; Italian Authority for Standardization: Milano, Italy, 2010.
39. UNI EN 15103. *Solid Biofuels, Determination of Bulk Density*; Italian Authority for Standardization: Milano, Italy, 2010.
40. UNI EN 14961-1. *Solid Biofuels, Determination of Particle Size Distribution. Part 1: Oscillating Screen Method Using Sieve Apertures of 1 mm and Above*; Italian Authority for Standardization: Milano, Italy, 2011.
41. ISO 3310-2. *Test Sieves e Technical Requirements and Testing-Part 1: Test Sieves of Metal Wire Cloth*; Italian Authority for Standardization: Milano, Italy, 2000.
42. Hartmann, H. *Handbuch Bioenergie-Kleinanlagen*; Sonderpublikation des Bundesministeriums für Verbraucherschutz, Ernährung und Landwirtschaft (BMVEL) und der Fachagentur Nachwachsende Rohstoffe (FNR): Gülzow, Germany, 2007.
43. Manzone, M.; Balsari, P.; Spinelli, R. Small-scale storage techniques for fuel chips from short rotation forestry. *Fuel* **2013**, *109*, 687–692. [CrossRef]
44. Pari, L.; Assirelli, A.; Acampora, A.; Del Giudice, A.; Santangelo, E. A new prototype for increasing the particle size of chopped *Arundo donax* (L.). *Biomass Bioenergy* **2015**, *74*, 288–295. [CrossRef]
45. Garcia-Galindo, D.; Gomez-Palmero, M.; Lopez, E.; Sebastián, F.; Jirjis, R.; Gebresenbet, G.; Germer, S.; Pari, L.; Suardi, A.; Lapeña, A.; et al. Agricultural pruning harvesting demostrationns in Germany, France, and Spain. Lessons learned reccommendations. In Proceedings of the 24th European Biomass Conference and Exhibition, Amsterdam, The Netherlands, 6–9 June 2016; pp. 1727–1733.
46. Zhou, L.P.; Wang, B.X.; Peng, X.F.; Du, X.Z.; Yang, Y.P. On the Specific Heat Capacity of CuO Nanofluid. *Adv. Mech. Eng.* **2010**, *1–4*, 172085. [CrossRef]
47. Tippayawong, N.; Tantakitti, S.; Thavornun, S. Energy efficiency improvements in longan drying practice. *Energy* **2008**, *33*, 1137–1143. [CrossRef]
48. Langsrud, Ø.; JØrgensen, K.; Ofstad, R.; Næs, T. Analyzing designed experiments with multiple responses. *J. Appl. Stat.* **2007**, *34*, 1275–1296. [CrossRef]
49. Jakob, K. Black locust (*Robinia pseudoacacia* L.). In *Energy Plant Species, Their Use And impact on Environment and Development*; James & James (Science Publishers) Ltd.: London, UK, 1998; pp. 98–103.
50. Kraszkiewicz, A. Chemical composition and selected energy proprieties of Black locust bark (*Robinia pseudoacacia* L.). *Agric. Eng.* **2016**, *20*, 117–124.
51. Hellrigl, B. *Elementi di Xiloenergetica. Definizioni, Formule e Tabelle*; AIEL-Italian Agriforestry Energy Association: Padova, Italy, 2006.
52. Civitarese, V.; Del Giudice, A.; Suardi, A.; Santangelo, E.; Pari, L. Study on the effect of a new rotor designed for chipping short rotation woody crops. *Croat. J. For. Eng.* **2015**, *36*, 101–108.
53. Anderson, J.-O.; Westerlund, L. Improved energy efficiency in sawmill drying system. *Appl. Energy* **2014**, *113*, 891–901. [CrossRef]
54. Johansson, L.; Westerlund, L. An open absorption system installed at a sawmill description of pilot plant used for timber and bio-fuel drying. *Energy* **2000**, *25*, 1067–1079. [CrossRef]
55. Chua, K.J.; Mujumdar, A.S.; Hawlader, M.N.A.; Chou, S.K.; Ho, J.C. Batch drying of banana pieces—Effect of stepwise change in drying air temperature on drying kinetics and product color. *Food Res. Int.* **2001**, *34*, 721–731. [CrossRef]
56. Ogura, H.; Yamamoto, T.; Otsubo, Y.; Ishida, H.; Kage, H.; Mujumdar, A.S. A control strategy for chemical heat pump dryer. *Dry. Technol.* **2005**, *23*, 1189–1203. [CrossRef]
57. De Fusco, L.; Jeanmart, H.; Blondeau, J. A modelling approach for the assessment of an air-dryer economic feasibility for small-scale biomass steam boilers. *Fuel Process. Technol.* **2015**, *134*, 251–258. [CrossRef]

Article

Influence of Oxidant Agent on Syngas Composition: Gasification of Hazelnut Shells through an Updraft Reactor

Francesco Gallucci [1,*], Raffaele Liberatore [2], Luca Sapegno [3], Edoardo Volponi [3], Paolo Venturini [3], Franco Rispoli [3], Enrico Paris [1], Monica Carnevale [1] and Andrea Colantoni [4]

1 Consiglio per la Ricerca in Agricoltura e l'analisi dell'economia Agraria (CREA)—Centro di Ricerca Ingegneria e Trasformazioni Agroalimentari (CREA-IT), Via della Pascolare 16, 00015 Monterotondo (Rome), Italy; enrico.paris@crea.gov.it (E.P.); monica.carnevale@crea.gov.it (M.C.)
2 ENEA-Casaccia Reserch Centre, Via Anguillarese, 301, 0123 Rome, Italy; raffaele.liberatore@enea.it
3 Department of Mechanical and Aerospace Engineering, Sapienza University of Rome, Via Eudossiana 18, 00184 Rome, Italy; lucasape@hotmail.it (L.S.); edoardo.volponi@gmail.com (E.V.); paolo.venturini@uniroma1.it (P.V.); franco.rispoli@uniroma1.it (F.R.)
4 Department of Agriculture and Forestry Science, Tuscia University, Via San Camillo de Lellis snc, 01100 Viterbo, Italy; colantoni@unitus.it
* Correspondence: francesco.gallucci@crea.gov.it; Tel.: +39-335-525-8415

Received: 12 November 2019; Accepted: 19 December 2019; Published: 24 December 2019

Abstract: This work aims to study the influence of an oxidant agent on syngas quality. A series of tests using air and steam as oxidant agents have been performed and the results compared with those of a pyrolysis test used as a reference. Tests were carried out at Sapienza University of Rome, using an updraft reactor. The reactor was fed with hazelnut shells, waste biomass commonly available in some parts of Italy. Temperature distribution, syngas composition and heating value, and producible energy were measured. Air and steam gasification tests produced about the same amount of syngas flow, but with a different quality. The energy flow in air gasification had the smallest measurement during the experiments. On the contrary, steam gasification produced a syngas flow with higher quality (13.1 MJ/Nm^3), leading to the best values of energy flow (about 5.4 MJ/s vs. 3.3 MJ/s in the case of air gasification). From the cold gas efficiency point of view, steam gasification is still the best solution, even considering the effect of the enthalpy associated with the steam injected within the gasification reactor.

Keywords: gasification; biomass; updraft; syngas; oxidizing agent

1. Introduction

In the last decades, because of increased interest in greenhouse gas emissions and related issues, biomass is becoming more attractive for several applications. Biomass gasification provides a syngas (also called producer gas) that can be used, for instance, in internal combustion engines after a cleaning process. It would widen the use of biomass and accordingly reduce greenhouse gas emissions. According to the last WEO (World Energy Outlook) by IEA (International Energy Agency) [1], biomass is the fourth most used energy source in the world, covering about 10% of the world's primary energy demand and 87% of it consists of solid (wooden) materials [2]. As reported in [3], in UE28 countries 13.2% of the gross inland energy consumption is covered by bioenergy, about 65% of which represents biomass, showing a slow but continuous increase over the last decade [3]. Italy is the fourth country in the UE28 in terms of gross inland energy consumption, representing 9% of the total, after the UK (12%), France (15%), and Germany (19%). Renewables in Italy represent 8% of the gross inland consumption, with about 50% provided by biomass and renewable wastes [3]. Biomass used for energy purposes comprises a wide range of substances, deriving from both dedicated crops and residues. However, well known energy-dedicated crops may collide with food-dedicated crops, subtracting lands and altering food prices and security. On the contrary, using residual biomass, that is biomass waste coming from production processes, is not expected to have any effect on food crops. Therefore, residual biomass should be considered a valuable option for fueling distributed energy systems [4] in all possible applications.

Biomass conversion processes can be classified into two main groups: Thermo-chemical (such as direct combustion, gasification, and pyrolysis), and biochemical (such as anaerobic digestion and alcoholic fermentation) [5]. Thermochemical processes are the most appropriate for wooden biomass, and therefore they are the most used.

Different biomass and reactors have been studied and used by researchers in the last decades. For instance (to cite some of the many papers published in the last years), Zainal et al. [6] tested wood chips and charcoal gasification in a downdraft reactor, analyzing the effect of the equivalence ratio on the gas composition and tar production. Wang et al. [7] performed steam gasification of municipal solid waste, varying steam to fuel ratio and gasification temperature to find optimal operative conditions for better gas yield. Borello et al. [8] performed an experimental and numerical campaign on gasification of olive pomace in an updraft reactor for a small scale CHP plant. Lucas et al. [9] tested gasification of wood pellets in an updraft gasifier using steam and air preheated up to 1250 °C as oxidant agents. Aghaalikhani et al. [10] and Ancona et al. [11] studied the use of poplar wood chips from phytoremediation as fuel for an updraft gasification reaction. Also, de Sales et al. [12] deepened the gasification of eucalyptus chips in a two-stage downdraft reactor.

Narvaez et al. [13] present the technical validation of novel, low complexity alternative remote, small-scale gasification facilities based on the inclusion of a new packed bed for improving performance. Mehta et al. [14] show a gasification experimental study of a top-lit updraft cook stove.

The use of a top-lit updraft gasifier is also analyzed by James et al. [15] for deepening the effect of woodchips physical properties. In addition, Huang et al. [16] carried out pilot-scale experiments with an updraft gasifier with the aim to remove and convert tar in syngas from woody biomass gasification.

Götz et al. [17] and Brunner et al. [18] explained the EU Horizon 2020 project HiEff-BioPower, which aims to develop a new, innovative, fuel-flexible, and highly efficient medium-scale biomass CHP (combined heat and power) technology for a capacity range of 1 to 10 MW total energy output. Other important recent research studies on the biomass gasification using updraft reactors include the effects of wood biomass type and airflow rate on the fuel and soil amendment properties of a solid by-product [19]; an experimental study of wood chips and grass waste gasification [20]; as well as lab-scale hydrogen production by supervised machine learning algorithms [21].

Yilmaz et al. presented a detailed thermodynamic performance assessment of an integrated system based on gasification for generation of cooling, heating, hydrogen, electricity and freshwater [22]. Bai et al. deepened kinetics and mechanisms of steam gasification from woody biomass treated with a hydrothermal process [23]. Huang et al. analyzed the effects of water content and particle size on pyrolysis and gasification of lignite chars [24]. Tian et al. carried out studies on the coupling of pyrolysis and gasification (CPG) process in the fluidized bed reactor to produce methane-rich syngas [25].

In the present paper, we focused our attention on the gasification of hazelnut shells. Hazelnut is one of the most important and investigated crops in Italy, located especially (but not only) in the central regions. The provinces of Rome, Avellino, and Naples have the most relevant hazelnut cultivations, but other significant spots are in Cuneo and Messina. These areas alone cover about 80% of the entire Italian production [26–29]. Hazelnut shells, as a waste byproduct, are currently used (in some cases) as fuel for household heating system. From a circular economy perspective, they would have better exploitation in more extensive and efficient plants.

Hazelnut shells are lignocellulosic matter; thus, they can be profitably used as fuel in thermo-chemical processes, such as direct combustion or gasification. The latter process in particular, produces a fuel gas that allows widening the applicability of this kind of substance. Gasification produces a syngas that is a mixture of gases mostly composed of CO, H_2, CO_2, CH_4, N_2, and tar (condensable hydrocarbons released during gasification). Syngas composition depends on the biomass used, in terms of lower heating value (LHV) and moisture, type of oxidant (air, pure oxygen, steam), and technologies and operative conditions, such as the equivalence ratio (ER), temperature, and pressure. The adoption of air as an oxidant agent produces syngas with the worst quality in terms of LHV (less than 6 MJ/Nm^3). It increases up to 9–10 MJ/Nm^3 in the case of oxygen, and 17–18 MJ/Nm^3 when using steam. The latter oxidant, usually employed in fluidized bed gasifiers, provides a syngas with high hydrogen content [30–34]. Syngas composition is also affected by the presence of impurities, including particulate, tar, and other elements (such as sulfur-based compounds), that make a cleaning process necessary before syngas is used [34–39]. Therefore, it is crucial to obtain syngas with a high LHV, and low particulate and tar content. Cerone et al. [40] analyzed the gasification of almonds and hazelnuts for syngas production through an updraft gasifier with a capacity up to 20–30 kg/h of biomass. Their experimental data were worked out by surface response analysis as a function of the equivalence ratios (ER) in relation to complete combustion and water reaction. By using only air at ER (O_2) 0.24 the ratio of H_2:CO in the syngas was 0.33; when adding steam at ER (H_2O) 0.28 the ratio reached a value of 1.0.

In the present paper, we aim to study the effect of an oxidant agent on syngas quality (namely its LHV), composition, producible energy, and cold gas efficiency. Moreover, the temperature distribution within the reactor is analyzed.

The adoption of an externally heated reactor also allows to analyze the effects of the system coupling with renewable energy sources such as the solar energy. The latter, for instance, would be possible through the use of a concentrating solar power (CSP) [35]. This system could supply the energy needed by the endothermic reactions of the gasification process, resulting in a better use of fuel energy content [36].

2. Materials and Methods

2.1. Experimental Setup

The gasification reactor used for the tests discussed in the present paper is an updraft reactor, which is the simplest type of reactor and, for this reason, it is also the most used one. The reactor (Figure 1, left) was installed at the laboratory of the Department of Mechanical and Aerospace Engineering, Sapienza University of Rome (Italy), and consisted of an AISI 310 stainless steel cylinder with a diameter of 8.3 cm and a height of 59.0 cm. It was equipped with an electric heating system (Figure 1, right), controlled by a Watlow 96 series temperature controller. The external surface was insulated with glass wool and aluminum in order to reduce thermal losses. At the top, a feeding cochlea transported the biomass within the reactor. In order to avoid any chocking of the feeding system and a syngas backflow from the reactor, a small flow of inert gas (namely N_2) was used to generate a slight overpressure in the feeding channel without altering the gasification reactions. Also, from the lower part, in the pyrolysis and hydrogasification tests, some nitrogen was introduced to allow the correct flow of the gases produced towards the reactor outlet. Within the reactor (at the bottom), a perforated metal plate worked as a support for the gasification bed which allowed the oxidant to flow through the bed itself. Underneath there was a "wind box" for preheating the oxidant agent, which was injected at the bottom. Oxidant temperature within the wind box can be regulated from the room temperature up to 450 °C. Syngas was extracted from the top, thus realizing a counter-current configuration. Downstream the reactor, syngas was cleaned first using a cyclone and a ceramic filter for particulate removal, and then through the passage in three bubblers containing isopropyl alcohol for tar removal. The bubblers were immersed in a thermal bath maintained at a constant temperature equal to−10 °C. Part of the syngas flow was then collected through a vacuum pump into the sampling box for the gas chromatography characterization using a VARIAN CP-4900 micro GC. The remaining syngas was sent to a flare for combustion. At different positions within the reactor, four K thermocouples (TH1–TH4, Figure 1, right) monitored the temperature evolution during the gasification process. Sensor TH4 was placed close to the syngas extraction duct, where the presence of the heating plates allowed for a constant temperature. The remnant sensors (TH1–TH3) were placed in the lower part of the reactor, at 2.0, 15.0, and 29.0 cm from the perforated metal plate, respectively. A sketch of the whole test plant is reported in Figure 2 which also shows a secondary reactor. It was used in another experimental campaign as a reformer to capture CO_2 and was filled with the catalyst/sorbent powder and thermally controlled [33], but was not employed here.

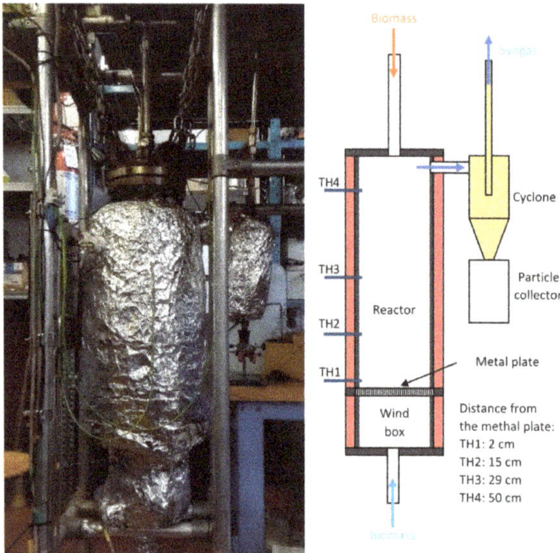

Figure 1. A gasification reactor installed at the Department of Mechanical and Aerospace Engineering Laboratory, Sapienza University of Rome (**left**). Detailed scheme of the reactor interior (**right**).

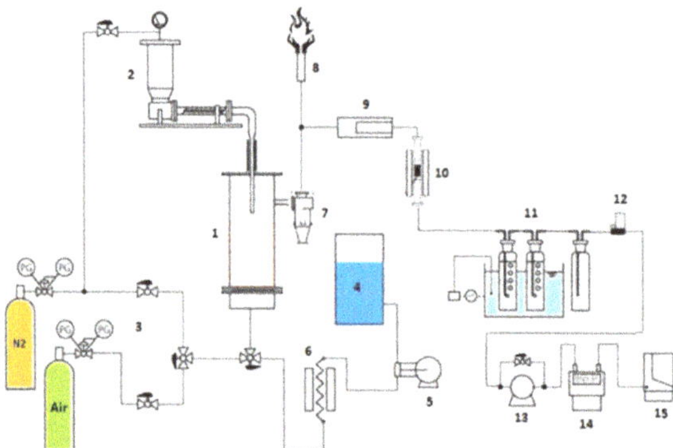

Figure 2. Scheme of the test plant. (1) updraft reactor; (2) feeding system; (3) air and nitrogen feeding systems; (4) steam generating unit; (5) pump to extract the distilled water; (6) water heater; (7) cyclone for particulate removal; (8) flare; (9) high temperature de-dusting system; (10) secondary reactor; (11) bubblers; (12) flow control system; (13) vacuum pump; (14) flowmeter; (15) gas-chromatograph.

2.2. *Characterization of the Gasification Bed*

The gasification bed is composed of biomass and inert material.

In the present experiments, we used olivine sand as inert material (Table 1 reports its main properties) and hazelnut shells as biomass. The latter came from the Soriano (VT) countryside, a small town in Central Italy. The only treatment the shells underwent before the test was shredding to about 2.5 mm. Table 2 reports proximate and elemental analysis of hazelnut shells. Data were obtained through a preliminary thermogravimetric analysis (using a Mettler–Toledo system) and thermal conductivity detection to measure element concentrations, after combusting and reducing a biomass sample (with the Leco 2000 CHN system). LHV was evaluated through the calorimetric bomb Parr 6200. Results show that hazelnut shells had a minimal moisture content (about 7.9 %) and a medium LHV. The main elements hazelnut shells are composed of are carbon and oxygen, covering together more than 85 % wt of the fuel. Accordingly, hazelnut shells raw formula is $C_4H_5O_2$ [33].

Table 1. Characterizationof the olivine sand.

Title	Physical Properties	
Material	**Density (kg/m^3)**	**Average Size (μm)**
Olivine	2640.0	351.0

Table 2. Physical and chemical properties of hazelnut shells.

Title	Physical Properties		Proximate Analysis			Elemental Analysis			
Biomass	**Density (kg/m^3)**	**Size (mm)**	**Moisture** (% wt)	**LHV (Lower Heating Value) (MJ/kg)**	**Ash** (% wt)	C (% wt)	H (% wt)	N (% wt)	O (% wt)
Hazelnut shells	945.0	2.5	7.90	17.23	1.16	46.65	5.55	3.04	38.74

2.3. Tests Description

Since the aim of the present paper is studying the influence of oxidant agents on syngas quality, we performed a series of tests using air (AG) or steam (SG), and another series with no oxidant (pyrolysis tests, PYR) as a reference. We fixed an arbitrary biomass feeding rate (namely 5.30 g/min) that was then kept constant during all the experiments. Moreover, to avoid possible clogging of the feeding channel and syngas backflow from the reactor, a small amount of nitrogen (0.4 Nl/min) was injected together with the biomass. N_2 was used because it is an inert gas; thus, it does not affect the gasification process itself. Of course, this amount of N_2 was then removed from the syngas composition, in order to not alter the measurements.

In order to compare operating conditions that provided about the same syngas flowrate while disregarding the oxidant agent used, the AG test showed that the reactor worked with an equivalence ratio (that is the ratio between the actual and stoichiometric airflow) equal to 8%, which means about 2.10 L/min air. On the other hand, SG tests were performed adopting a steam-to-biomass ratio equal to 0.5, corresponding to 2.70×10^{-3} L/min steam. The temperature of the electric heater was 845 °C for PYR and AG, and 900 °C for SG. In the latter case, we had to increase the temperature to compensate the smaller amount of heat produced by an exothermal reaction occurring within the fuel bed. Table 3 summarizes the test conditions.

Table 3. Test conditions.

Test Conditions	PYR	AG	SG
Biomass flow rate (g/min)	5.30	5.30	5.30
Oxidant flow rate (L/min)	0.0	2.10 (air)	2.70×10^{-3} (steam)
Heating system temperature (°C)	845	845	900

Before each test the reactor, connection tubes, cyclone filter, and ceramic filter were cleaned in order to avoid any char/ash deposition from previous tests or tar condensation.

Since in real applications tar content has to be removed because it is dangerous for the environment and people as well as for combustion devices [41], in our test plant we placed some bubblers before the gas-chromatograph inlet (Figure 2) in order to remove it. We are not interested in discussing tar composition in this study—this will be the focus of a further publication.

Syngas composition was studied through gas-chromatography analysis. In order to have significant values, eight measurements were performed for each test with a maximum error of ±1.8%, and averaged values were computed. Results were corrected to not account for the nitrogen used to help the biomass feeding. Since tar was removed from the gas stream, syngas composition was measured considering only nitrogen, methane, hydrogen, carbon monoxide, and carbon dioxide. Other trace gasses were neglected.

3. Results and Discussion

In this section, the results obtained in the above experimental campaign are presented together with a discussion of the motivations that carried out each of these outputs. The selected outcomes concern the temperature distribution within the reactor, the syngas quality in terms of its composition, the LHV, the cold gas efficiency, and the producible energy per unit time. They are indeed the main characteristics to evaluate the influence of the oxidant agent on syngas produced through biomass gasification, also in view of coupling with an external energy source.

During all tests, the biomass flow rate was kept constant while the oxidant agent varied.

3.1. Temperature Distribution within the Reactor

Reactor characteristics strongly affected the temperature. At the bottom, where exothermic reactions occur and where one would expect a higher temperature, there was actually a lower temperature. This was mainly due to the temperature at which the oxidant was injected within the reactor. As reported in Section 2.1, the oxidant was heated up to 450 °C, thus affecting the temperature within the reactor at the bottom. To have an idea of the cooling effect of the oxidant, we performed a preliminary test in which we set the temperature of the electric heating system to 845 °C (controlling TH2, see Figure 1) and injected air at 450 °C. In this case, the temperature measured by TH1, which was just 13 cm below TH2, was about 580 °C.

Figure 3 reports the average temperature distributions within the reactor in the three tests performed. Most of the works in the literature [34,35,37,38,42] show a close link between the amount of oxidant injected within the reactor and the temperature reached inside it. The temperature depends on the energy developed by the combustion reactions, which in turn depends on the amount of oxidant available for the reactions. In the present case, while the temperature measured by TH1 during AG and PYR tests varied in 555–712 °C, the temperature measured by the other thermocouples was mostly constant, slightly decreasing going toward the top. This means that despite the heat supplied by the electric heating system being rather large, the temperature in the oxidation region is still strongly influenced by exothermic reactions, as is also demonstrated in [41]. As previously stated, in steam gasification (SG) because of the strong endothermic reactions occurring within the fuel bed, and because of the huge amount of steam at 450 °C entering the reactor, to obtain an acceptable temperature in the oxidation zone we had to set a larger control temperature (about 900 °C). This resulted in a temperature at TH1 very close to that obtained during AG experiments, but a nearly constant temperature gap remained according to measurements by the other thermocouples. This is an indirect confirmation that reactions mostly occur at the bottom of the gasification reactor.

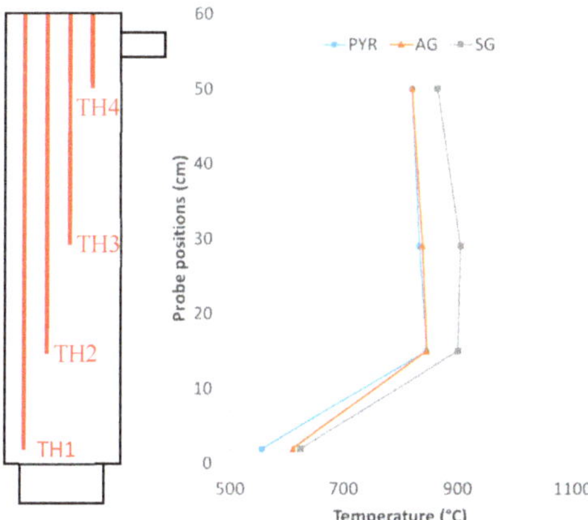

Figure 3. Temperature distribution within the reactor during PYR, AG and SG tests.

3.2. Syngas Composition

Figures 4 and 5 show the volume flow rate and composition of the syngas which was produced during the three tests. It is clear that both the flow rate and composition depend on the amount of oxidant injected. The PYR test produced about 2880.1 mL/min (0.0433 Nm^3/h) syngas, in which CO was the most relevant compound (about 42%). Content of CO, H_2, and CH_4 were similar, about 18–21%. During AG and SG tests the syngas flow rate was very similar (0.0885 vs. 0.0875 Nm^3/h), but the compositions were not. The producer gas in AG tests contained 50.0% nitrogen, while methane and hydrogen content were quite small, covering together about 13% of the total volume. CO equaled 25.3% and CO_2 11.6%. Syngas in the case of SG experiments was mostly composed of H_2 (35.5%) and CO (31.8%), then CO_2 (18.1%) and CH_4 (14.6%). In SG tests, N_2 content was negligible.

By removing N_2 (which is an inert) from the gas stream, we can better understand the effect of oxidant on syngas composition. Since nitrogen content in hazelnut shells is negligible, nitrogen in the syngas could only be due to the air injected which alters the concentration of different species composing the syngas. Figure 6 shows the N_2-free syngas composition. By comparing PYR and AG it is clear that oxygen injected with the air is mostly used in partial oxidation of C. Indeed, in the PYR test where the only oxygen available for oxidations was that contained in the biomass, we found a large amount of CH_4 (21.1%), 42.5% CO, 18.0% CO_2, and 18.5% H_2. Injecting air (AG tests), and thus increasing the amount of oxygen available for reactions, resulted in a reduction of CH_4 content (17.4%) and an increase of both CO (43.1%) and CO_2 (22.6%). H_2 content decreased a bit (16.9%) due to its partial oxidation. The amount of these components is interconnected by gas-phase reactions [41] and by those of carbon oxidation in the heterogeneous phase.

Figure 4. Volume flow rate (in ml/min) of syngas compounds.

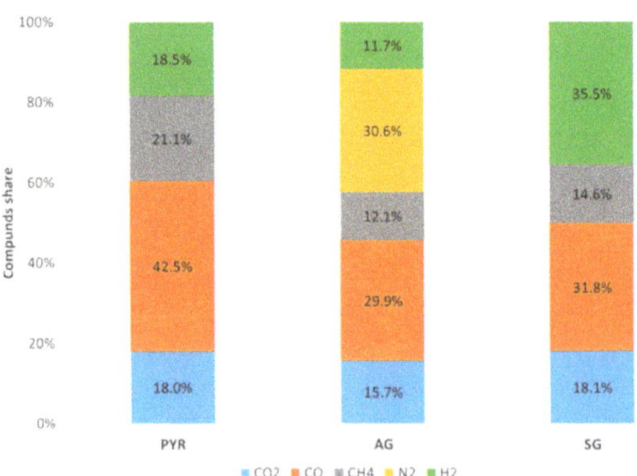

Figure 5. Syngas composition (vol %).

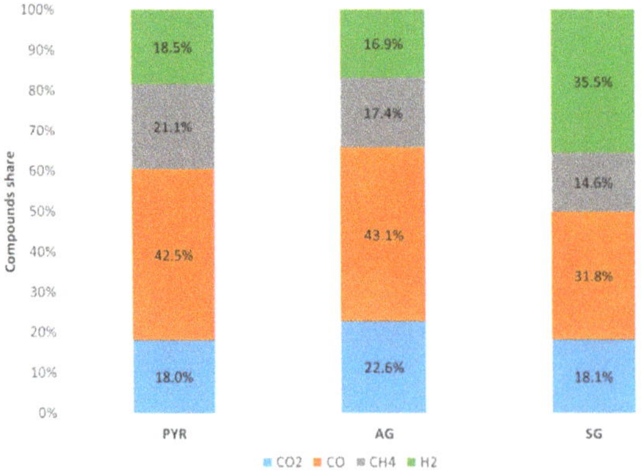

Figure 6. Syngas composition (vol %, N_2-free basis).

$$H_2 + \frac{1}{2}O_2 \rightarrow H_2O \quad (1)\ \text{Combustion}$$
$$CO + \frac{1}{2}O_2 \rightarrow CO_2 \quad (2)\ \text{Combustion}$$
$$CH_4 + \frac{1}{2}O_2 \rightarrow CO + 2H_2 \quad (3)\ \text{Combustion}$$
$$CH_4 + CO_2 \rightarrow 2CO + 2H_2 \quad (4)\ \text{Dry reforming}$$
$$CH_4 + H_2O \rightarrow CO + 3H_2 \quad (5)\ \text{Steam reforming}$$
$$CO + H_2O \rightarrow CO_2 + H_2 \quad (6)\ \text{Water gas-shift}$$

In SG tests, reactions (1)–(3) are very limited by the small amount of oxygen injected within the reactor. On the contrary, reactions (4)–(6) are enhanced. By comparing syngas composition in PYR and SG tests, it is evident that the use of steam as an oxidant results in a more significant amount of H_2. This mainly comes from the water gas-shift reaction (6), transforming CO into CO_2 and H_2, but also from dry and steam reforming (Equations (4) and (5)). CO_2 content is about the same in PYR and SB, meaning CO_2 produced by reactions (6) and destroyed by reaction (4) are almost in equilibrium. CO content decreases going from PYR to SG, which means that the water gas-shift reaction (6) is more active than reactions (4) and (5).

3.3. Syngas Heating Value and Energy Produced

LHV of the syngas depends on its content of fuel gases, namely H_2, CO, and CH_4. Since other fuel gases are negligible, and the tar content is captured bythe bubblers, here we can compute LHV according to Equation (7) [37]. Figure 7 reports the LHVs of the dry syngas from different tests.

$$LHV = 107.98H_2 + 126.36CO + 358CH_4 \quad (7)$$

The PYR test is the one producing the best LHV, while AG is the worst, which is related to the amount of nitrogen present within the reactor. In the case of the AG test, we were injecting air, and thus nitrogen, which was reflected in the decrease of LHV. Syngas from SG test showed an LHV closer to the PYR test. This is due to the composition (Figure 5). In the SG test, H_2 content is more than in the PYR test. On the contrary, CO and CH_4 decrease, thus resulting in a slightly smaller LHV.

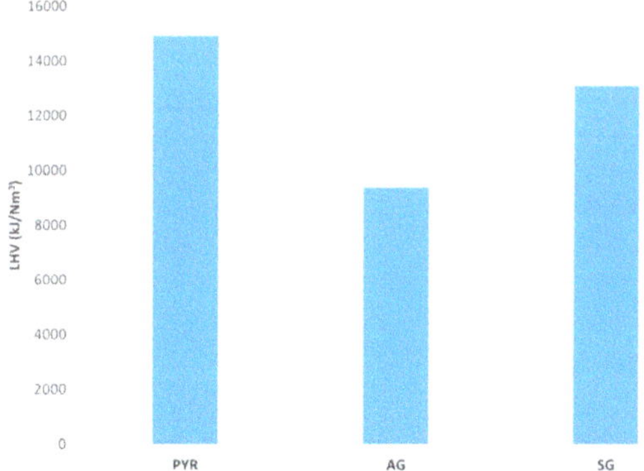

Figure 7. LHV (kJ/Nm^3) of syngas produced during the tests.

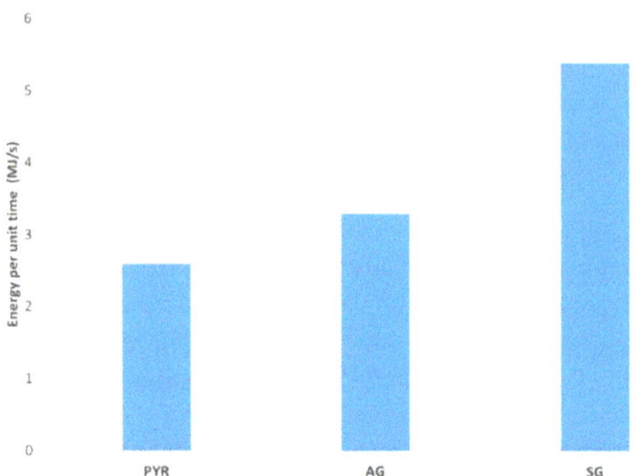

Figure 8. Energy per unit time (MJ/s) of syngas produced during the tests.

By multiplying LHV by the syngas flow rate (Figure 4), we evaluated the syngas energy producible per unit of time by the same amount of biomass but at different test conditions. Figure 8 reports this quantity. It is clear that the large amount of steam entering the reactor during SG tests results in a larger gas flow, which in turn results in a larger energy producible per unit time. On the contrary, the PYR test produces smaller gas flow, resulting in a smaller energy producible per unit of time. Lastly, the combination of a good LHV and high gas flow produce a high energy per unit time in the case of the SG test.

3.4. Cold Gas Efficiency

Cold gas efficiency is a measurement of how efficient the gasification reaction is, being the ratio between the energy content of the syngas and that of the original biomass. According to Cao et al. [43], the cold gas efficiency η_{CG} can be written asEquation (8).

$$\eta_{CG} = \frac{\text{LHV of fuel gas}\left(\text{in}\frac{\text{kJ}}{\text{Nm}^3}\right) \times \text{fuel gas production}\left(\text{in}\frac{\text{kJ}}{\text{kg}}\right)}{\text{LHV of the biomass}\left(\text{in}\frac{\text{kJ}}{\text{kg}}\right)} \tag{8}$$

In the case of SG, the modified denominator in Equation (8) accounts for the enthalpy owned by the steam at 450 °C (3382.0 kJ/kg) entering the reactor. Therefore, the denominator is given by the sum of the LHV of the biomass used and the enthalpy of the steam injected per kg of biomass. Cold gas efficiency computed in the three test sets is reported in Table 4. As shown, despite the contribution given by the enthalpy of the steam, the cold gas efficiency is larger in the case of SG which is due to the larger amount of H_2 present in the syngas.

Table 4. Cold gas efficiency.

Cold Gas Efficiency	PYR	AG	SG
η_{CG}	0.11	0.14	0.18

4. Conclusions

In the present paper, we reported an analysis of the influence of the oxidant agent on syngas produced through biomass gasification. In particular, we used waste biomass quite common in Italy, namely hazelnut shells, which are used from time to time as a fuel in small domestic combustion plants. The idea is to increase the energy content of the biomass profitably used by adopting it as a fuel in a gasification process. The effect of the oxidant agent is measured in terms of temperature distribution within the reactor, syngas quality in terms of its composition and LHV, as well as producible energy per unit of time.

Based on the experiments, AG and SB tests produce about the same syngas flow but have different quality in terms of composition, LHV, and cold gas efficiency.

The producer gas in AG tests contains 50.0% nitrogen (negligible in SG tests), while methane and hydrogen cover together about 13% of the total volume. CO equals 25.3% and CO_2 11.6%. On the contrary, in the SG experiments the producer gas is mostly composed of H_2 (35.5%) and CO (31.8%), then CO_2 (18.1%) and CH_4 (14.6%).

Due to the nitrogen content of the air, the AG syngas' LHV appears very small, thus resulting in a small amount of energy producible per unit of time. On the contrary, the syngas flow produced during SG tests had a very high LHV (about 13.1 MJ/Nm^3) because of its large content of hydrogen, CO, and CH_4. This leads to a larger energy flow producible through syngas, and a better cold gas efficiency (0.18 versus 0.14 of the AG and 0.11 of the PYR), that can be seen as a measure of the reaction gasification efficiency. It indicates that steam gasification of biomass would be a possibility coupling the steam production with renewable sources (i.e., solar energy).

This analysis can be useful for external heat source exploitation, such as solar energy (i.e., a CSP plant) in order to supply part of the heat necessary for the gasification. Thus, the adoption of an external heat source allows the ER optimization and/or steam usage as an oxidant agent, with the aim of maximizing both the syngas flow rate and its quality (LHV).

Author Contributions: Data curation, E.V., F.R. and E.P.; Formal analysis, R.L. and L.S.; Project administration, F.G.; Supervision, F.G. and A.C.; Writing—review & editing, P.V. and M.C. All authors have read and agreed to the published version of the manuscript.

Funding: This research received no external funding.

Conflicts of Interest: The authors declare no conflict of interest.

References

1. International Energy Agency. World Energy Outlook 2018 Report, 2018 Edition. Available online: https://www.iea.org/reports/world-energy-outlook-2018 (accessed on 12 November 2019).
2. World Biomass Association. WBA Global Bioenergy Statistics 2018 Report, 2018 Edition. Available online: https://worldbioenergy.org/uploads/181017%20WBA%20GBS%202018_Summary_hq.pdf (accessed on 12 November 2019).
3. Bioenergy Europe. Statistical Report, 2018 Edition. Available online: https://bioenergyeurope.org/statistical-report.html (accessed on 12 November 2019).
4. Asadullah, M. Barriers of commercial power generation using biomass gasification gas: A review. *Renew. Sustain. Energy Rev.* **2014**, *29*, 201–215. [CrossRef]
5. Dincer, I. Green methods for hydrogen production. *Int. J. Hydrogen Energy* **2012**, *37*, 1954–1971. [CrossRef]
6. Zainal, Z.A.; Rifau, A.; Quadir, G.A.; Seetharamu, K.N. Experimental investigation of a downdraft biomass gasifier. *Biomass Bioenergy* **2002**, *23*, 283–289. [CrossRef]
7. Wang, J.; Cheng, G.; You, Y.; Xiao, B.; Liu, S.; He, P.; Guo, D.; Guo, X.; Zhang, G. Hydrogen-rich gas production by steam gasification of municipal solid waste (MSW) using NiO supported on modified dolomite. *Int. J. Hydrogen Energy* **2012**, *37*, 6503–6510. [CrossRef]
8. Borello, D.; Pantaleo, A.M.; Caucci, M.; De Caprariis, B.; De Filippis, P.; Shah, N. Modeling and experimental study of a small scale olive pomace gasifier for cogeneration: Energy and profitability analysis. *Energies* **2017**, *10*, 1930. [CrossRef]
9. Lucas, C.; Szewczyk, D.; Blasiak, W.; Mochida, S. High-temperature air and steam gasification of densified biofuels. *Biomass Bioenergy* **2004**, *27*, 563–575. [CrossRef]
10. Aghaalikhani, A.; Savuto, E.; Di Carlo, A.; Borello, D. Poplar from phytoremediation as a renewable energy source: Gasification properties and pollution analysis. *Energy Procedia* **2017**, *142*, 924–931. [CrossRef]

11. Ancona, V.; Barra Caracciolo, A.; Campanale, C.; De Caprariis, B.; Grenni, P.; Uricchio, V.F.; Borello, D. Gasification treatment of poplar biomass produced in a contaminated area restored using plant assisted bioremediation. *J. Environ. Manag.* **2019**, *239*, 137–141. [CrossRef]
12. De Sales, C.A.; Maya, D.M.; Lora, E.E.; Jaén, R.L.; Reyes, A.M.; González, A.M.; Andrade, R.V.; Martínez, J.D. Experimental study on biomass (eucalyptus spp.) gasification in a two-stage downdraft reactor by using mixtures of air, saturated steam and oxygen as gasifying agents. *Energy Convers. Manag.* **2017**, *145*, 314–323. [CrossRef]
13. Narvaez, R.A.; Blanchard, R.; Dixon, R.; Ramirez, V.; Chulde, D. Low-Cost Syngas Shifting for Remote Gasifiers: Combination of CO_2 Adsorption and Catalyst Addition in a Novel and Simplified Packed Structure. *Energies* **2018**, *11*, 311. [CrossRef]
14. Mehta, Y.; Richards, C. Gasification performance of a top-lit updraft cook stove. *Energies* **2017**, *10*, 1529. [CrossRef]
15. James, R.A.M.; Yuan, W.; Boyette, M.D. The Effect of Biomass Physical Properties on Top-Lit Updraft Gasification of Woodchips. *Energies* **2016**, *9*, 283. [CrossRef]
16. Huang, J.; Schmidt, K.G.; Bian, Z. Removal and conversion of tar in Syngas from woody biomass gasification for power utilization using catalytic Hydrocracking. *Energies* **2011**, *4*, 1163–1177. [CrossRef]
17. Götz, T.; Saurat, M.; Kaselofsky, J.; Obernberger, I.; Brunner, T.; Weiss, G.; Bellostas, B.C.; Moretti, C. First stage environmental impact assessment of a new highly efficient and fuel flexible medium-scale chp technology based on fixed-bed updraft biomass gasification and a SOFC. In Proceedings of the 27th European Biomass Conference and Exhibition, Lisbon, Portugal, 27–30 May 2019; pp. 1586–1594.
18. Brunner, T.; Biedermann, F.; Obernberger, I.; Hirscher, S.; Schöch, M.; Milito, C.; Leibold, H.; Sitzmann, J.; Megel, S.; Hauth, M.; et al. Development of a new highly efficient and fuel flexible medium-scale CHP technology based on fixed-bed updraft biomass gasification and a SOFC. In Proceedings of the 26th European Biomass Conference and Exhibition, Copenhagen, Denmark, 14–17 May 2018; pp. 470–477.
19. Diez, H.E.; Perez, J.F. Effects of wood biomass type and airflow rate on fuel and soil amendment properties of biochar produced in a top-lit updraft gasifier. *Environ. Prog. Sustain. Energy* **2019**, *38*, 13105. [CrossRef]
20. Palencia Diaz, A.; Aguillon Martinez, J.E. Experimental study of forestry waste gasification: Pinewood chips-grass mixtures. *J. Renew. Sustain. Energy* **2019**, *11*, 044701. [CrossRef]
21. Ozbas, E.E.; Aksu, D.; Ongen, A.; Aydin, M.A.; Ozcan, H.K. Hydrogen production via biomass gasification, and modeling by supervised machine learning algorithms. *Int. J. Hydrogen Energy* **2019**, *44*, 17260–17268. [CrossRef]
22. Yilmaz, F.; Ozturk, M.; Selbas, R. Design and thermodynamic analysis of coal-gasification assisted multigeneration system with hydrogen production and liquefaction. *Energy Convers. Manag.* **2019**, *186*, 229–240. [CrossRef]
23. Bai, L.; Kudo, S.; Norinaga, K.; Wang, Y.G.; Hayashi, J.I. Kinetics and mechanism of steam gasification of char from hydrothermally treated woody biomass. *Energy Fuels* **2014**, *28*, 7133–7139. [CrossRef]
24. Huang, Y.; Wang, Y.; Zhou, H.; Gao, Y.; Xu, D.; Bai, L.; Zhang, S. Effects of water content and particle size on yield and reactivity of lignite chars derived from pyrolysis and gasification. *Molecules* **2018**, *23*, 2717. [CrossRef]
25. Tian, B.; Qiao, Y.; Fan, J.; Bai, L.; Tian, Y. Coupling Pyrolysis and Gasification Processes for Methane-Rich Syngas Production: Fundamental Studies on Pyrolysis Behavior and Kinetics of a Calcium-Rich High-Volatile Bituminous Coal. *Energy Fuels* **2017**, *31*, 10665–10673. [CrossRef]
26. Franco, S.; Pancino, B.; Cristofori, V. Hazelnut production and local development in Italy. *Acta Hortic.* **2014**, *1052*, 347–352. [CrossRef]
27. Speranza, S.; Bucini, D.; Paparatti, B. New observations on biology of European shot-hole borer [Xyleborus dispar (F.)] on hazelnut in Northern Latium (Central Italy). *Acta Hortic.* **2009**, *845*, 539–542. [CrossRef]
28. Paparatti, B.; Speranza, S. Biological control of hazelnut weevil (*Curculio nucum* L., Coleoptera, Curculionidae) using the entomopathogenic fungus Beauveria bassiana (Balsamo) Vuill. (Deuteromycotina, Hyphomycetes). *Acta Hortic.* **2005**, *686*, 407–412. [CrossRef]
29. Balestra, G.M.; Bucini, D.; Paparatti, B.; Speranza, S.; Proietti Zolla, C.; Pucci, C.; Varvaro, L. Bio-etology of Anisandrus dispar F. and its possible involvment in dieback (Moria) diseases of hazelnut (*Corylus avellana* L.) plants in central Italy. *Acta Hortic.* **2005**, *686*, 435–444.

30. McKendry, P. Energy production from biomass (part 2): Conversion technologies. *Bioresour. Technol.* **2002**, *83*, 47–54. [CrossRef]
31. McKendry, P. Energy production from biomass (part 3): Gasification technologies. *Bioresour. Technol.* **2002**, *83*, 55–63. [CrossRef]
32. Couto, N.; Ruoboa, A.; Silva, V.; Monteiro, E.; Bouziane, K. Influence of the biomass gasification pro-cesses on the final composition of syngas. *Energy Procedia* **2013**, *36*, 596–606. [CrossRef]
33. Sisinni, M.; Di Carlo, A.; Bocci, E.; Micangeli, A.; Naso, V. Hydrogen-rich gas production by sorption enhanced steam reforming of wood gas containing tar over a commercial Ni catalyst and calcined dolomite as co2 sorbent. *Energies* **2013**, *6*, 3167–3181. [CrossRef]
34. Dabai, F.; Paterson, N.; Millan, M.; Fennell, P.; Kandiyoti, R. Tar formation and destruction in a fixed bed reactor simulating downdraft gasification: Effect of reaction conditions on tar cracking products. *Energy Fuels* **2014**, *28*, 1970–1982. [CrossRef]
35. Gallucci, F.; Liberatore, R.; Sapegno, L.; Volponi, E.; Venturini, P.; Paris, E.; Carnevale, M.; Rispoli, F. Biomass gasification: The effect of equivalence ratio on syngas quality in the case of externally heated reactor. In Proceedings of the 27th European Biomass Conference and Exhibition, Lisbon, Portugal, 27–30 May 2019.
36. Liberatore, R.; Crescenzi, T.; Sapegno, L.; Volponi, E.; Venturini, P.; Rispoli, F.; Paris, E.; Carnevale, M.; Gallucci, F. Analysis on the coupling of biomass gasification process with a parabolic trough concentrating solar plant. In Proceedings of the European Biomass Conference and Exhibition, Lisbon, Portugal, 27–30 May 2019.
37. Guo, F.; Dong, Y.; Dong, L.; Guo, C. Effect of design and operating parameters on the gasification process of biomass in a downdraft fixed bed: An experimental study. *Int. J. Hydrogen Energy* **2014**, *39*, 5625–5633. [CrossRef]
38. Perez, J.F.; Melgar, A.; Nel Benjumea, P. Effect of operating and design parameters on the gasification/combustion process of waste biomass in fixed bed downdraft reactors: An experimental study. *Fuel* **2012**, *96*, 487–496. [CrossRef]
39. Molino, A.; Chianese, S.; Musmarra, D. Biomass gasification technology: The state of the art overview. *J. Energy Chem.* **2016**, *25*, 10–25. [CrossRef]
40. Cerone, N.; Zimbardi, F. Gasification of Agroresidues for syngas production. *Energies* **2018**, *11*, 1280. [CrossRef]
41. Parthasarathy, P.; Narayanan, K.S. Hydrogen production from steam gasification of biomass: Influence of process parameters on hydrogen yield—A review. *Renew. Energy* **2014**, *66*, 570–579. [CrossRef]
42. Houben, M.P.; de Lange, H.C.; van Steenhoven, A.A. Tar reduction through partial combustion of fuel gas. *Fuel* **2005**, *84*, 817–824. [CrossRef]
43. Cao, Y.; Wang, Y.; Riley, J.T.; Pan, W.P. A novel biomass air gasification process for producing tar-free higher heating value fuel gas. *Fuel Process. Technol.* **2006**, *87*, 343–353. [CrossRef]

Article

Sustainability Assessment of Alternative Strip Clear Cutting Operations for Wood Chip Production in Renaturalization Management of Pine Stands

Janine Schweier [1,2,*], Boško Blagojević [3], Rachele Venanzi [4], Francesco Latterini [4] and Rodolfo Picchio [4]

[1] Swiss Federal Institute for Forest, Snow and Landscape Research (WSL), Zürcherstrasse 111, 8903 Birmensdorf, Switzerland
[2] Chair of Forest Engineering, University of Freiburg, Werthmannstrasse 6, 79085 Freiburg, Germany
[3] Department of Forest Biomaterials and Technology, Swedish University of Agricultural Sciences (SLU), Skogsmarksgränd, 901 83 Umeå, Sweden
[4] Department of Agriculture and Forest Sciences (DAFNE), Tuscia University, Via S. Camillo de Lellis, 01100 Viterbo, Italy
* Correspondence: janine.schweier@wsl.ch; Tel.: +41-44-739-2478

Received: 19 July 2019; Accepted: 19 August 2019; Published: 27 August 2019

Abstract: In Mediterranean regions, afforested areas were planted to ensure the permanence of land cover, and to protect against erosion and to initiate the vegetation processes. For those purposes, pine species were mainly used; however, many of these stands, without silvicultural treatments for over fifty-sixty years, were in a poor state from physical and biological perspective, and therefore, clear-cutting on strips was conducted as silvicultural operation with the aim to eliminate 50% of the pine trees and to favor the affirmation of indigenous broadleaves seedlings. At the same time, the high and increasing demand of the forest based sector for wood biomass related to energy production, needs to be supplied. In a modern and multifunctional forestry, in which society is asking for sustainable forestry and naturalistic forest management, forestry operations should ideally be carried out in a sustainable manner, thus support the concept of sustainable forest management. All these aspects are also related to the innovation in forestry sector for an effective energetic sustainability. Three different forest wood chains were applied in pine plantations, all differing in the extraction system (animal, forestry-fitted farm tractor with winch, and double drum cable yarder). The method of the sustainability impact assessment was used in order to assess potential impacts of these alternative management options, and a set of 12 indicators covering economic, environmental, and social dimensions was analyzed. Further, to support decision makers in taking informed decisions, multi-criteria decision analysis was conducted. Decision makers gave weight towards the indicators natural tree regeneration and soil biological quality to support the achievement of the forest management goal. Results showed that first ranked alternative was case 2, in which extraction was conducted by a tractor with a winch. The main reason for that lies in the fact that this alternative had best performance for 80% of the analyzed criteria.

Keywords: horse skidding; winch skidding; cable yarder; life cycle assessment; societal assessment; economic assessment; multi-criteria decision analysis; sustainable forest management

---

1. Introduction

Mediterranean pines play a key role in the vegetation dynamics of the Mediterranean regions [1]. This group of species includes *Pinus nigra* Arnold, *Pinus brutia* Ten., *Pinus halepensis* Mill., and others such as *Pinus pinaster* Aiton, as the main representatives. These trees are well adapted to the fire regime that characterizes the area; they have a rapid and early growth and a general colonizing capacity; which all might explain why they have been traditionally used for afforestation projects and today often form extensive plantations in the overall Mediterranean basin. Afforestation was conducted mainly since the second half of the 19th century, aiming to improve protection functions (e.g., catchment hydrology and soil erosion) and socioeconomic functions [2–4] after centuries of forest exploitation and conversion to agricultural areas [5]. The total surface occupied by these pine plantations is estimated at 13 million ha, or 25% of the total forest area of the Mediterranean basin.

As many afforestation efforts lacked any kind of management, today, two main problems can be observed: First, many stands are in a poor physical and biological state with no dynamic processes [6]. This is due to several factors (i.e., biotic and environmental adversity and the inadequate treatment). As one consequence, forest health will decline, the stability of forests will be reduced [7], thus the permanence of land cover cannot be ensured.

Second, from a management perspective the pine-dominated vegetation is an intermediate step in succession to a climax state dominated by broadleaved trees [3]. However, due to climate change, many stands expand far beyond the limits of their natural ranges [7]. These changes are accompanied by a loss of biodiversity, a shift to non-site adapted tree species and a reduction of the resistance against climate inducted fluctuations, such as droughts, storms, insects and fungi [7–9], and an active forest management is urgently necessary.

Consequently, in order to redirect plantations toward more natural densities, there is a strong need for silvicultural treatments, such as thinning [10–12]. Different strategies exist to manage the pine plantations. They are mainly linked to renaturalization of artificial pine stands and consist generally in medium-high intensity thinning followed by a clear-cut after the affirmation of indigenous broadleaves seedlings [13–15].

In all case, the thinning approach should be chosen carefully as, e.g., a selective thinning might cause a higher risk of crown fire when overtopped trees remain untouched [16,17].

Additionally, forest operations (FO) to implement this renaturalization strategy could have important impacts on environmental, economic, and/or social performances, hence on all pillars of sustainability [18]. Forest Operations might affect carbon dioxide efflux [19], porosity, bulk density, shear strength [20], tree growth rate [21] soil horizon mixing and topsoil removal [22], and mineral soil respiration [23].

In particular, extraction processes, such as forwarding and skidding, have a high potential for soil compaction [24–26]. Further, damages to remaining stands might occur [27] and lead to negative impacts on regeneration [28].

It is well-known that fuel consumption is the most relevant contributor to greenhouse gas (GHG) emissions which cause global warming and thus, should be reduced [29,30]. Potential impacts in other environmental categories, such as the eutrophication potential (EP) and the acidification potential (AP) might be of particular interest when stands are included within natural reserves and underlie special conservation rules.

Further, with regard to social aspects, FO, especially when deployed on a low level of mechanization, is associated with a high risk of fatal accidents [31,32], particularly in felling and extracting operations.

Although in recent times there have been significant technological innovations in FO [33], felling and extracting in Italy, also in many other countries of Europe, are often deployed by traditional methods; i.e., motor-manual felling with chainsaws and the use of mules and/or agricultural tractors for extraction (e.g., [34,35]). Cable yarding systems might be another suitable extraction method [36].

To conclude, from a management perspective, there is an urgent need to apply silvicultural management strategies that support vegetation dynamics and enhance stand ecology, like the renaturalization concept. At the same time, the high and increasing demand of the forest based sector for wood biomass, related to energy productions, needs to be supplied.

The increasing global energy demand; the increasing fuel prices; the environmental impacts and the limited availability of fossil fuels; the aim to reduce emissions of greenhouse gases and to become more independent from fossil fuels, are some of the drivers why biomass resources are increasingly demanded for the production of renewable and sustainable energy. In contrast to wind and solar, biomass can provide base load capacity to the grid. In particular, the versatility of wood chips allows its flexible use in large heating plants, small combustion units and domestic boilers.

In a modern and multifunctional forestry, in which society is asking for sustainable forestry and naturalistic forest management [7], FO should ideally be carried out in a sustainable manner and thus, support the concept of Sustainable Forest Management (SFM) [18,37,38]. It aims to improve economic, but also environmental and social performances of forest processes, products and/or ecosystem services. All these aspects are also related to the innovation in forestry sector for an effective energetic sustainability. Indeed, renewable energy sources and the rational use of energy represent an important forestry resource in a local and global context against climate change.

It is a major challenge for decision makers (DMs) to consider the manifold consequences of decisions and to estimate the economic, environmental, and social performances of different alternatives before an action is carried out. Different indicators might have conflicting results and potential consequences should be known and taken into account in order to improve the silvicultural management strategies and respective methods of FO.

Therefore, the aim of this study was to assess possible impacts on sustainability that are related to FOs and resulting forest wood chains supporting the renaturalization strategy in typical afforested pine plantations in the Mediterranean basin. To be more concrete, we aimed to (i) identify alternative FOs that are suitable silvicultural actions for renaturalization of the pine stands, thereby putting a special emphasis to the extraction process; (ii) assess the potential impacts on all three pillars of sustainability; and (iii) make comprehensive evaluations of the alternative forest wood chains in order to support DMs. To do so, the method of sustainability impact assessment (SIA) was used. It supports assessing economic, environmental and social dimensions of forest processes, products and/or ecosystem services aiming to improve them [39].

In addition, a multi-criteria decision analysis (MCDA) was applied to support DMs. Forestry decision making is a very complex issue that requires consideration of trade-offs among different criteria (or indicators) [40]. MCDA is described by Belton and Stewart 2002 [41] "as an umbrella term to describe a collection of formal approaches, which seek to take explicit account of multiple criteria in helping individuals or groups explore decisions that matter." In other words, MCDA handles the process of making decisions in the presence of multiple, usually conflicting, criteria and provides a formal model to compare a finite number of alternatives on a one-dimensional preference scale [42]. MCDA has been widely used as decision-support tool in forest management [40,43] and FO [44,45].

2. Materials and Methods

2.1. Case Study Site

The Abruzzo region in Italy accounts for about 4% of the entire Italian forest surface. This region is quite representative for Mediterranean regions. There are about 19,000 ha of coniferous plantations, and of these, black pine (*Pinus nigra* Arnold subsp. *nigra* var. *italica* Villetta Barrea) afforestation covers approximately 13,000 ha [46]. The afforestation was conducted aiming to provide soil protection and to initiate new vegetation processes.

In this typical Mediterranean region, a case study fostering thinning operations supporting the renaturalization of the stands was carried out. Comparative trials were conducted in a 60 year old black pine (*Pinus nigra* Arnold subsp. *nigra* var. *italica* Villetta Barrea) plantation located near Passo delle Capannelle Municipality of Pizzoli (AQ 42°26′49, N 13°20′1) in the Abruzzo region. The studied afforestation covers about 27 ha along the middle mountain slope [46].

Planting was carried out with bare root black pine transplants at a distance of 1 m in the step [46]. It was a homogeneous and pure stand, with poor social differentiation and a high slenderness ratio. The degree of coverage was 90–100%. No significant meteoric damage had occurred, there were no obvious signs of fungal and insect attacks; dead wood snags were substantially absent, and logs were not consistent. None thinning had been applied since the establishment. Thus, the results of this study refer to operations being carried out at the stand age of 60 years. The average tree diameter at breast height was between 18 cm and 24 cm and the average tree height was between 13.2 m and 14.4 m. Further information regarding site characteristics, temperature and precipitation values were reported in Picchio et al. [46].

2.2. System Description

The forest management goal was to ensure the partial permanence of land cover, with the gradual replacement of pine with late successional tree species that are typical of more mature stages of evolution. Clear-cutting (dismantling cutting) on strips was conducted as silvicultural operation with the aim to eliminate 50% of the surface of the pine plantations.

Three different forest wood chains were applied (Figure 1), which are called case 1, case 2 and case 3 hereafter. In all cases, trees were felled motor-manually by using a chainsaw. Felling was conducted by a team of two workers; the first operated with the chainsaw and the second supported directing trees and cleaned stumps before cutting.

After felling, different extraction processes were applied: In case 1, extraction was conducted by animal (heavy rapid skidding horse, TPR-horse) (Figure 2a); in case 2 extraction was conducted by forestry-fitted farm tractor with a winch (Figure 2b); and in case 3 extraction was conducted by double drum cable yarder (Figure 2c).

Depending on the extraction, felling also differed: Since for the areas extracted by horses no directional felling was required (traditional practice) a simple directional felling (in winching area) and herringbone directional felling (in yarding area) was performed.

Transport and further processing did not differ between the cases: After extraction, trees were transported by using the same tractor with winch that was used earlier during extraction. The average transport distance was 400 m (SD ± 38 m). At the landing site trees were chipped for energy purposes by using a mobile chipper. It is common that the total harvesting material of these stands is chipped and used as biofuel. Table 1 shows assessment relevant machinery and animal data.

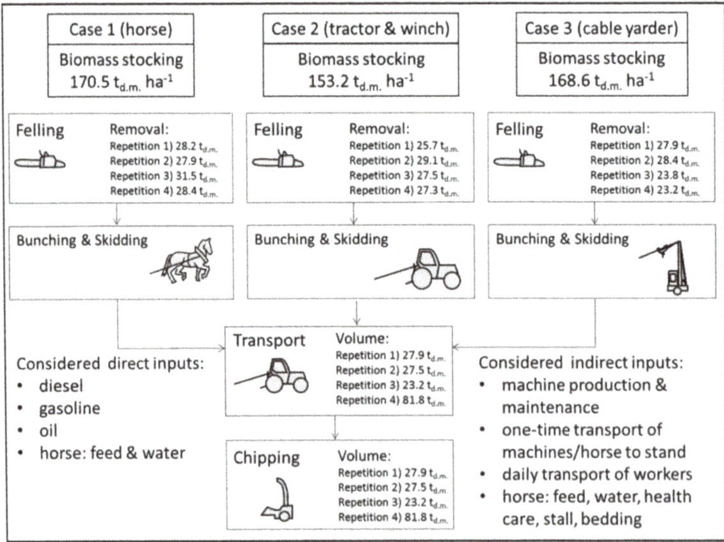

Figure 1. Schematic presentation of the four analyzed forest wood chains that differ in the extraction processes and amount of biomass removal, in tons dry matter ($t_{d.m.}$).

Figure 2. Extraction of whole trees in Mediterranean black pine plantations by (**a**) heavy rapid skidding horse; (**b**) forestry-fitted farm tractor with winch; and (**c**) double drum cable yarder.

Table 1. Inventory relevant machinery and animal data.

Machine	Brand	Type	Power [kW]	Lifetime [h]	Mass [kg]
Chainsaw	Stihl	MS441 C-M	4.2	2000	6.6
Horse	TPR	n.a.	n.a.	18,000 [a]	1200
Tractor	New Holland	88–85 M	62.5	10,000	4000
Winch	Farmi	7 tons	n.a. [b]	10,000	600
Cable yarder	Valentini	V600/3	175	17,000	12,000
Chipper	Pezzolato	PTH 700/660	129	14,000	8200

Notes: n.a. = not applicable; [a] = referring to productive working hours; [b] = depending on tractor power take-off; TPR = heavy rapid skidding horse.

2.3. Experimental Design of the Trials

As reported in Picchio et al. [46], the plantation was divided into two experimental blocks (replicates) being located on the southeastern slope in the altitudinal range 1200–1300 m a.s.l. (above sea level), with an average slope of about 50%. The first experimental block, which was at an altitude of 1200 m a.s.l. (east–southeast), consisted of 12 strips that were 100 m long (according to the lines of maximum slope) and 15 m wide. This block was surrounded on all sides, excluding the track, with a protection buffer that was a minimum of 20 m wide. The second block, with similar characteristics, was realized slightly lower, at an altitude of 1100 m a.s.l. (southeast).

The experimental design of the study considered three alternatives, derived from the three extractions methods (Figure 2). A randomized block design was assigned for the extraction methods, while the silvicultural operations were systematically assigned (one uncut strip and one clear-cut strip). Each extraction method was replicated two times in each block, thus, four times in total.

2.4. Methodological Approach

The method of SIA was used to assess the impact of the three alternatives on sustainability. It´s unique feature is that the economic, energetic, environmental and social dimensions of forest processes, products and/or ecosystem services can be addressed, thus it is a powerful concept to implement SFM. This method was proposed by [39,47] who suggest the following rules: (i) Supply chains are described as a set of processes; (ii) each process is characterized by a set of sustainability indicators; (iii) the total amount of material flowing through the processes is the basis for assessing the overall sustainability impact and (iv) an analysis of trade-offs between the characteristics is carried out to assess holistically the impact of changes between proposed alternatives.

2.5. Modelling and System Boundaries

The three alternatives were modelled as forest wood chains using the software Umberto (v 5.6), developed by IFU Hamburg GmbH. With Umberto, material flow networks are created allowing to model material and energy flows occurring in the system. The so-called "cradle-to-gate" approach was applied, meaning that the analysis was restricted to a selected life cycle stage [48]. In our case, the study concentrated on the felling, extracting, transporting and chipping of trees, as shown in Figure 1.

According to the modelling rules [47] in each process, the wood material changes its appearance and/or moves to another location. Thus, the SIA builds on the conceptual representation of forest wood chains as chains of value-adding production processes [49]. System boundaries were designated to be from where machines/animals, personnel and equipment were brought to the working sites to where the produced wood chips were at landing (Figure 1). For all processes, impacts due to direct (e.g., fuels) and indirect inputs (e.g., machinery) were considered.

The transportation of the chips to the final destination was not considered. Further, the building of roads and road maintenance, the disposal of machines and horse manure, the CO_2 uptake due to tree growth, and its release to the environment after biomass oxidation at the end of the life cycle, were not considered. Neither were changes in the soil organic matter stocks—all due to rare data.

2.6. Selection of Sustainablility Indicators

The sustainability indicators (SIs) selected for the calculation were relevant and balanced with regard to economic, environmental, and social sustainability, as well as feasibility in terms of data availability and quality [50]. A set of 12 SIs was chosen (Table 2) to be analyzed based on existing indicator sets (e.g., [51,52]).

The most relevant economic SIs are (#1) productivity (PROD), (#2) costs (COST), and (#3) working delays (DELAY). Productivity was described as machine performance per productive machine hour; production costs include personnel costs and fix and variable machine costs; and delays express nonproductive working times caused by mechanical, personal or operational issues.

As the environmental SI, the (#4) cumulated energy demand (CED) of fossil energy was calculated. Further, impacts in the well-known category (#5) global warming potential (GWP) were assessed, as well as in the following environmental impacts categories: (#6) Eutrophication potential (EP) and (#7) acidification potential (AP). All of them are important categories for biomass cultivation and distribution and are highly influenced by nitrous and carbon oxides, which are of special interest to coastal pine plantations along the Tyrrenian coast and generally in Central Italy, where most such stands are included within natural reserves, under special conservation rules (e.g., Gran Sasso and Monti della Laga National Park, Abruzzo National Park, and Majella National Park).

When it comes to social SI, attention was put on (#8) employment (EMP). The amount of fatal accidents was not included due to missing reliable data. Statistical data are neither available for the accidents occurring during the thinning of Italian coastal pine plantations, nor for working accidents in Italian forestry in general, since the Italian work accident statistics lump forestry and agriculture together.

As the provision of ecosystem services, in this case, the prevention against erosion and the initiation of the vegetation processes, and the increase in tree biodiversity, directly impact societal and living conditions; the SI (#9) tree regeneration density (TRD), (#10) tree species diversity (TSD), (#11) soil biological quality (QBS-ar), and (#12) soil microarthropod community density (SMD) were considered as social indicators. Tree regeneration was estimated according to the phytosociological method applied by Pourbabaei et al. and Picchio et al. [53,54]. The Shannon index was used to estimate floristic biodiversity [55]. It is a model that measures species diversity and the degree of homogeneity in species abundance. It is sensitive to changes in rare species, it clearly discriminates, and is well represented in the literature [56,57]. To analyze the impact on soil and short-term recovery the arthropod-based soil biological quality index (known as QBS-ar index) was used (e.g., [58,59]). It is a valuable tool in ecosystem restoration programs for monitoring the development of soil functions and biodiversity and is based on the following concept: The higher soil quality, the higher the number of microarthropod groups well adapted to soil habitats will be [1]. The organisms belonging to each biological taxon were counted in order to estimate their density at the sampled depth (0–10 cm) and ratio of the number of individuals (IND), and the sample area to 1 dm^2 of the surface (IND dm^{-2}) [60,61].

This indicator, called soil microarthropod community density, it has been validly applied as a further quantitative biological soil index by [28,59].

2.7. Indicator Calculation

Machine costs referred to Euros (€) per productive machine hour (PMH_{15}), meaning that delays up to 15 min were included. Costs (#2) were calculated according to Picchio [62]. Delay (#3) time was reported separately in order to calculate delay factors [63]; i.e., the ratio of delay time to productive working time. Data related to time input and machine productivity (#1) were determined with a time study. Data about utilization and maintenance of machines and value recovery were obtained directly from the machine owners and from the consultation of machine data sheets.

The analysis of the CED (#4) as well as environmental impacts in the categories GWP (#5), EP (#6) and AP (#7) focused on technical aspects of the alternative FO and followed the ISO 14040-44 guidelines which prescribes the inclusion of direct (e.g., use of fuel) and indirect (e.g., use of machines) impacts. Respective data of direct fuel inputs were shown in Table 3. Fuel consumption was determined by measurements during FO. In particular, data with regard to fuel and oil use were collected for all machines involved.

The feed and water requirements of the horse belong to both categories, direct and indirect inputs. According to Engel et al. [64], the lifespan for a horse was set at 20 years. It can be assumed that their training requires 5 years. For the residual 15 years, a constant work performance of 1200 productive working hours (PWH) per year was assumed, which is equal to 7 PWH per day on 171 days per year [64]. The feed and water requirements on these 2565 working days (171 days per year × 15 years) were considered as direct inputs. Data refer to a daily feedstuff of 72 kg water, 7 kg hay, 5 kg straw, and 9 kg barley [64,65]. Barley was used for calculation instead of oats due to missing emission data of oats in the database. The feed and water requirements for the first 5 years of life (365 days × 5 years = 1825 days) as well as for the non-working days (194 days per year × 15 years = 2910 days) were considered as indirect inputs.

The production and maintenance of the chainsaw and the harvesting machines belong to indirect inputs, too. Data represent an average value and were taken from literature [66–68], including a repair factor of 50%.

Further, the transportation of the machines and the horse to the forest stand and the daily transportation of the forest workers to the stand were considered. The machines and the horse stayed in the forest during the overall FO. The transport distance of the horse and of all machines to the forest stand was 40 km for one way, except the yarder, where it was 350 km per way. The forest workers used a car to get to the stand every day and the transport distance was 35 km per way.

The modelling software Umberto [69] and the database Ecoinvent (vs. 2.3) [70] were used to conduct the life cycle inventory. In Ecoinvent, emission data for several materials (e.g., oil) can be found. They were connected to the material's specific use (e.g., required diesel in a process) and then in the life cycle impact assessment linked to the contributing environmental categories (e.g., CO_2 to GWP).

The effect on EMP was calculated from the productivity data observed in the study, considering 1500 h per year as full employment of one worker unit, according to Italian National Collective Agreement for FO.

The TRD was assessed via systematically accounting for each species according to literature [60,71]. The Shannon index was calculated as reported in Picchio et al. [46]. The QBS-ar index was calculated according to Venanzi et al. [59] and the SMD was assessed as reported in Marchi et al. [60]. Both, the impact of the silvicultural management on natural tree regeneration and on soil have been analyzed in a previous study; methods were reported in detail in Picchio et al. [46].

Indicator results were reported per ton dry matter ($t_{d.m.}$) of wood chips and on a per hectare basis. Total indicator results refer to 27 ha. However, the studied area region, there are about 19,158 ha of coniferous plantations, and of these, black pine afforestation covers about 13,000 ha [72].

Table 2. Applied sustainability indicators.

#Nr.	Indicator	Abbreviation	Unit	Description
#1	Productivity	PROD	PMH_{15} $t_{d.m.}^{-1}$	Rate of product output per unit of time for a production system including delays up to 15 min. A productivity ratio may also be calculated for resources other than time.
#2	Costs	COST	€ $t_{d.m.}^{-1}$	Sum of production costs (fixed costs accruing regardless the rate of activity inclusive personnel costs as well as variable costs that vary with quantity of production).
#3	Delays	DELAY	Minutes $t_{d.m.}^{-1}$	Interruptions of the work process that can be related back to the organization of the work; commonly subdivided into the categories mechanical (e.g., repair), personal (e.g., rest breaks) and operational delays (e.g., waiting times).
#4	Cumulated energy demand of fossil energy	CED	MJ $t_{d.m.}^{-1}$	The cumulative energy demand of fossil energy investigates the energy use throughout the overall life cycle, including the use of direct and indirect consumption of energy.
#5	Global warming potential	GWP	kg CO_2-eq. $t_{d.m.}^{-1}$	The potential of global warming is mainly caused by the release of greenhouse gas emissions due to anthropogenic activities such as fossil fuel combustion and transportation.
#6	Eutrophication potential	EP	kg PO_4-eq. $t_{d.m.}^{-1}$	Potential eutrophication due to some substances, calculated through the conversion factor of phosphorous and nitrogen compounds into phosphorous equivalents.
#7	Acidification potential	AP	kg SO_2-eq. $t_{d.m.}^{-1}$	Potential acidification due to atmospheric deposition of sulfur and nitrogen.
#8	Employment	EMP	FTE 1000 $t_{d.m.}^{-1}$	Rate of full-time employments related to forest operations.
#9	Tree regeneration density	TRD	n° $t_{d.m.}^{-1}$	Number of individuals of tree seedlings per area referred to harvested biomass.
#10	Tree species diversity	TSD	Shannon Index	Degree of uncertainty of predicting the species of a random sample is related to the diversity of a community and is based on measuring uncertainty.
#11	Soil biological quality	QBS-ar	QBS-ar index	Ecological index which joins the biodiversity of soil microarthropods community with the degree of soil vulnerability.
#12	Soil microarthropod community density	SMD	n° ind t $t_{d.m.}^{-1}$	Quantitative biological indicator of soil microarthropod community, expressed as number of individuals per area, and referred to harvested biomass.

Note: gt = green tonne (fresh weight); Min = minutes; MJ = megajoule; EE = energy efficiency; CO_2 = carbon dioxide; PO_4 = phosphate; SO_2 = sulfur dioxide; FTE = full-time equivalent.

Table 3. Inventory data of direct inputs, in kg per ton $_{d.m}$.

Case No.	Process	Felling		Skidding & Bunching						Transport		Chipping	
	Input Material	Gasoline	Oil	Diesel	Oil	Hay	Straw	Barley	Water	Diesel	Oil	Diesel	Oil
Case 1	Repetition 1	0.73	0.24	0.91	0.06	0.97	0.70	1.25	29.54	2.12	0.15	2.25	0.15
	Repetition 2	0.76	0.25	0.98	0.07	0.78	0.56	1.00	36.91	2.12	0.15	2.25	0.15
	Repetition 3	0.69	0.23	0.73	0.05	1.00	0.72	1.29	28.61	2.12	0.15	2.25	0.15
	Repetition 4	0.61	0.20	0.64	0.04	1.01	0.72	1.30	28.52	2.12	0.15	2.25	0.15
	Average	0.70	0.23	0.82	0.06	0.94	0.67	1.21	30.90	2.12	0.15	2.25	0.15
Case 2	Repetition 1	0.63	0.21	1.89	0.13	0.00	0.00	0.00	0.00	2.12	0.15	2.25	0.15
	Repetition 2	0.72	0.24	1.92	0.13	0.00	0.00	0.00	0.00	2.12	0.15	2.25	0.15
	Repetition 3	0.80	0.26	1.40	0.10	0.00	0.00	0.00	0.00	2.12	0.15	2.25	0.15
	Repetition 4	0.70	0.23	1.47	0.10	0.00	0.00	0.00	0.00	2.12	0.15	2.25	0.15
	Average	0.71	0.23	1.67	0.11	0.00	0.00	0.00	0.00	2.12	0.15	2.25	0.15
Case 3	Repetition 1	0.62	0.20	1.92	0.13	0.00	0.00	0.00	0.00	2.12	0.15	2.25	0.15
	Repetition 2	0.68	0.23	2.13	0.15	0.00	0.00	0.00	0.00	2.12	0.15	2.25	0.15
	Repetition 3	0.60	0.20	1.68	0.12	0.00	0.00	0.00	0.00	2.12	0.15	2.25	0.15
	Repetition 4	0.58	0.19	1.63	0.11	0.00	0.00	0.00	0.00	2.12	0.15	2.25	0.15
	Average	0.62	0.20	1.84	0.13	0.00	0.00	0.00	0.00	2.12	0.15	2.25	0.15

2.8. Multi-Criteria Decision Analysis

For ranking the presented three alternatives two fundamental MCDA methods were used: Multi-attribute utility theory (MAUT) [73] and the PROMETHEE method [74].

MAUT belongs to Value Measurement group of methods [75]. MAUT is compensatory and thereby produces complete rankings of alternatives. In MAUT [73], the preferences of DMs are represented by sub-utility function for each criterion. This sub-function (s) must be constructed by the DM(s). In that way, different criteria (e.g., employment, tree species diversity, etc.) are transformed into one common utility scale (with range 0–10) [76]. Summing the products of the sub-utilities multiplied with the corresponding weights of the criteria—which are defined by DMs—the final utility of each alternative is obtained. The alternative with the highest utility value is the first ranked alternative (the best one). A detailed description of MAUT can be found in Keeney and Raiffa [73]. In this paper, MAUT analysis was done with Simple Value Tree software.

In contrast, the PROMETHEE I and PROMETHEE II methods belong to group of outranking methods. They are based on the pairwise comparison of alternatives for every selected criterion using preference function which translates this comparison into one common scale (from zero to one) [77]. Brans et al. [74] proposed six criteria functions (usual, U-shaped, V-shaped, level, linear and Gaussian). In the PROMETHEE method, DMs need to define: (i) weights of criteria; and (ii) the shapes of preference functions and corresponding indifference, and/or preference thresholds. After that, positive and negative preference flows for each alternative are calculated using previously obtained values. PROMETHEE I will produce full ranking of alternatives only in situations when one alternative is better than another with respect to both positive and negative flow, otherwise they are incomparable. In PROMETHEE II, the difference between positive and negative flow (net flow) is used; therefore, results will always be complete ranking of alternatives [45,78,79]. A thorough description of PROMETHEE is given in Brans et al. [74]. In this paper, PROMETHEE analysis was done with Visual PROMETHEE software.

In this study, 10 relevant criteria (previously described as SI in Section 2.6) were used to rank the three alternatives. To avoid double counting, two criteria (productivity and delays) were excluded from the MCDA because they were included in costs criterion. Weights of criteria were obtained by two experts (or DMs) from forestry using the DIRECT method. In the DIRECT method, the DM allocates points to each criterion. For example, the DM is asked to distribute 100 points among the criteria. The DM is also allowed to distribute more (or less) than 100 points. The final weights are the points of each criterion divided by the sum of all points. The selection of utility functions, preference functions and thresholds for this study was based on previous studies [44,80], as well as the authors' judgment. In MAUT method we used linear utility-function for all criteria while for PROMETHEE method, a V-shape preference function has been applied. The preference threshold (for V-shape preference function) was set to be 10% of the highest value for each SI [44].

3. Results

3.1. Economic Indicator Results

The average productivity of the process felling varied from 4.35 ± 0.52 $t_{d.m.}$ PMH_{15}^{-1} (case 2) to 4.69 ± 1.09 $t_{d.m.}$ PMH_{15}^{-1} (case 3). In case 1, results were slightly higher than in case 2, but showed higher standard deviation (4.40 ± 2.2 $t_{d.m.}$ PMH_{15}^{-1}) (Figure 3).

In all cases, the most time-consuming process was bunching and skidding (Figure 3). On average, it reached highest productivities in case 2 (2.26 ± 0.31 $t_{d.m.}$ PMH_{15}^{-1}), followed by case 1 (1.08 ± 0.12 $t_{d.m.}$ PMH_{15}^{-1}), and case 3 (0.81 ± 0.10 $t_{d.m.}$ PMH_{15}^{-1}).

Transport and chipping operations were carried out independently from the felling and extraction processes and did not differ between the cases. On average, the productivity of transport was 2.54 ± 0.37 $t_{d.m.}$ PMH_{15}^{-1}. In case of chipping, it was 15.14 ± 3.57 $t_{d.m.}$ PMH_{15}^{-1}.

The resulting average system productivity ranged from 0.52 ± 0.05 $t_{d.m.}$ PMH_{15}^{-1} (case 3) to 0.88 ± 0.05 $t_{d.m.}$ PMH_{15}^{-1} (case 2). In case 1, it was 0.62 ± 0.07 $t_{d.m.}$ PMH_{15}^{-1} on average. In all cases, a team consisting of 2 workers was necessary.

A more detailed look into the distribution of the net working time of the process felling showed that cutting was the most time-consuming working step (55.9% of the working time in case 1, 50.4% in case 2, and 51.4% in case 3; Table 4). The working step movement was significantly less time-consuming in case 1 (18.7%) compared to case 2 (27.8%) and case 3 (29.2%) (Table 4). It differed in the process bunching and skidding: in case 1, the empty movement was most time-consuming working step (41.2%), while it was bunching extraction in case 2 (38.8%) and hooking in case 3 (32.9%) (Table 4).

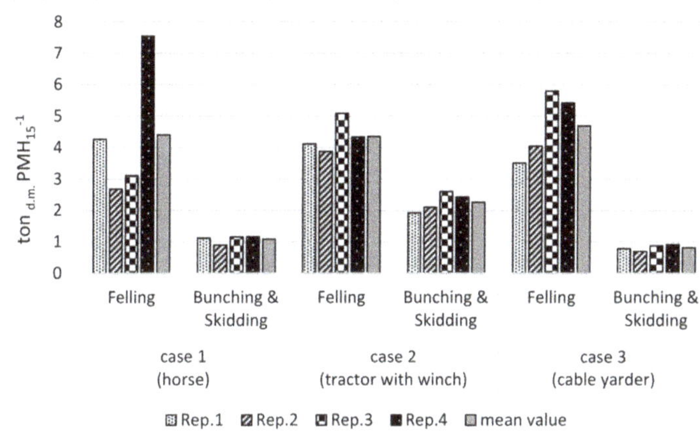

Figure 3. Resulting working productivity for the processes felling and bunching and skidding, per case and in $t_{d.m.}$ PMH_{15}^{-1}. Note: productivity results were shown in $t_{d.m.}$ PMH_{15}^{-1}; and not in the functional unit (PMH_{15} $t_{d.m.}^{-1}$) in order to make findings comparable to other studies.

Table 4. Resulting average distribution of net working time per working step of the processes felling and bunching and skidding.

Process	Working Step	Case 1 (Horse)		Case 2 (Winch)		Case 3 (Yarder)	
		Min $t_{d.m.}^{-1}$	SD	Min $t_{d.m.}^{-1}$	SD	Min $t_{d.m.}^{-1}$	SD
Felling	Movement	1.07	±0.10	2.11	±0.30	1.91	±0.18
	Preparation	1.08	±0.10	1.66	±0.18	0.95	±0.21
	Cutting	3.20	±0.26	3.83	±0.43	3.36	±0.57
	Tree grounding	0.37	±0.21	0.00	±0.00	0.32	±0.08
Bunching & Skidding	Empty Movement	14.11	±2.24	5.08	±1.03	5.69	±0.96
	Hooking	4.40	±1.29	2.96	±0.30	14.09	±2.12
	Bunching extraction	13.04	±3.05	7.28	±1.16	11.76	±2.39
	Unhooking	2.71	±0.45	3.44	±0.99	11.31	±1.95

Resulting costs followed the same pattern we the system productivity (Table 5): lowest felling costs were reached in case 3 (€3.40 ± €0.81 $t_{d.m.}^{-1}$) and lowest bunching and skidding costs were reached in case 2 (€12.05 ± €1.64 $t_{d.m.}^{-1}$). The average transport costs were €8.41 ± €1.14 $t_{d.m.}^{-1}$ and average chipping costs were €7.24 ± €1.69 $t_{d.m.}^{-1}$. In sum, case 2 was cheapest (€31.34 ± €4.68 $t_{d.m.}^{-1}$), while case 3 was most expensive one (€76.98 ± €10.50 $t_{d.m.}^{-1}$).

The highest share of DELAY occurred in the motor-manual felling operations. On average, delay time was 51.3% (45.4–58.4%) of the total felling time. In bunching and extraction processes, the average delay time was 30.4% when extraction was conducted by using the tractor with a winch, 39.4% when using the horse, and 42.9% when using the cable yarder. Average delays accounted for 17.1% in transportation processes and 9.9% in chipping processes.

Resulting delay factors for the overall forest wood chains were on average 18.2 for the alternative forest wood chains in which extraction was conducted by using the tractor with the winch, 18.6 when extraction was conducted by using the cable yarder, and 19.9 in cases when extraction was conducted by horse.

Table 5. Resulting production €s per process and case, in € per ton $_{d.m.}$.

Case	Process	€ $t._{d.m.}{}^{-1}$	SD
Case 1 (extraction by horse)	Felling	4.33	±1.72
	Bunching & Skidding	16.65	±2.17
	Transport	8.41	±1.14
	Chipping	7.24	±1.69
	Sum	36.63	±5.66
Case 2 (extraction by tractor with winch)	Felling	3.64	±0.41
	Bunching & Skidding	12.05	±1.64
	Transport	8.41	±1.14
	Chipping	7.24	±1.69
	Sum	31.34	±4.68
Case 3 (extraction by cable yarder)	Felling	3.40	±0.81
	Bunching & Skidding	57.93	±7.31
	Transport	8.41	±1.14
	Chipping	7.24	±1.69
	Sum	76.98	±10.50

3.2. *Envionmental Indicator Results*

The total CED varied between 423 ± 20 MJ $t_{d.m.}{}^{-1}$ (case 1) and 499 ± 25 MJ $t_{d.m.}{}^{-1}$ (case 3) (Table 6). The result in case 3 was mainly caused due to the more intensive energy requirement of the process bunching and skidding (169 ± 21 MJ $t_{d.m.}{}^{-1}$). The process felling contributed 11.0% to 14.7% to the total CED (case 3 and case 1, respectively); bunching and skidding with 20.2% to 33.8% (case 1 and case 3, respectively); transport with 28.1% to 33.1% (case 3 and case 1, respectively) and chipping with 27.1% to 32.0% (case 3 and case 1, respectively).

The total GWP varied between 6.66 ± 0.27 kg CO_2 $t_{d.m.}{}^{-1}$ (case 2) and 9.10 ± 0.76 kg CO_2 $t_{d.m.}{}^{-1}$ (case 1) (Table 6). The process felling contributed with 11.1–16.7% to the total GWP (case 3 and case 2, respectively); bunching and skidding with 29.5% to 49.5% (case 2 and case 3, respectively); transport with 23.3–31.7% (case 1 and case 3, then case 2, respectively) and chipping with 16.2–22.1% (case 1 and case 3, then case 2, respectively).

The total EP varied between 0.0113 ± 0.0005 kg PO_4-eq. $t_{d.m}.^{-1}$ (case 2) and 0.0494 ± 0.0056 kg PO_4-eq. $t_{d.m}.^{-1}$ (case 1) (Table 6). The process felling contributed with 3.0–12.6% to the total EP (case 1 and case 2, respectively); bunching and skidding with 31.4%–84.2% (case 2 and case 1, respectively); transport with 7.7–33.3% (case 1 and case 2, respectively) and chipping with 5.2–15.0% (case 1 and case 3, respectively).

The total AP varied between 0.0527 ± 0.0018 kg SO_2-eq. $t_{d.m}.^{-1}$ (case 2) and 0.701 ± 0.0049 kg SO_2-eq. $t_{d.m}.^{-1}$ (case 1) (Table 6). The process felling contributed with 11.5–15.2% to the total AP (case 1 and case 2, respectively); bunching and skidding with 26.5–44.6% (case 2 and case 1, respectively); transport with 23.3–30.9% (case 1 and case 2, respectively) and chipping with 20.6–27.4% (case 1 and case 2, respectively).

Table 6. Results table of life cycle impact assessment, in kg per ton $_{d.m}$.

Input t d.m. -1	Process	Felling				Extraction				Transport				Chipping				Sum Impact			
	Case no./IC	CED	GWP	EP	AP	CED	GWP	EP	AP	CED	GWP	EP	AP	CED	GWP	EP	AP	CED	GWP	EP	AP
Diesel	case 1	0.0000	0.0000	0.0000	0.0000	43.8454	0.4178	0.0007	0.0046	114.0733	1.0869	0.0019	0.0119	120.8435	1.1515	0.0020	0.0126	278.7622	2.6562	0.0046	0.0291
	case 2	0.0000	0.0000	0.0000	0.0000	89.6789	0.8545	0.0015	0.0094	114.0733	1.0869	0.0019	0.0119	120.8435	1.1515	0.0020	0.0126	324.5956	3.0929	0.0053	0.0339
	case 3	0.0000	0.0000	0.0000	0.0000	98.7596	0.9410	0.0016	0.0103	114.0733	1.0869	0.0019	0.0119	120.8435	1.1515	0.0020	0.0126	333.6764	3.1794	0.0055	0.0348
Gasoline	case 1	39.4986	0.4974	0.0007	0.0050	0.0000	0.0000	0.0000	0.0000	0.0000	0.0000	0.0000	0.0000	0.0000	0.0000	0.0000	0.0000	39.4986	0.4974	0.0007	0.0050
	case 2	40.2070	0.5063	0.0007	0.0051	0.0000	0.0000	0.0000	0.0000	0.0000	0.0000	0.0000	0.0000	0.0000	0.0000	0.0000	0.0000	40.2070	0.5063	0.0007	0.0051
	case 3	34.9650	0.4403	0.0006	0.0045	0.0000	0.0000	0.0000	0.0000	0.0000	0.0000	0.0000	0.0000	0.0000	0.0000	0.0000	0.0000	34.9650	0.4403	0.0006	0.0045
Oil	case 1	17.1688	0.2041	0.0003	0.0018	4.1757	0.0496	0.0001	0.0004	10.8121	0.1285	0.0002	0.0011	11.4832	0.1365	0.0002	0.0012	43.6398	0.5188	0.0009	0.0045
	case 2	17.4484	0.2074	0.0003	0.0018	8.5005	0.1010	0.0002	0.0009	10.8121	0.1285	0.0002	0.0011	11.4832	0.1365	0.0002	0.0012	48.2442	0.5735	0.0010	0.0049
	case 3	15.2115	0.1808	0.0003	0.0016	9.4140	0.1119	0.0002	0.0010	10.8121	0.1285	0.0002	0.0011	11.4832	0.1365	0.0002	0.0012	46.9207	0.5578	0.0009	0.0048
Fodder	case 1	0.0000	0.0000	0.0000	0.0000	5.3405	0.7395	0.0120	0.0064	0.0000	0.0000	0.0000	0.0000	0.0000	0.0000	0.0000	0.0000	5.3405	0.7395	0.0120	0.0064
	case 2	0.0000	0.0000	0.0000	0.0000	0.0000	0.0000	0.0000	0.0000	0.0000	0.0000	0.0000	0.0000	0.0000	0.0000	0.0000	0.0000	0.0000	0.0000	0.0000	0.0000
	case 3	0.0000	0.0000	0.0000	0.0000	0.0000	0.0000	0.0000	0.0000	0.0000	0.0000	0.0000	0.0000	0.0000	0.0000	0.0000	0.0000	0.0000	0.0000	0.0000	0.0000
impact machine/horse	case 1	0.7972	0.1257	0.0002	0.0001	13.6026	1.8488	0.0280	0.0155	7.4732	0.3592	0.0014	0.0015	1.5953	0.0767	0.0003	0.0003	23.4682	2.4104	0.0298	0.0175
	case 2	0.6975	0.1100	0.0002	0.0001	8.4980	0.4085	0.0016	0.0017	7.4732	0.3592	0.0014	0.0015	1.5953	0.0767	0.0003	0.0003	18.2640	0.9544	0.0034	0.0037
	case 3	0.6681	0.1053	0.0002	0.0001	36.3619	1.7480	0.0067	0.0023	7.4732	0.3592	0.0014	0.0015	1.5953	0.0767	0.0003	0.0003	46.0985	2.2892	0.0086	0.0042
daily transport	case 1	4.7233	0.3317	0.0002	0.0011	16.7391	1.1755	0.0007	0.0039	7.0099	0.4923	0.0003	0.0016	1.1753	0.0825	0.0000	0.0003	29.6476	2.0819	0.0012	0.0070
	case 2	4.1330	0.2902	0.0002	0.0010	7.9712	0.5598	0.0003	0.0019	7.0099	0.4923	0.0003	0.0016	1.1753	0.0825	0.0000	0.0003	20.2893	1.4248	0.0008	0.0048
	case 3	3.9584	0.2780	0.0002	0.0009	22.2269	1.5608	0.0009	0.0052	7.0099	0.4923	0.0003	0.0016	1.1753	0.0825	0.0000	0.0003	34.3704	2.4136	0.0014	0.0081
one-time transport machine/horse	case 1	0.0000	0.0000	0.0000	0.0000	1.9177	0.1263	0.0001	0.0004	0.6185	0.0437	0.0000	0.0001	0.3092	0.0218	0.0000	0.0001	2.8454	0.1918	0.0001	0.0006
	case 2	0.0000	0.0000	0.0000	0.0000	0.6185	0.0437	0.0000	0.0001	0.6185	0.0437	0.0000	0.0001	0.3092	0.0218	0.0000	0.0001	1.5462	0.1092	0.0001	0.0003
	case 3	0.0000	0.0000	0.0000	0.0000	1.8040	0.1274	0.0001	0.0004	0.6185	0.0437	0.0000	0.0001	0.3092	0.0218	0.0000	0.0001	2.7317	0.1930	0.0001	0.0006
SUM	case 1	62.1879	1.1588	0.0015	0.0080	85.6210	4.3576	0.0416	0.0313	139.9869	2.1107	0.0038	0.0163	135.4065	1.4690	0.0026	0.0145	423.2023	9.0961	0.0494	0.0700
	case 2	62.4860	1.1139	0.0014	0.0080	115.2671	1.9675	0.0036	0.0139	139.9869	2.1107	0.0038	0.0163	135.4065	1.4690	0.0026	0.0145	453.1464	6.6611	0.0113	0.0527
	case 3	54.8030	1.0044	0.0013	0.0071	168.5663	4.4892	0.0095	0.0192	139.9869	2.1107	0.0038	0.0163	135.4065	1.4690	0.0026	0.0145	498.7627	9.0733	0.0172	0.0570

Note: IC = impact category; CED = cumulative energy demand (of fossil energy), reported in MJ ton $_{d.m.}^{-1}$; GWP = global warming potential, reported in kg CO_2-eq. ton $_{d.m.}^{-1}$; EP = eutrophication potential, reported in kg PO_4-eq. ton $_{d.m.}^{-1}$; AP = acidification potential, reported in kg SO_2-eq. ton $_{d.m.}^{-1}$. Indirect inputs are marked in grey color.

Indirect inputs included the (i) production and maintenance of machines; (ii) transport of the machines to the forest stand; and (iii) daily transport of the forest workers to the stand. With regard to CED of fossil energy, on average, the share of indirect inputs was 13.2% in case 1; 8.8% in case 2 and 16.7% in case 3 (Figure 4). In the category GWP, on average, the share of indirect inputs was 51.5% in case 1, 37.4% in case 2, and 54.0% in case 3 (Figure 4). In the category EP, on average, the share of indirect inputs was 63.2% in case 1, 38.2% in case 2, and 59.0% in case 3 (Figure 4). In the category AP, on average, the share of indirect inputs was 35.8% in case 1, 16.6% in case 2, and 22.7% in case 3 (Figure 4). Among the indirect inputs, the daily transport of the workers to the stand contributed highest to that value. On average it was as follows: With regard to CED of fossil energy, it varied between 41.0% (case 3) and 53.6% (case 1); in the category GWP, it varied between 44.4% (case 1) and 57.1% (case 2); in the category EP it varied between 3.8% (case 1) and 18.5% (case 2); and in the category AP it varied between 27.9% (case 1) and 62.6% (case 3). It is worth mentioning that the process bunching and skidding caused high shares of indirect emissions in two cases. In case 3 (extraction by cable yarder), indirect emissions had an average share of 35.8% in CED; 76.6% in GWP; 81.0% in EP; and 41.3% in AP—mainly caused by the production and maintenance of the heavy yarder. In case 2 (extraction by horse), indirect emissions had an average share of 37.7% in CED; 72.3% in GWP; 69.1% in EP; and 63.6% in AP—mainly caused by the daily care for the horse (e.g., fodder).

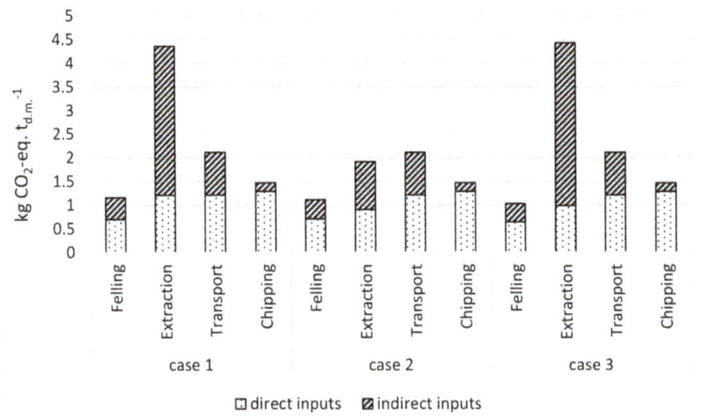

Figure 4. Resulting environmental impacts per process and case in the category global warming potential, distributed with regard to direct and indirect inputs, in kg CO_2-eq. $t_{d.m.}{}^{-1}$.

3.3. *Socio-Ecological Indicator Results*

The total EMP was highest in case 3 (2.64 ± 0.26 FTE 1000 $t_{d.m.}{}^{-1}$), followed by case 1 (2.60 ± 0.08 FTE 1000 $t_{d.m.}{}^{-1}$) and case 2 (1.58 ± 0.11 FTE 1000 $t_{d.m.}{}^{-1}$) (Figure 5). The process bunching and skidding differed most among the cases. On average, it was 1.66 ± 0.21 FTE 1000 $t_{d.m.}{}^{-1}$ for extraction by yarder (case 3); 1.57 ± 0.20 FTE 1000 $t_{d.m.}{}^{-1}$ for extraction by horse (case 1); and 0.60 ± 0.08 FTE 1000 $t_{d.m.}{}^{-1}$ for extraction by tractor with winch (case 1) (Figure 5).

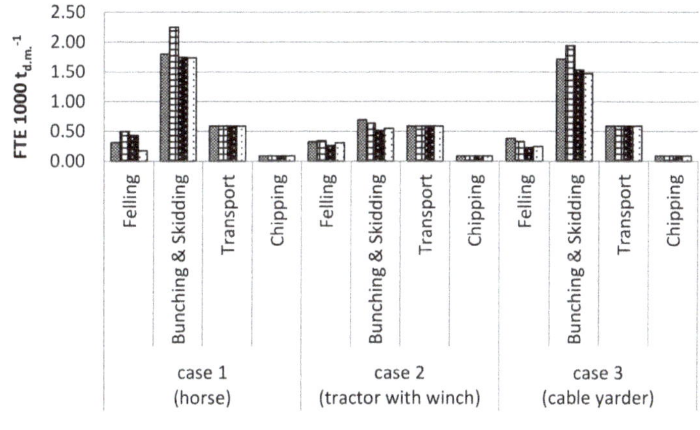

Figure 5. Resulting employment rate per process and case, in FTE 1000 $t_{d.m.}{}^{-1}$. Note: FTE = full time equivalent (1500 h year^{-1} for workers and 1200 h year^{-1} for horse).

After a period of three years after the FO, the highest natural tree regeneration density was found in case 2 with an average of 21,018 ± 1399 trees ha^{-1} (Table 7). Case 1 and case 3 were on a similar level (18,737 ± 1204 trees ha^{-1} and 18,729 ± 1236 trees ha^{-1}, respectively). These values were about twice as high compared to control areas (stand without FO) (data not shown). In a temporal trend of three years after the FO, case 1 had a considerable increase, while case 2 and case 3 showed slight decreases (data not shown).

After harvesting, a constant increase in species richness was determined [46]. In case 3, where extraction was conducted by using a cable yarder, the highest diversity was found. In particular, the applied extraction system more positively influenced the richness (data not shown) and marginally influenced the diversity. The cases 2 and cases 3 had higher richness values than case 1. These indexes were only marginally different respect to control areas (stand without FO) (data not shown) and their trend was positive during the three years after FO.

With regard to the QBS-ar index it turned out that case 1 showed higher, thus better, values than the other cases: 228 ± 9.1 in case 1, compared to 199 ± 13.4 in case 2, and 179 ± 13.2 in case 3 (Table 7).

However, as shown in Picchio et al. [46], the QBS-ar index showed significant differences only among treatments and years, with a positive trend during the three years after FO, but with values still lower than the control for case 2 and case 3 (data not shown).

Soil microarthropod community density showed statistically significant differences among treatments and years too [46]. In particular, in the harvested strips, the density values were lower than in the control, but the trends were positive. Three years after the FO, the density varied between 100 ± 6.6 million n° ha^{-1} (case 3), 124 ± 6.3 million n° ha^{-1} (case 1), and 161 ± 8.8 million n° ha^{-1} (case 2) (Table 7).

All indicator values were converted to the functional unit hectare, too, as this unit is more relevant for forest management (Table 8). The results per hectare were the basis for the subsequent MCDA.

Table 7. Resulting indicator values of tree regeneration density (TRD), floristic diversity (Shannon index), Soil Biological Quality (QBS-ar index) and soil microarthropod community density (SMD), per ha.

Case	Rep. No.	TRD	Shanon Index	QBS-ar Index	SMD
		[n° ha^{-1}]			[Million n° ind ha^{-1}]
Case 1 (horse)	Rep. 1	19,950	1.55	233	132
	Rep. 2	17,650	1.69	223	126
	Rep. 3	17,756	1.49	238	120
	Rep. 4	19,592	1.59	218	118
	average	18,737	1.58	228	124
Case 2 (tractor & winch)	Rep. 1	22,600	1.89	213	163
	Rep. 2	19,600	1.87	187	170
	Rep. 3	20,114	1.83	188	149
	Rep. 4	21,756	1.89	208	162
	average	21,018	1.87	199	161
Case 3 (cable yarder)	Rep. 1	20,100	1.59	165	91
	Rep. 2	17,500	1.59	172	105
	Rep. 3	17,889	1.53	195	105
	Rep. 4	19,425	1.61	184	99
	average	18,729	1.58	179	100

Note: Rep. = Repetition; TRD = Tree Regeneration Density; SMD = Soil Microarthropod community Density.

Table 8. Resulting average indicator values per hectare.

SI	Unit	Case	Resulting Value
Productivity	PMH_{15} ha^{-1}	Case 1	7.4123
		Case 2	6.3102
		Case 3	7.2864
COST	€ ha^{-1}	Case 1	6244.5625
		Case 2	4800.9050
		Case 3	12,978.4065
Delay	Minutes ha^{-1}	Case 1	63.15
		Case 2	55.04
		Case 3	62.77
CED	MJ-eq. ha^{-1}	Case 1	72,155.9898
		Case 2	69,422.0291
		Case 3	84,091.3891
GWP	Kg CO_2-eq. ha^{-1}	Case 1	1550.8824
		Case 2	1020.4863
		Case 3	1529.7583
EP	Kg PO_4-eq. ha^{-1}	Case 1	8.4206
		Case 2	1.7384
		Case 3	2.8931
AP	Kg SO_2-eq. ha^{-1}	Case 1	11.9496
		Case 2	8.0750
		Case 3	9.6135
EMP	FTE ha^{-1}	Case 1	0.4430
		Case 2	0.2426
		Case 3	0.4447
TRD	n° ha^{-1}	Case 1	18,737.0000
		Case 2	21,017.5000
		Case 3	18,728.5000
TSD	Shannon-Index	Case 1	1.5800
		Case 2	1.8700
		Case 3	1.5800

Table 8. *Cont.*

SI	Unit	Case	Resulting Value
		Case 1	228.0000
QBS-ar	QBS-ar-Index	Case 2	199.0000
		Case 3	179.0000
		Case 1	124.0000
SMD	Million n° ha^{-1}	Case 2	161.0000
		Case 3	100.0000

3.4. Multi-Criteria Decision Analysis

Table 9 shows input data for MCDA. According to methodological approach, five criteria should be minimized and five should be maximized. Weights of criteria were obtained with DIRECT method and it can be seen that #9 (TRD) and #11 (QBS) were the most important criteria (0.200), while the least important criteria were #2 (COST), #8 (EMP), #10 (TSD), and #12 (SMD), with weights of 0.050. This decision was related to DMs intention to give higher priority to environmental criteria in mountain areas, closely related to land cover and soil biological quality.

Table 9. Input data (decision matrix) for multi-criteria decision analysis (MCDA).

Criteria	COST	CED	GWP	EP	AP	EMP	TRD	TSD	QBS	SMD
Min/Max	Min	Min	Min	Min	Min	Max	Max	Max	Max	Max
Shape of Function	V	V	V	V	V	V	V	V	V	V
Preference threshold (p)	1298	8409	155	0.842	1.19	0.044	2102	0.187	2.28	16.1
Weights of criteria	0.050	0.100	0.100	0.100	0.100	0.050	0.200	0.050	0.200	0.050
Case 1 (horse)	6245	72,156	1551	8.42	11.95	0.443	18,737	1.580	228	124
Case 2 (tractor with winch)	4801	69,422	1020	1.74	8.07	0.243	21,018	1.870	199	161
Case 3 (cable yarder)	12,978	84,091	1530	2.89	9.61	0.445	18,729	1.580	179	100

Table 10 presents results of the MCDA when applying the MAUT method. When considering the DMs´ weighting of indicators, case 2 was the first ranked alternative, case 1 was second while last ranked alternative was case 3. It should be noticed that case 1 had utility of 8.3 (out of 10), meaning that this alternative was very dominant in comparison to others. Identical rankings were obtained when the different methods PROMETHEE I and II were applied (Table 11, not all data shown).

Table 10. MCDA results for application of MAUT method.

Alternatives	Utility	Ranks
Case 1 (horse)	3.9	2
Case 2 (tractor with winch)	8.3	1
Case 3 (cable yarder)	2	3

Table 11. MCDA results for application of PROMETHEE method.

Alternatives	Phi	Phi+	Phi−	Ranks
Case 1 (horse)	−0.124	0.325	0.449	2
Case 2 (tractor with winch)	0.666	0.816	0.150	1
Case 3 (cable yarder)	−0.543	0.133	0.675	3

Case 2 had the best performance for eight (out of 10) criteria. Only for two criteria (#8 EMP and #11 QBS-ar), other alternatives had better performances. A sensitive analysis was conducted in order to analyze how much one need to change (increase) weights of these two criteria in order to change the first ranked alternative. When the weight of #8 (EMP) became higher than 0.34, case 1 became first ranked alternative instead of case 2 (when using MAUT method) (Figure 6a). For criterion #11 (QBS-ar), the value was even higher. It was necessary to increase weight of #11 to 0.54 in order to change first ranked alternative (Figure 6b). For PROMETHEE method results were similar. Case 2 was the first ranked alternative in the range from 0 to 0.427 (for #8 EMP, (Figure 7a) and in the range from 0 to 0.511 (for #11 QBS-ar, Figure 7b).

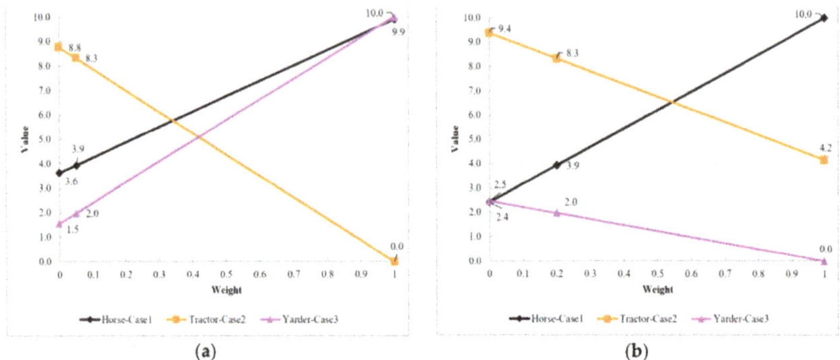

Figure 6. Results of sensitivity analysis using MAUT method when changing weight of #8 employment (EMP) (**a**) and #11 soil biological quality (QBS-ar) (**b**).

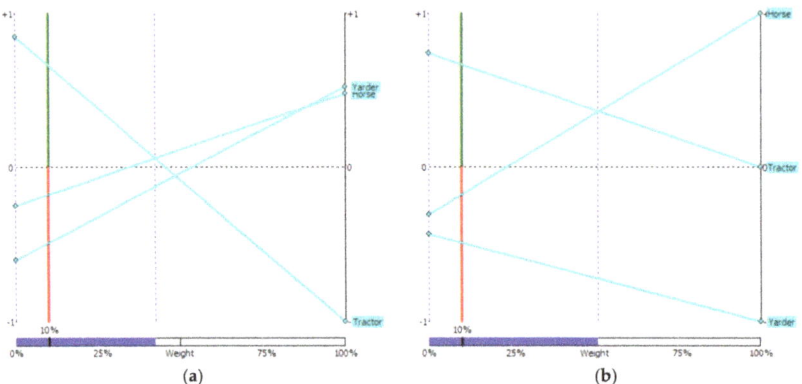

Figure 7. Results of sensitivity analysis using PROMETHEE method when changing weight of #8 EMP (0–0.427) (**a**) and #11 QBS-ar (0–0.511) (**b**).

4. Discussion

This study showed results from a case study that was carried out in a 60 year old black pine stand in the Abruzzo region in Italy. The mainstream silvicultural prescription for these stands is two to four thinning operations, followed by clear-cutting and replanting or renaturalization.

In this case, the forest management goal was to ensure the partial permanence of land cover, with the gradual replacement of pine with late successional tree species that are typical of more mature stages of evolution. Clear-cutting on strips was conducted as silvicultural operation with the aim to eliminate 50% of the surface of the plantations, and thereby to support natural renaturalization.

Thinning operations can be carried out by using many harvesting systems. The most popular are cut-to-length and whole-tree harvesting. The latter was applied in this study. However, different extraction processes were conducted (Figures 1 and 2): In case 1, extraction was conducted by animal (heavy rapid skidding horse, TPR-horse), in case 2 extraction was conducted by, forestry-fitted farm tractor with winch, and in case 3 extraction was conducted by double drum cable yarder. After extraction, trees were transported to the landing, where the trees were chipped for energy purposes, which is a common procedure. All forest wood chains were repeated four times.

One of the most challenging tasks in forest management is to consider the consequences of different strategies or FO and to estimate the economic, environmental and social performance of each alternative before an action is carried out. It is important to consider different pillars of sustainability and to link environmental impacts to socio-economic activities in order to guide DMs in their actions and to ensure that the impacts of their decisions are measured.

Therefore, the aim of the study was to conduct a SIA aiming to assess potential impacts on sustainability that are related to FO being applied to support the renaturalization strategy in typical afforested pine plantations in the Mediterranean basin.

The system boundaries included all processes necessary for turning standing trees in the forest into whole-tree chips loaded on trucks and ready for delivery to the mill. Twelve indicators were considered to be important and feasible with regard to data collection.

Input data were gathered from field studies (as reported in [46]) and respective indicator values were calculated by the use of different tools; e.g., potential environmental impacts of exhaust gases under the use of the Ecoinvent database and Umberto, a tool for LCA.

Recent studies have shown that that there are few studies related to FO considering all pillars of sustainability [38]. Most studies are focused only on either environmental or on economic and environmental aspects. However, the use of several indicators and the combination of different methods to calculate indicator values leads to a strong analytical power for embracing financial, technological, environmental, and other aspects of a production system [30].

Different software tools exist to conduct a SIA [81], including within the context of forestry (e.g., TOSIA, as presented by [39,47,82]). We decided to use a LCA software tool for modelling and analysis, because the method of LCA was used to determine potential environmental impacts, and in the software used, SI other than environmental ones can be defined and added, too.

In all three cases, a team consisting of two workers was necessary to conduct the working processes. Resulting system productivity was highest in case 2. Felling was always conducted motor-manually; there were differences between the cases: The average cutting productivity was low in case 2, when trees could be felled non-accurate orientation. The reason was that the two workers struggled less as the trees were felled based on their natural inclination, but this led to a maze of crossed trees on the ground or situations of hanging trees. Therefore, materially, their proceedings were very often difficult and confusing. The result was higher working times than the oriented felling.

It turned out that the most time-consuming process was bunching and skidding. It reached the highest productivities in case 2, followed by case 1 and case 3 (Table 4). When considering average tree diameters (18–24 cm) and steepness of the terrain (50%) the delays in case 1 might be explained (39% of the total working time of bunching and skidding). In case 3, a high share of the total working of bunching and skidding time was spent on hooking and unhooking (34%) and the share of delays was quite high, too (43%). This could be related to the average tree low dimensions, that for yarder extraction needed mainly one chain for tree, with consequential hooking and unhooking time increasing. In contrast, using a tractor with a winch is a common method to extract trees in the case study region. Thus, operators were experienced and spent less working time on hooking and unhooking (24%) and had fewer delays (30%). More training with a yarder would probably lead to an increase in productivity, too.

Costs followed the same pattern and were almost 2.5 times lower in case 2 than in case 3. These figures are quite impressive when indicator results were scaled up to hectares (Table 8), and when considering that there are about 13,000 ha black pine plantations growing in the studied area. To give an example: Managing all plantations with the harvesting systems and machines presented in case 2 would result in total costs of million € 62.4 while it results in million 168.7 when choosing the harvesting systems and machines presented in case 3.

It has to be noted that, in contrast to productivity and costs, the employment (#8 EMP) was highest in case 3 due to the above-mentioned reasons, followed by case 1 and case 2 (Figure 5). Decision makers should have in mind (i) which infrastructure is given in a specific region (e.g., would a yarder be available?); (ii) that it is increasingly difficult to find skilled labor; and (iii) consideration for the question of which possibilities for rural development of an area there are.

LCA results showed that the cumulated energy demand of fossil energy was lowest in case 1, followed by case 2 and case 3. This fact can be explained by the amount of fuels required by machines. However, surprisingly, the share of indirect emission was quite high (Table 6, Figure 4). For example, it was 38% in the process bunching and skidding in case 1, mainly caused by the daily transport of workers (35 km/way) and the "impact" of the horse on non-working days. It was also high in case 3 (36%) due to the production and maintenance of the yarder.

The potential impacts in the environmental categories global warming potential, eutrophication potential, and acidification potential all followed the same trend (Table 6): Extraction by tractor with a winch resulted in lowest impacts. As inputs were not exclusively, but mainly, fuels we can ascribe to facilitating high productivity and thus, lower fuel consumption reached in case 2.

In mountains areas, the plantations and treatment operations related to re-forestation, had strong and variable effects on plant species occurrence and diversity due to the alteration of ecological processes [83]. However, these plantations contributed to biodiversity conservation in various ways, as found by Poorbabaei and Poorrahmati [84]: A high similarity in species composition between plantation and the adjacent natural forest, which is the main source of seed in plantations, was present. The actual necessity of an active management of pine plantations could have strong and variable effects on plant species' occurrence and diversity due to treatment operations and canopy cover changes.

As found by Picchio et al. [46], both silvicultural treatment and FO applied in this research, showed changes on density, richness, and biodiversity of tree species in only three years after harvesting. The good density and richness of tree species in this pine plantation indicate the high potential reached by the stand for biodiversity restoration, following what was found in other studies [85,86].

Referring to stand regeneration, different taxonomic compositions of the tree forest community among the cases are shown, in particular in the percentage of distribution, showing a simplification in case 1 with respect to the others. In general, in the cases 1 and 2 (ground-based logging) allowed for the presence of *Robinia pseudoacacia* and only marginal *Pinus nigra* regeneration.

The treatments applied showed a positive effect to the SI tree regeneration density, with greater consistency in the cases 2 and 3; compared to the control, they showed increases of 85% and 72%, respectively. The case 1 showed a positive trend, with an increase of about 69% compared to the control.

Other important ecological aspects were assessed, such as the tree richness and diversity of tree species; in particular the tree species diversity was chosen. The case 2 had higher richness values than the control and the cases 1 and 3. However, it is important to note that the data presented so far concern a limited period of time; more time is needed to further evaluate whether the cutting effect on biodiversity will last long [87,88].

Indicator values of the SI soil biological quality showed for the three cases an impact, and the observed variation is explained by the different degrees of soil compaction and the abundance of litter associated with sudden stand removal [59,89]. The QBS-ar values were lowest in the case 3, followed by the case 1. The best situation was found for case 2. In addition, the SI soil microarthropod community density was assessed, and, as can be observed from the data gathered, it was impacted by FOs. Case 2 had higher values than the cases 1 and 3.

To help DMs judging these results, a MCDA was conducted. Weights of criteria were obtained using the DIRECT method. As shown in Table 9, the SI tree regeneration density and soil biological quality were set as most important criteria, because they support the achievement of the forest management goal, followed by the environmental criteria cumulated energy demand, global warming potential, eutrophication potential and acidification potential, while the least important criteria were tree species diversity, soil microarthropod community density, employment, and COSTs (with a weight of 0.050). This decision was related to DMs' intention to give higher priority to environmental criteria. The two SIs productivity and delay were excluded from the MCDA because they were included in other SI, e.g., in Costs.

For the ranking of alternatives (cases), two different MCDA methods were applied, namely MAUT and PROMETHEE. They have different philosophies, and therefore often produce different results (rankings), but here, this was not the case. The main reason for that lies in a fact that first ranked alternative (case 2) had best performance for eight (out of 10) criteria. Only for the two criteria, employment and soil biological quality, other alternatives had better performances. Because of that, a sensitive analysis was carried out aiming to estimate how much one needs to change (increase) the weights of those two criteria—employment and soil biological quality—in order to change first ranked alternative (Figures 6 and 7). From the results of the sensitive analysis, we can conclude that case 2 is a very stable first ranked alternative and can be selected as the best one for this case study. Worthy of mention is that the results of the MCDA process were presented to participating individuals; i.e., DMs. No significant complaints by DMs were made about the ranks of analyzed alternatives. Overall, presented approach can improve (and simplify) decision making process and may help experts (or DMs) to select the best alternative for given context.

5. Conclusions

In the preceding years, several changes were ongoing the forest world; for example, the growing interest in sustainability, due to the new awareness of people about the importance of forests from environmental and social points of view, which increased the need of having strong and reliable instruments for decision makers (DMs) to optimize choices in order to satisfy all forests' stakeholders and interests.

From this perspective, this paper was born with the aim to assess possible impacts on sustainability that were related to FO and the resulting forest wood chains to support the renaturalization strategy in typical. afforested pine plantations in the Mediterranean basin. In detail, three main topics were studied in order to: (i) Identify alternative FO concerning silvicultural actions suitable for renaturalization of the pine stands, thereby putting a special emphasis to the extraction process; (ii) assess the potential impacts on all three pillars of sustainability; and (iii) make comprehensive evaluations of the alternative forest wood chains in order to support DMs.

In order to reach aim the first aim, a SIA and a MCDA were conducted for three different extraction methods in pine stands thinning operations, considering Mediterranean setting. In particular, the analyzed extraction systems were: TPR horse, forestry-fitted farm tractor with a winch, and double drum cable yarder. Obtained results showed that a tractor with a winch was clearly the best alternative, since it showed the best performance for eight out of 10 investigated variables. Thus, it can be said that a forestry-fitted farm tractor with a winch was the best alternative from an economic, environmental, and social point of view. This result was reached setting the SIA and MCDA with particular attention to the environmental aspects, considering that study area is located in a Natural Reserve and that the most important aim of the silvicultural intervention was not economic gain but renaturalization.

The specific result focused on the second aim showed a detailed assessment of FO consequences on all three pillars of sustainability. From economic point of view, only cable yarder showed no positive results, more related to the silvicultural treatment applied. TPR-horses and a tractor with a winch, instead, reached good economic performance. About environmental pillar, all FO applied in this research showed changes on density, richness, and biodiversity of tree species in only three years after harvesting. Indicator values of the QBS-ar showed an impact for the three cases, so one might say that soil ecosystem restoration, in this case, is slower than forest stand one. However, for all these parameters, tractors with winches showed the best values. Concerning the social point of view, it can be said that all three extraction methods had the same labor requirements. In central Italy's context a TPR-horse and the tractor with a winch are the best-known extraction methods, and this partially explains the cheap results of a cable yarder from economic point of view. In this context, an improvement in cable yarder use, linked to workers' proper formation, should be recommended; however, that should consider the conditions of high slopes and lack of viability of central Italy forest, in particular, pine stands.

These are important results that fit with one of the major challenges of forest management, regarding the consequences of different management strategies or FOs, by assessing the economic, environmental, and social performance of each individual option before an action is carried out.

Focusing around the third aim, it was possible to affirm that tractor with winch resulted to have the best performance from all point of views, and it represented the best choice for pine stands renaturalization interventions. In fact, it combined good productivity and so quite low costs, contained environmental impacts and good recovery capacity of pre-intervention conditions, and optimum knowledge of its functioning and safety rules of work by central Italy forest workers.

In relation to cable yarder it was important to underline how the poor performances were mainly linked to the silvicultural treatment design (strips of 100 m length were a limit for this equipment).

On the other hand, obtained results confirmed what detected in other previous studies about extraction with animals. The general performances of this extraction methodology were often worse than mechanical ones, not only related to productivity aspects but also to environmental impacts. Even though in this study a TPR-horse resulted to be a good alternative to cable yarder.

Finally, it was possible to say that SIA and MCDA showed satisfying performance in analyzing FO alternatives and thus they resulted to be strong instruments to support DM; and this is very important in the perspective of reaching a sustainable forest management, which leads to satisfy all three pillars of sustainability.

Author Contributions: R.P., R.V. and F.L. performed the field experiments and calculated economic and social indicators; J.S. conducted the LCA; B.B. ran the MCDA; J.S. wrote the manuscript with contributions from R.P. and B.B.

Funding: This research received no external funding.

Acknowledgments: This work was supported by the Italian Ministry for education, University and Research (MIUR) for financial support (Law 232/2016, Italian University Departments of excellence)—UNITUS-DAFNE WP3. While working at the University of Freiburg, J.S. was supported by the European Social Fund and by the Ministry of Science, Research and Arts Baden-Württemberg.

Conflicts of Interest: The authors declare no conflict of interest. The founding sponsors had no role in the design of the study; the collection, analyses, or interpretation of data; the writing of the manuscript, or in the decision to publish the results.

Abbreviations

Sustainability impact assessment (SIA), sustainable forest management (SFM), sustainable development (SD), sustainability indicators (SI), life cycle assessment (LCA), forest operations (FO), heavy rapid skidding horse (TPR-horse), productive machine hours (PMH), productive working hours (PWH), productive machine hours (PMH), greenhouse gas (GHG), global warming potential (GWP), eutrophication potential (EP), acidification potential (AP), employment (EMP), tree regeneration density (TRD), tree species diversity (TSD), soil biological quality index (QBS-ar), soil microarthropod density (SMD), multi-criteria decision analysis (MCDA), multi-attribute utility theory (MAUT), decision makers (DMs), individuals (IND).

References

1. Tapias, R.; Climent, J.; Pardos, J.A.; Gil, L. Life histories of Mediterranean pines. *Plant Ecol.* **2004**, *171*, 53–68. [CrossRef]
2. Peñuelas, J.L.; Ocaña, L. *Cultivo de Plantas Forestales en Contenedor*; Pesca y Alimentación & Mundi-Prensa: Madrid, Spain, 1996.
3. Barbéro, M.; Loisel, R.; Quézel, P.; Richardson, D.M.; Romane, F. Pines of the Mediterranean Basin. In *Ecology and Biogeography of Pinus*; Richardson, D.M., Ed.; Cambridge University Press: Cambridge, UK, 2000; pp. 153–170.
4. Maestre, F.T.; Cortina, J. Are Pinus halepensis plantations useful as a restoration tool in semiarid Mediterranean areas? *For. Ecol. Manag.* **2004**, *198*, 303–317. [CrossRef]

5. Pausas, J.G.; Bladé, C.; Valdecantos, A.; Seva, J.P.; Fuentes, D.; Alloza, J.A.; Vilagrosa, A.; Bautista, S.; Cortina, J.; Vallejo, R. Pines and oaks in the restoration of Mediterranean landscapes of Spain: New perspectives for an old practice—A review. *Plant Ecol.* **2004**, *171*, 209–220. [CrossRef]
6. Marchi, M.; Paletto, A.; Cantiani, P.; Bianchetto, E.; de Meo, I. Comparing thinning system effects on ecosystem services provision in artificial black pine (Pinus nigra J. F. Arnold). *Forests* **2018**, *9*, 188. [CrossRef]
7. Spiecker, H. Silvicultural management in maintaining biodiversity and resistance of forests in Europe—Temperate zone. *J. Environ. Manag.* **2003**, *67*, 55–65. [CrossRef]
8. Moriondo, M.; Good, P.; Durao, R.; Bindi, M.; Giannakopoulos, C.; Côrte-Real, J. Potential impact of climate change on fire risk in the Mediterranean area. *Clim. Res.* **2006**, *31*, 85–95. [CrossRef]
9. Sarris, D.; Christodoulakis, D.; Körner, C. Impact of recent climatic change on growth of low elevation eastern Mediterranean forest trees. *Clim. Chang.* **2011**, *106*, 203–223. [CrossRef]
10. Gómez-Aparicio, L.; Zavala, M.A.; Bonet, F.J.; Zamora, R. Are pine plantations valid tools for restoring Mediterranean forests? An assessment along gradients of climatic conditions, stand density and distance to seed sources. *Ecol. Appl.* **2009**, *19*, 2124–2141. [CrossRef]
11. Jiménez, M.N.; Spotswood, E.N.; Cañadas, E.M.; Navarro, F.B. Stand management to reduce fire risk promotes understorey plant diversity and biomass in a semi-arid Pinus halepensis plantation. *Appl. Veg. Sci.* **2015**, *18*, 467–480. [CrossRef]
12. Navarro-Cerrillo, R.M.; Sánchez-Salguero, R.; Rodriguez, C.; Duque Lazo, J.; Moreno-Rojas, J.M.; Palacios-Rodriguez, G.; Camarero, J.J. Is thinning an alternative when trees could die in response to drought? The case of planted Pinus nigra and P. Sylvestris stands in southern Spain. *For. Ecol. Manag.* **2019**, *433*, 313–324. [CrossRef]
13. Ciancio, O.; Mercurio, R.; Nocentini, S. First thinnings of young conifer stands: Strategy and perspectives. In Proceedings of the IUFRO XX World Congress, Tampere, Finland, 6–12 August 1995; pp. 61–66.
14. De Meo, I.; Cantiani, P.; Becagli, C.; Bianchetto, E.; Cazau, C.; Mocali, S.; Salerni, E. Thinnings to enanche biodiversity in black pine stands: A case study in Italian Appennine. In Proceedings of the Forestry: Bridge to the Future Conference, Sofia, Bulgaria, 6–9 May 2015.
15. Muscolo, A.; Settineri, G.; Bagnato, S.; Mercurio, R.; Sidari, M. Use of canopy gap openings to restore coniferous stands in Mediterranean environment. *iForest Biogeosci. For.* **2017**, *10*, 322–327. [CrossRef]
16. Pollet, J.; Omi, P.N. Effect of thinning and prescribed burning on crown fire severity in ponderosa pine forests. *Int. J. Wildland Fire* **2002**, *11*. [CrossRef]
17. Crecente-Campo, F.; Pommerening, A.; Rodríguez-Soalleiro, R. Impacts of thinning on structure, growth and risk of crown fire in a Pinus sylvestris L. plantation in northern Spain. *For. Ecol. Manag.* **2009**, *257*, 1945–1954. [CrossRef]
18. Marchi, E.; Chung, W.; Visser, R.; Abbas, D.; Nordfjell, T.; Mederski, P.S.; McEwan, A.; Brink, M.; Laschi, A. Sustainable Forest Operations (SFO): A new paradigm in a changing world and climate. *Sci. Total Environ.* **2018**, *634*, 1385–1397. [CrossRef] [PubMed]
19. Olajuyigbe, S.; Tobin, B.; Saunders, M.; Nieuwenhuis, M. Forest thinning and soil respiration in a Sitka spruce forest in Ireland. *Agric. For. Meteorol.* **2012**, *157*, 86–95. [CrossRef]
20. Cambi, M.; Certini, G.; Neri, F.; Marchi, E. The impact of heavy traffic on forest soils: A review. *For. Ecol. Manag.* **2015**, *338*, 124–138. [CrossRef]
21. Grigal, D.F. Effects of extensive forest management on soil productivity. *For. Ecol. Manag.* **2000**, *138*, 167–185. [CrossRef]
22. Korb, J.E.; Fulé, P.Z.; Gideon, B. Different Restoration Thinning Treatments Affect Level of Soil Disturbance in Ponderosa Pine Forests of Northern Arizona, USA. *Ecol. Restor.* **2007**, *25*, 43–49. [CrossRef]
23. Petrenko, C.L.; Friedland, A.J. Mineral soil carbon pool responses to forest clearing in Northeastern hardwood forests. *GCB Bioenergy* **2015**, *7*, 1283–1293. [CrossRef]
24. Frey, B.; Niklaus, P.A.; Kremer, J.; Lüscher, P.; Zimmermann, S. Heavy-Machinery Traffic Impacts Methane Emissions as Well as Methanogen Abundance and Community Structure in Oxic Forest Soils. *Appl. Environ. Microbiol.* **2011**, *77*, 6060–6068. [CrossRef]
25. Picchio, R.; Neri, F.; Petrini, E.; Verani, S.; Marchi, E.; Certini, G. Machinery-induced soil compaction in thinning two pine stands in central Italy. *For. Ecol. Manag.* **2012**, *285*, 38–43. [CrossRef]
26. Tavankar, F.; Bonyad, A.E.; Majnounian, B. Affective factors on residual tree damage during selection cutting and cable-skidder logging in the Caspian forests, Northern Iran. *Ecol. Eng.* **2015**, *83*, 505–512. [CrossRef]

27. Picchio, R.; Magagnotti, N.; Sirna, A.; Spinelli, R. Improved winching technique to reduce logging damage. *Ecol. Eng.* **2012**, *47*, 83–86. [CrossRef]
28. Cambi, M.; Hoshika, Y.; Mariotti, B.; Paoletti, E.; Picchio, R.; Venanzi, R.; Marchi, E. Compaction by a forest machine affects soil quality and Quercus robur L. seedling performance in an experimental field. *For. Ecol. Manag.* **2017**, *384*, 406–414. [CrossRef]
29. Đuka, A.; Vusić, D.; Horvat, D.; Šušnjar, M.; Pandur, Z.; Papa, I. LCA Studies in Forestry—Stagnation or Progress? *Croat. J. For. Eng.* **2017**, *38*, 311–326.
30. Schweier, J.; Molina-Herrera, S.; Ghirardo, A.; Grote, R.; Díaz-Pinés, E.; Kreuzwieser, J.; Haas, E.; Rennenberg, H.; Schnitzler, J.-P.; Becker, G.; et al. Environmental impacts of bioenergy wood production from poplar short-rotation coppice grown at a marginal agricultural site in Germany. *GCB Bioenergy* **2017**, *9*, 1207–1221. [CrossRef]
31. Albizu-Urionabarrenetxea, P.M.; Tolosana-Esteban, E.; Roman-Jordan, E. Safety and health in forest harvesting operations. Diagnosis and preventive actions. A review. *For. Syst.* **2013**, *22*, 392–400. [CrossRef]
32. Laschi, A.; Marchi, E.; González-García, S. Forest operations in coppice: Environmental assessment of two different logging methods. *Sci. Total Environ.* **2016**, *562*, 493–503. [CrossRef]
33. Lindroos, O.; La Hera, P.; Häggström, C. Drivers of Advances in Mechanized Timber Harvesting—A Selective Review of Technological Innovation. *Croat. J. For. Eng.* **2017**, *38*, 243–258.
34. Laschi, A.; Marchi, E.; Foderi, C.; Neri, F. Identifying causes, dynamics and consequences of work accidents in forest operations in an alpine context. *Saf. Sci.* **2016**, *89*, 28–35. [CrossRef]
35. Spinelli, R.; Magagnotti, N.; Schweier, J. Trends and perspectives in coppice harvesting. *Croat. J. For. Eng.* **2017**, *38*, 219–230.
36. Enache, A.; Kühmaier, M.; Visser, R.; Stampfer, K. Forestry Operations in the European Mountains: A study of current practices and efficiency gaps. *Scand. J. For. Res.* **2016**, *31*, 1–39. [CrossRef]
37. Siry, J.P.; Cubbage, F.W.; Potter, K.M.; McGinley, K. Current Perspectives on Sustainable Forest Management: North America. *Curr. For. Rep.* **2018**, *4*, 138–149. [CrossRef]
38. Schweier, J.; Magagnotti, N.; Labelle, E.R.; Athanassiadis, D. Sustainability Impact Assessment of Forest Operations: A Review. *Curr. For. Rep.* **2019**, 1–13. [CrossRef]
39. Lindner, M.; Suominen, T.; Palosuo, T.; Garcia-Gonzalo, J.; Verweij, P.; Zudin, S.; Päivinen, R. ToSIA—A tool for sustainability impact assessment of forest-wood-chains. *Ecol. Model.* **2010**, *221*, 2197–2205. [CrossRef]
40. Diaz-Balteiro, L.; Romero, C. Making forestry decisions with multiple criteria: A review and an assessment. *For. Ecol. Manag.* **2008**, *255*, 3222–3241. [CrossRef]
41. Belton, V.; Stewart, T. *Multiple Criteria Decision Analysis: An Integrated Approach*; Springer: Norwell, MA, USA, 2002; p. 372.
42. Triantaphyllou, E.; Shu, B.; Nieto Sanchez, S.; Ray, T. Multi-Criteria Decision Making: An Operations Research Approach. In *Encyclopedia of Electrical and Electronics Engineering*; John Wiley & Sons: New York, NY, USA, 2015; Volume 15, pp. 175–186.
43. Acosta, M.; Corral, S. Multicriteria Decision Analysis and Participatory Decision Support Systems in Forest Management. *Forests* **2017**, *8*, 116. [CrossRef]
44. Schweier, J.; Spinelli, R.; Magagnotti, N.; Wolfslehner, B.; Lexer, M.J. Sustainability Assessment of Alternative Thinning Operations in Mediterranean Softwood Plantations. *Forests* **2018**, *9*, 375. [CrossRef]
45. Blagojević, B.; Jonsson, R.; Björheden, R.; Nordström, E.-M.; Lindroos, O. Multi-Criteria Decision Analysis (MCDA) in Forest Operations—An Introductional Review. *Croat. J. For. Eng.* **2019**, *40*, 191–205.
46. Picchio, R.; Mercurio, R.; Venanzi, R.; Gratani, L.; Giallonardo, T.; Monaco, A.L.; Frattaroli, A.R. Strip Clear-Cutting Application and Logging Typologies for Renaturalization of Pine Afforestation—A Case Study. *Forests* **2018**, *9*, 366. [CrossRef]
47. Päivinen, R.; Lindner, M.; Rosén, K.; Lexer, M.J. A concept for assessing sustainability impacts of forestry-wood chains. *Eur. J. For. Res.* **2012**, *131*, 7–19. [CrossRef]
48. European Commission; Joint Research Centre; Institute for Environment and Sustainability (Eds.) *International Reference Life Cycle Data System (ILCD) Handbook—General Guide for Life Cycle Assessment—Detailed Guidance*, 1st ed.; Publications Office of the European Union: Luxembourg, 2010.
49. Päivinen, R.; Lindner, M. *Assessment of Sustainability of Forest Wood Chains*; Cesaro, L., Gatto, P., Pettenella, D., Eds.; European Forest Institute: Joensuu, Finland, 2008; pp. 153–160.

50. Pülzl, H.; Prokofieva, I.; Berg, S.; Rametsteiner, E.; Aggestam, F.; Wolfslehner, B. Indicator development in sustainability impact assessment: Balancing theory and practice. *Eur. J. For. Res.* **2012**, *131*, 35–46. [CrossRef]
51. OECD. *Environmental Indicators. Towards Sustainable Development*; OECD Publications: Paris, France, 2001.
52. MCPFE. *Improved Pan-European Indicators for Sustainable Forest Management*; MCPFE Liaison Unit: Vienna, Austria, 2003; Available online: http://www.foresteurope.org/documentos/improved_indicators.pdf (accessed on 18 October 2017).
53. Pourbabaei, H.; Haddadi-Moghaddam, H.; Begyom-Faghir, M.; Abedi, T. Theinfluence of gap size on plant species diversity and composition in beech (Fagus orientalis) forests, Ramsar, Mazandaran Province, North of Iran. *Biodiversitas* **2013**, *14*, 89–94. [CrossRef]
54. Picchio, R.; Tavankar, F.; Venanzi, R.; Lo Monaco, A.; Nikooy, M. Study of forest road effect on tree community and stand structure in three Italian and Iranian temperate forests. *Croat. J. For. Eng.* **2018**, *39*, 57–70.
55. Pielou, E. The measurement of diversity in different types of biological collections. *J. Theor. Boil.* **1966**, *13*, 131–144. [CrossRef]
56. Burton, P.J.; Balisky, A.C.; Coward, L.P.; Kneeshaw, D.D.; Cumming, S.G. The value of managing for biodiversity. *For. Chron.* **1992**, *68*, 225–237. [CrossRef]
57. Begehold, H.; Rzanny, M.; Winter, S. Patch patterns of lowland beech forests in a gradient of management intensity. *For. Ecol. Manag.* **2016**, *360*, 69–79. [CrossRef]
58. Parisi, V.; Menta, C.; Gardi, C.; Jacomini, C.; Mozzanica, E. Microarthropod communities as a tool to assess soil quality and biodiversity: A new approach in Italy. *Agric. Ecosyst. Environ.* **2005**, *105*, 323–333. [CrossRef]
59. Venanzi, R.; Picchio, R.; Piovesan, G. Silvicultural and logging impact on soil characteristics in Chestnut (Castanea sativa Mill.) Mediterranean coppice. *Ecol. Eng.* **2016**, *92*, 82–89. [CrossRef]
60. Marchi, E.; Picchio, R.; Mederski, P.S.; Vusi´c, D.; Perugini, M.; Venanzi, R. Impact of silvicultural treatment and forest operation on soil and regeneration in Mediterranean Turkey oak (*Quercus cerris* L.) coppice with standards. *Ecol. Eng.* **2016**, *95*, 475–484. [CrossRef]
61. Blasi, S.; Menta, C.; Balducci, L.; Delia Conti, F.; Petrini, E.; Piovesan, G. Soil microarthropod communities from Mediterranean forest ecosystems in Central Italy under different disturbances. *Environ. Monit. Assess.* **2013**, *185*, 1637–1655. [CrossRef] [PubMed]
62. Picchio, R.; Spina, R.; Maesano, M.; Carbone, F.; Monaco, A.L.; Marchi, E. Stumpage value in the short wood system for the conversion into high forest of an oak coppice. *For. Stud. China* **2011**, *13*, 252–262. [CrossRef]
63. Spinelli, R.; Visser, R. Analyzing and Estimating Delays in Harvester Operations. *Int. J. For. Eng.* **2008**, *19*, 36–41. [CrossRef]
64. Engel, A.-M.; Wegener, J.; Lange, M. Greenhouse gas emissions of two mechanised wood harvesting methods in comparison with the use of draft horses for logging. *Eur. J. For. Res.* **2012**, *131*, 1139–1149. [CrossRef]
65. Picchio, R.; Maesano, M.; Savelli, S.; Marchi, E. Productivity and energy balance in conversion of a Quercus cerris L. coppice stand into high forest in Central Italy. *Croat. J. For. Eng.* **2009**, *30*, 15–26.
66. Knechtle, N. *Materialprofile von Holzerntesystemen—Analyse Ausgewählter Beispiele als Grundlage für ein Forsttechnisches Ökoinventar*; ETH Zürich: Zürich, Switzerland, 1997.
67. Athanassiadis, D.; Lidestav, G.; Nordfjell, T. Energy use and emissions due to the manufacture of a forwarder. *Resour. Conserv. Recycl.* **2002**, *34*, 149–160. [CrossRef]
68. Timberjack. *Green Forests Machines for Sustainable Development. Environmental Declaration*; Timberjack: Tampere, Finland, 2002.
69. Institut für Umweltinformatik. *Umberto*; Institut für Umweltinformatik: Hamburg, Germany, 2011.
70. Ecoinvent. Swiss Center for Life Cycle Inventories: St Gallen, Switzerland. 2010. Available online: https://www.ecoinvent.org/ (accessed on 20 May 2019).
71. Tavankar, F.; Majnounian, B.; Bonyad, A. Felling and skidding damage to residual trees following selection cutting in Caspian forests of Iran. *J. For. Sci.* **2013**, *59*, 196–203. [CrossRef]
72. INFC. *Inventario Nazionale Delle Foreste E Dei Serbatoi Forestali Di Carbonio [Forest Area Estimates 2005—Section 1. MiPA—Corpo Forestale Dello Stato—Ispettorato Generale, CRA-ISAFA, Trento, Italy. Italian National Forest Inventory, INFC]*; INFC: Trento, Italy, 2007; pp. 1–413. (In Italian)
73. Keeney, R.; Raiffa, H. *Decision with Multiple Objectives: Preferences and Value Tradeoffs*; Wiley: New York, NY, USA, 1976; p. 569.
74. Brans, J.; Vincke, P.; Mareschal, B. How to select and how to rank projects: The Promethee method. *Eur. J. Oper. Res.* **1986**, *24*, 228–238. [CrossRef]

75. Ishizaka, A.; Nemery, P. *Multi-Criteria Decision Analysis: Methods and Software*; John Wiley & Sons: Chichester, UK, 2013; p. 310.
76. Linkov, I.; Varghese, A.; Jamil, S.; Seager, T.P.; Kiker, G.M.; Bridges, T. Multi-criteria decision analysis: A framework for structuring remedial decisions at contaminated sites. In *Comparative Risk Assessment and Environmental Decision Making*; Springer: Dordrecht, The Netherlands, 2004; pp. 15–54.
77. Behzadian, M.; Kazemzadeh, R.; Albadvi, A.; Aghdasi, M. PROMETHEE: A comprehensive literature review on methodologies and applications. *Eur. J. Oper. Res.* **2010**, *200*, 198–215. [CrossRef]
78. Hokkanen, J.; Salminen, P. Choosing a solid waste management system using multicriteria decision analysis. *Eur. J. Oper. Res.* **1997**, *98*, 19–36. [CrossRef]
79. Kangas, A.; Kurttila, M.; Hujala, T.; Eyvindson, K.; Kangas, J. *Decision Support for Forest Management*, 2nd ed.; Springer: Berlin, Germany, 2015; p. 307.
80. Sultana, A.; Kumar, A. Ranking of biomass pellets by integration of economic, environmental and technical factors. *Biomass Bioenergy* **2012**, *39*, 344–355. [CrossRef]
81. Taisch, M.; Sadr, V.; May, G.; Stahl, B. Sustainability Assessment Tools—State of Research and Gap Analysis. In *Advances in Production Management Systems. Sustainable Production and Service Supply Chains*; Prabhu, V., Taisch, M., Kiritsis, D., Eds.; Springer: Berlin/Heidelberg, Germany, 2013; pp. 426–434.
82. Rosén, K.; Lindner, M.; Nabuurs, G.J.; Paschalis-Jakubowicz, P. Challenges in implementing sustainability impact assessment of forest wood chains. *Eur. J. For. Res.* **2012**, *131*, 1–5. [CrossRef]
83. Brosofske, K.; Chen, J.; Crow, T. Understory vegetation and site factors: Implications for a managed Wisconsin landscape. *For. Ecol. Manag.* **2001**, *146*, 75–87. [CrossRef]
84. Poorbabaei, H.; Poorrahmati, G. Plant species diversity in loblolly pine (Pinus taeda L.) and sugi (Cryptomeria japonica D. Don.) plantations in the western Guilan, Iran. *Int. J. Biodvers. Conserv.* **2009**, *1*, 38–44.
85. Carnevale, N.J.; Montagnini, F. Facilitating regeneration of secondary forests with the use of mixed and pure plantations of indigenous tree species. *For. Ecol. Manag.* **2002**, *163*, 217–227. [CrossRef]
86. Gratani, L.; Frattaroli, A.R.; Console, C. Regeneration of the undergrowth in reafforested areas with Pinus nigra, in the High Aterno Valley (Italy). *Belg. J. Bot.* **1994**, *127*, 61–66.
87. Lust, N.; Muys, B.; Nachtergale, L. Increase of biodiversity in homogeneous Scots pine stands by an ecologically diversified management. *Biodivers. Conserv.* **1998**, *7*, 249–260. [CrossRef]
88. Götmark, F.; Paltto, H.; Nordén, B.; Götmark, E. Evaluating partial cutting in broadleaved temperate forest under strong experimental control: Short-term effects on herbaceous plants. *For. Ecol. Manag.* **2005**, *214*, 124–141. [CrossRef]
89. Picchio, R.; Spina, R.; Calienno, L.; Venanzi, R.; Lo Monaco, A. Forest operations for implementing silvicultural treatments for multiple purposes. *Ital. J. Agron.* **2016**, *11*, 156–161.

Article

Production of Wood Pellets from Poplar Trees Managed as Coppices with Different Harvesting Cycles

Vincenzo Civitarese [1], Andrea Acampora [1], Giulio Sperandio [1], Alberto Assirelli [1] and Rodolfo Picchio [2,*]

1 Consiglio per la ricerca in agricoltura e l'analisi dell'economia agraria (Crea), Via Po n° 14, 00198 Rome, Italy
2 Department of Agriculture and Forest Sciences (DAFNE), University of Tuscia, Via San Camillo de Lellis, 01100 Viterbo, Italy
* Correspondence: r.picchio@unitus.it; Tel.: +39-0761-357-400

Received: 20 June 2019; Accepted: 30 July 2019; Published: 1 August 2019

Abstract: High-density biomass plantations have played a key role in the national energy landscape in Italy since the 1990s but, to date, an inversion of tendency and a significant reduction of cultivated areas has been noted. Despite this, the existing plantations have seen their coppicing rotation become significantly lengthened, resulting in large quantities of biomass per hectare. This study aimed to identify the best raw material suitable for pellet production using whole trees or stems without branches from poplar plantations at the end of the third, sixth and ninth year of age. All types of pellets made reach the requirements of class A1 for diameter, length, moisture content, ash melting point, lower heating value, as well as nitrogen (N), sulfur (S), and heavy metals. None of the theses satisfied the bulk density parameters while for ashes and mechanical durability, a great variability was observed according to the different raw materials used. An improvement in terms of heating value was observed by transforming the poplar wood chips refined into pellets. The pelletizing process using high density poplar plantation as a raw material highlights the possibility of obtaining a product that meets many of the quality standards required on the market. These aspects are closely related to the innovation carried out in the agro-forestry sector for effective energetic sustainability.

Keywords: chipping; pellet; poplar; SRWC; pelletization; biomass quality; energy quality

1. Introduction

The critical issues related to the decreasing availability of energy sources of fossil origin, as well as their geographical distribution in politically unstable areas, together with huge environmental problems at a global scale, have led to an increased focus on the search for alternative energy sources. The use of renewable energy has increased steadily over time due to the need to mitigate climate change by reducing the use of fossil fuels [1,2], which are responsible for the constant increase in the concentration of greenhouse gases (GHG) in the atmosphere [3].

The European Commission, through the 2020 climate and energy package, has drafted a set of binding rules to ensure that the EU achieves its climate and energy targets by 2020, providing for the cutting of 20% of greenhouse gas emissions (compared to the levels of 1990), 20% of energy requirements needing to be derived from renewable sources, and a 20% improvement in energy efficiency [4].

Although biomass has been a subject of great interest in terms of power generation, its use at the industrial level has attracted less attention; the various factors to explain this lack of attention are mainly the low mass and energy density, the dispersion of the raw material and its availability in less convenient forms, and the high transport costs [5].

One way to overcome the limitations resulting from the low bulk density and high transport costs is to use densification processes (pelletization and/or briquetting) before using this material for energy purposes, in order to exploit a homogeneous and easy-to-use solid biofuel, which is also characterized by a higher energy density [6,7].

For these reasons the pellet sector, unlike biomasses from Short Rotation Woody Crops (SRWC), has seen important developments both in terms of production and in terms of the number of installed transformation plants, with a market price exceeding €300/tons in 2018 [8].

The pellet production process is an extrusion process that consists of subjecting the very fine dry biomass to a high pressure and high temperatures, compressing it through a hole of a few millimeters and producing small cylinders that are cut to the desired length and cooled [9,10].

The bioenergy policies implemented by the individual Member States of the European Union are characterized by tax exemptions, mandatory targets to be achieved, subsidies, and biomass sustainability policies that stimulate the growth of the imported wood pellets [11]. The global annual production of wood pellets was recently estimated by the International Energy Agency (IEA) to be about 6 to 8 million tons, with a net potential of about 13 million tons [12]. Pellet production grew from 1.7 million tons in the year 2000 to 28 million tons in 2015 [13], showing an annual increase of 14% compared to 2011 [14]. It is estimated that the demand for pellets from 2020 will be about 50 million tons per year [15] and the consumption of industrial pellets will grow steadily at a rate of 21% year^{-1}, whereas the increment in the consumption of domestic pellets will reach 8.5% year^{-1} [7].

Europe is the major pellet producer and consumer, followed by the USA and then the rest of the world. Europe is also a global net importer of wood pellets [13,16]: the highest consumption of pellets is recorded in the United Kingdom, Finland and Sweden, where they are mainly used to produce electricity and heating; other important pellet consumers in the EU are Belgium, the Netherlands, Denmark and Italy [17]. On the production side, Portugal and Latvia are Europe's biggest exporters of pellets, followed by Germany, Lithuania, Estonia, Finland and Sweden [18,19]. In 2016 the thermal energy obtained from biomasses in Italy amounted to about 7.06 Mtep [20].

In this context the Council for Agricultural Research and Economics (CREA) (AGROENER project Energy from agriculture: sustainable innovations for the bio-economy, financing MiPAAF D.D. n. 26329 of 4 April 2016 [21]) and the Department of Agriculture and Forest Sciences (DAFNE-University of Tuscia, Italy) started an experimental activity aimed to enhance the value of different lignocellulosic materials through the promotion of a demonstration model of pellet production on a company scale.

In Italy high-density biomass plantations have played a key role in the national energy landscape since the 1990s [22,23] but, to date, the interest in this type of crop seems to be disappeared, indicating an inversion of tendency and a significant reduction of cultivated areas [24]. Despite this, the existing plantations have seen their coppicing rotation significantly lengthened, reaching even 5 to 6 years to over 9 to 10 years and this results in larger quantities of biomass per hectare, with much better qualitative characteristics than those found in the classic two- or three-year cycles.

The market price is a key factor for the development of pellet manufacturing [13] and the shredded material from SRWC (Short Rotation Wood Coppice) can then be valorized through a pelletization process, thereby allocating the transformed material to much more profitable markets.

The purpose of the work was to evaluate the possibility of enhancing the SRWC of poplar, verifying the ability for this type of biomass to be transformed into pellet, starting from different coppicing intervals (third, sixth and ninth year of vegetation) and different fractions (whole trees and stems without branches). The dendrometric characteristics of the raw wood material and the qualitative characteristics of the pellets produced were assessed. Some parameters, such as the calorific and moisture content, ashes and heavy metals were determined before and after pelleting in order to identify any differences directly related to the transformation process.

2. Materials and Methods

The activity, developed within the CREA farm (Lat. 42° 06′ 07″ N, Long. 12° 37′ 39″ E), involved the plantation of Short Rotation Coppice of poplar (clone AF6-Populus x Euroamericana), with a density of 7142 stumps per hectare, divided into sectors characterized by different coppicing intervals, with shoots aged between 3 and 11 years [25]. The experimentation was carried out using different treatments with harvesting cycle of: 3 years (roots of 11 years and stems of 3 years), 6 years (roots of 11 years and stems of 6 years), and 9 years (roots and stems of 9 years). The experimental design involved a total of 30 sample trees, 10 for each crop cycle, taken by following the indications suggested by Mitchell et al. [26]. The dimensional analyses concerned the basal diameter of the stem, the total height, the percentage by weight of the branches and the relative dimensions (diameter, length and number).

The sample trees cut down on the 20 February 2018 (time T0), were left in a storage yard for two months (time T1), subdivided by crop cycle and type of product:

- five three-year-old whole trees,
- five six-year-old whole trees,
- five nine-year-old whole trees,
- five three-year-old stems without branches,
- five six-year-old stems without branches,
- five nine-year-old stems without branches.

The biomass was subsequently chipped, refined and positioned within six different bins. The dehydration process of the material was enhanced by periodically exposing it to sunlight for further 40 days (time T2). The pelletization was finally carried out the following day (time T3). The chipping, refining and pelletizing were carried out, respectively, with a Farmi Forest CH260 forestry chipper, a BL-100 shredder, and a 4 kW Bianco line pelletizer. For the reining process, a 6 mm grid was used according to the method provided by Bergstrom et al. [27].

The moisture content [28] of the biomass was monitored from the date of the felling of trees to the time of final pelleting, by detecting the parameters in the 4 intervals of time mentioned above (T0, T1, T2, T3). During the storage in the field, the main meteorological data (monthly precipitation, rainy days, average monthly temperatures) were also recorded by the local weather station, in order to provide information about the climatic trend of the area.

The initial moisture content (T0) was detected on samples taken from other trees of the same crop cycle cut down on the same day. Specifically, basal, median and apical stem wheels were used (thickness of 1.5 cm), as well as portions of different diameters and lengths of branches and treetops, in order not to compromise the integrity of the stored material.

The moisture content of the biomass in the three intervals of time T1, T2, and T3 was determined by taking a total of 90 samples of 500 g of chips, refined and pelletized material (5 for each crop cycle, fraction, type of transformation and reference period considered). After the storage, a further reduction of moisture content may occur during the production process because the biomass is subjected to high pressures with a significant increase in temperature. For this reason, particular attention was paid to verify the possible differences of moisture content before (T2) and after the passage in the pellet machine (T3).

The characterization of the biomass was carried out at the CREA and DAFNE laboratories and concerned the ash content and ash melting point, the heating value, the metals, nitrogen and sulfur, the bulk density, the pellet dimensions, the mechanical durability, and the moisture content. Five samples were used for each parameter, except for pellet sizes (50 samples).

The ash content was calculated according to EN ISO 18122 [29]. Samples of about 1 g were placed in the Lenton EF11/8B muffle furnace and heated to 250 °C for one hour. Subsequently, the temperature was raised to 550 °C for two hours. The determination of ash content was calculated considering the weight loss of the sample before and after the heating process.

The ash samples subsequently underwent a granulometry reduction process in order to obtain a fine and homogeneous powder that was able to guarantee the preparation of the samples for the fusion analysis with a more regular form. Subsequently, the procedure provided for the preparation of a cylindrical ash sample to be introduced into the Sylab SHV-IF 1500 analyzer, according to the CEN/TS 15370-1 [30]. The fusibility analysis is based on the identification of the temperature which corresponds to the start of sample deformation, monitored by a camera connected to a computer (image analysis).

The most important parameter to characterize a substance as fuel is the heating value, determined according to EN ISO 18125 [31]. A sample of dried wood chips was first ground by a knife mill Retsch SM 100, and secondly by a centrifuge mill Retsch ZM 200. The Higher Heating Value (HHV) was determined using the calorimeter Anton Paar 6400 while the Lower Heating Value (LHV) was determined using a logarithmic formula. Four samples of shredded wood were prepared by means of the pellet mill Pellet Press 2810 to produce tablets, weighing 1 g each. Before every single analysis, the instrument was calibrated with benzoic acid.

The determination of heavy metals, which directly influence the formation of aerosols and fly ash during the combustion of wood material [32], was performed using Agilent 7700 ICP-MS according to the provisions of EN ISO 16968 [33]. An aliquot of each sample (about 500 mg), was transferred into special Teflon containers and subjected to acid attack (HNO_3 and H_2O_2) using a microwave digester (Start D, Milestone S.r.l., Sorisole, Italy), and the solutions obtained were diluted and subjected to analysis.

The content of nitrogen (N) and sulfur (S) were measured according to EN ISO 16948 [34] and EN ISO 16994 [35] using an elemental analyzer CHNS-O Costech ECS 4010. The tin capsules with approx. 1 mg of sample were inserted through the autosampler into the analyzer's combustion oven.

The bulk density was evaluated in accordance with EN ISO 17828 [36]. A standard container was filled with a certain amount of shredded material of a given size and shape and subsequently weighed. The bulk density was calculated from the net weight per standard volume and reported with the determined moisture content. The bulk density of the pellet was calculated using a metal cylinder of known volume (0.005 m^3), filled to the rim and weighed using a field dynamometer. The measurement was replicated 8 times.

The pellet dimensions were determined according to EN ISO 17829 [37] by measuring the length and diameter of 50 individual pellets randomly selected per sample. Average values for diameter and length were calculated.

The mechanical durability was analyzed by means of a mechanical durability tester (Andritz Sprout rotation pellet testing apparatus) according to EN ISO 17831-1 [38].

The moisture content was determined according to EN ISO 18134-1 [39], using a Memmert UFP800 drying oven. The samples were taken to the laboratory where they were oven-dried at 103 ± 2 °C, until a constant weight was achieved (weight variation not exceeding 0.2% during a further drying period of 60 minutes). The determination of moisture content was calculated as a percentage of weight loss before and after the drying process.

Data were statistically analyzed using PAST and Statistics software. Morphometric parameters were examined by using One-Way Anova, Welch F test and Kruskal-Wallis. A T-test and Welch F test for moisture content, bulk density and for the comparison of physical-chemical parameters between refined and pelletized wood, were performed.

3. Results

3.1. Size Characteristics of the Crops

Figure 1 shows the main morphometric parameters of the poplar detected in the three cultivation cycles considered. Trees of 3 years were characterized by a diameter, a height, and a weight equal to 8.32 cm, 6.97 m, and 19.62 kg. The length and diameter of branches insertion were slightly less than 1 m and 1 cm while the average length of the treetops was 1.79 m, with a diameter of about 2 cm.

Significantly larger sizes were recorded for the poplars of 6 and 9 years of age: 16.09 and 21.39 cm in diameter, 13.07 and 17.70 m in height, as well as 79.63 and 183.92 kg in weight. The 9-year-old branches were almost double the size of the 6-year-old material, both in terms of length and diameter of insertion on the stem. The treetops, on the other hand, showed very similar dimensions with an average length of more than 3 m and a diameter of about 5 cm.

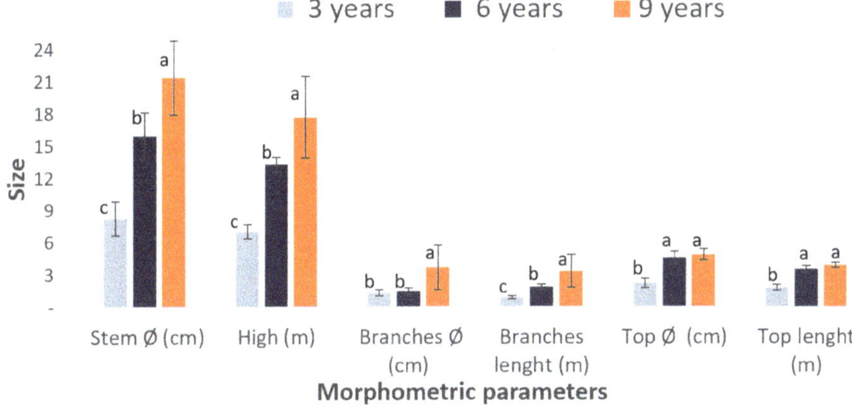

Figure 1. Main morphometric parameters (mean ± St. dev.) of SRWC of poplar at the third, sixth and ninth year of vegetation. Stem diameter (One-Way Anova and Tuckey post hoc test): F = 66.97, df = 27, $p < 0.001$; height (Welch F test and Tuckey post hoc test): F = 253.1, df = 16.15, $p < 0.001$; branches diameter (Welch F test and Tuckey post hoc test): F = 8.997, df = 16.1, $p < 0.01$; branches length (Welch F test and Tuckey post hoc test): F = 71.99, df = 15.39, $p < 0.001$; treetops diameter (One-Way Anova and Tuckey post hoc test): F = 82.41, df = 27, $p < 0.001$; treetops length (Kruskal-Wallis and Mann-Whitney pairwise comparisons Bonferroni corrected): H = 21.83, Hc = 21.87, $p < 0.001$. Different letters indicate statistically significant differences between the groups obtained by post hoc tests applied ($p < 0.05$).

Statistical analysis showed no significant differences with respect to the diameter of the 3- and 6-year-old branches and to the diameter and length of the 6- and 9-year-old treetops. In all other cases, the differences were significant.

The percentage by weight of branches and treetops increased progressively with the increase of the age of the shoots, with the poplar of 3 years that was characterized by the lower average value (Table 1). The one-way Anova highlighted the existence of significant differences between material of 3 years of age and that of 6 and 9 years of age ($p < 0.05$). However, there were no statistically significant differences between the poplars of 6 and 9 years. In the latter case, the values measured were very similar to each other, with about 80% of the weight represented by the stem.

Table 1. Percentage weight distribution of stems, branches and tops (mean ± St. dev.). One-Way Anova: F = 3.603, df = 27, $p < 0.05$. Tuckey post hoc test: different letters indicate statistically significant differences ($p < 0.05$). Percentage values were previously transformed into a square root of the arcosine.

Harvesting Cycle	3 Years	6 Years	9 Years
Branches and tops	17.65% (bc)	19.41% (ac)	20.36% (a)
Stems	82.35%	80.59%	79.64%

3.2. Dehydration Process and Moisture Content of Different Types of Biomass

During the outdoor storage of the wood material, 365 mm of rain were recorded in 33 days, mainly concentrated in the month of March. Average temperatures were 5.72, 11.48 and 15.1 °C, respectively, in the last week of February, in March and in the first two weeks of April. At the time the trees were cut down, they had an average moisture content of 54.74% (T0) reaching, after 2 months of storage, an average value of 46% (T1). Subsequently, the refined biomass was conserved under the roof and periodically exposed to sunlight, attested on average values of just under 10% (T2).

The reduction of the moisture content after pelleting, which was assessed comparing the values for T2 and T3 periods, ranged between 7.61 and 18.53 percentage points. The t-test performed showed, in all the examined cases, statistically significant differences (Table 2).

Table 2. Biomass moisture content in the 4 reference periods and loss of moisture during palletization (mean ± St. dev.). T test columns Δ T3/T2. The asterisks indicated different levels of statistical significance: $p < 0.05$ (*), $p < 0.01$ (**), $p < 0.001$ (***). Percentage values were previously transformed into a square root of the arcosine.

Type of Biomass Source	Trees T_0 20 February	Chips T_1 20 April	Refined T_2 31 May	Pellet T_3 1 June	Δ T_3/T_2
3 years old whole tree	52.28% (±0.96)	46.05% (±4.07)	9.53% (±0.38)	8.32% (±0.37)	−12.72% (±2.42) ***
3 years old stem	52.72% (±0.73)	46.69% (±3.07)	10.21% (±0.43)	8.46% (±0.35)	−17.04% (±4.77) ***
6 years old whole tree	53.98% (±0.87)	45.77% (±4.39)	10.05% (±0.55)	8.30% (±0.23)	−17.25% (±4.72) ***
6 years old stem	53.88% (±1.36)	48.03% (±3.43)	9.43% (±0.33)	8.71% (±0.31)	−7.61% (±4.96) **
9 years old whole tree	56.27% (±1.81)	47.95% (±3.93)	9.32% (±0.63)	8.35% (±0.31)	−10.24% (±4.54) *
9 years old stem	56.85% (±2.09)	50.78% (±4.53)	10.40% (±0.39)	8.47% (±0.37)	−18.53% (±4.94) ***
Average	**54.74%**	**46.05%**	**9.76%**	**8.50%**	**−12.62%**

3.3. Length, Diameter and Bulk Density of the Pellets

The average diameter of the cylinders was just over 6 mm, with a length between 15.13 and 17.93 mm for the whole material and 17.35 and 20.08 mm for the material without branches. At the same crop cycle, the pellets obtained from whole trees were characterized by a lower average length: −7.48%, −5.84% and −32.65% for the cycle of 3, 6 and 9 years, even if this lower average length was statistically significant exclusively for the poplar of a 9-year-old (One-way Anova and Tuckey post hoc test. Df: 144, MS: 32.7052, F: 2.51, $p < 0.05$).

Bulk density, on the other hand, varied between 576 and 584 kg·m^{-3} for trees without branches and 553 and 556 kg·m^{-3} for whole trees. The use of the branches and tree-tops leaded to a reduction in the average bulk density values, confirming the trend observed for the length of the material produced. The reductions recorded stood at −4.32%, −5.53% and −3.67% for the poplar of 3, 6 and 9 years. The statistical analysis, in this case, showed significant differences between the whole material and that without branches in all the cycles considered (Figure 2).

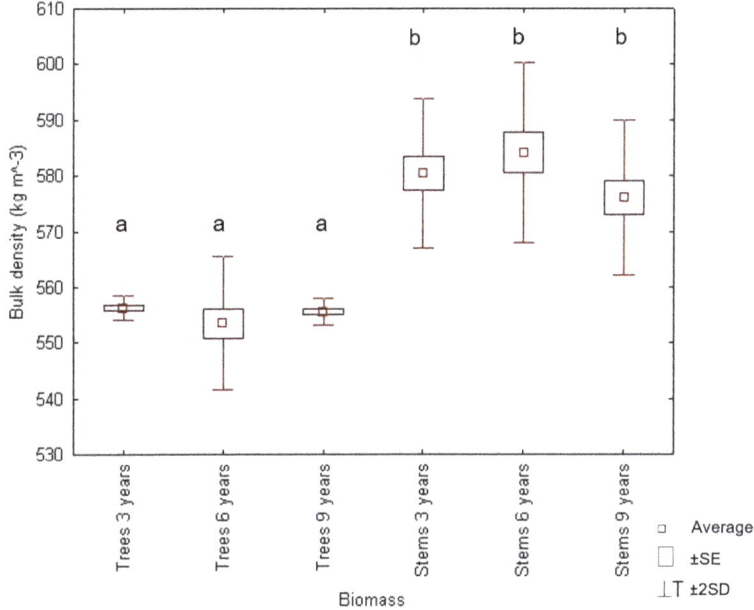

Figure 2. Bulk density of the pellets (average, standard error, standard deviation). Welch F test and Tuckey post hoc test): F = 26.12, df = 10.47, $p < 0.001$). Different letters indicate statistically significant differences between the groups obtained by Tukey post hoc test ($p < 0.001$).

3.4. Heating Value, Ash Content, Ash Melting Point and Heavy Metals of the Pellets

Tables 3 and 4 show the average values of lower heating value, ash content, ash melting point and heavy metals detected in the various types of pellet produced.

Table 3. Heating value, ash content and ash melting point of the different types of pelletized biomass (mean ± St. dev.). LHV: Kruskal Wallis, H = 3.134, Hc = 3.136, $p > 0.05$; Ash content: Welch F test and Tuckey post hoc test, F = 94.56, df = 10.77, $p < 0.001$; Ash melting point: Welch F test and Tuckey post hoc test, F = 189.8, df = 9.981, $p < 0.001$. Different letters indicate statistically significant differences between the groups obtained by Tukey post hoc test ($p < 0.01$).

Biomass Source after Pelletization	LHV ($MJ{\cdot}kg^{-1}$)	Ash (%)	Ash Melting Point (°C)
3 years old whole tree	17.67 [a] ±0.27	2.78 [a] ±0.09	1433.4 [c] ±4.4
3 years old stem	17.85 [a] ±0.38	2.71 [a] ±0.31	1455.8 [d] ±5.8
6 years old whole tree	17.85 [a] ±0.11	1.82 [c] ±0.23	1404.6 [c] ±17.2
6 years old stem	17.68 [a] ±0.26	1.89 [c] ±0.08	1433.2 [c] ±4.3
9 years old whole tree	17.72 [a] ±0.28	2.27 [b] ±0.06	1479.8 [a] ±1.1
9 years old stem	17.55 [a] ±0.62	1.87 [c] ±0.04	1457.6 [b] ±4.6

Table 4. Heavy metals of the different types of pelletized biomass (mean ± St. dev.). As: One-Way Anova, F = 11.07, df = 24, $p < 0.001$. Cd, Cr, Cu, Pb, Ni, Zn: Kruskal-Wallis and Mann-Whitney pairwise comparisons Bonferroni corrected, H = 26.03, 27.05, 27.85, 28.23, 28.23, 25.39 respectively, Hc = 26.18, 27.58, 27.92, 28.39, 28.33, 25.94, $p < 0.001$ Different letters indicate statistically significant differences between the groups obtained by Tukey post hoc test ($p < 0.05$).

Biomass Source after Pelletization	As (mg·kg^{-1})	Cd (mg·kg^{-1})	Cr (mg·kg^{-1})	Cu (mg·kg^{-1})	Pb (mg·kg^{-1})	Ni (mg·kg^{-1})	Zn (mg·kg^{-1})
3 years old whole tree	0.0050 ^{b,c} ±0.0007	0.2282 ^{a} ±0.0011	0.1562 ^{b} ±0.0015	0.9330 ^{c} ±0.0071	0.0352 ^{e} ±0.0004	0.2406 ^{c} ±0.0009	12.1950 ^{c} ±0.1897
3 years old stem	0.0056 ^{a,c} ±0.0009	0.2318 ^{a} ±0.0056	0.1440 ^{c} ±0.0007	0.8874 ^{e} ±0.0017	0.0464 ^{c} ±0.0009	0.2222 ^{d} ±0.0019	11.4421 ^{d} ±0.2845
6 years old whole tree	0.0046 ^{b,c} ±0.0009	0.1946 ^{c,d} ±0.0082	0.1558 ^{b} ±0.0024	0.8306 ^{f} ±0.0138	0.0320 ^{f} ±0.0012	0.2022 ^{e} ±0.0013	12.5868 ^{c} ±0.2566
6 years old stem	0.0028 ^{d} ±0.0011	0.1966 ^{c} ±0.0018	0.1058 ^{e} ±0.0043	0.9950 ^{a} ±0.0007	0.0418 ^{d} ±0.0013	0.1742 ^{f} ±0.0011	12.3915 ^{c} ±0.2021
9 years old whole tree	0.0068 ^{a} ±0.0011	0.2028 ^{b} ±0.0011	0.2462 ^{a} ±0.0055	0.9470 ^{b} ±0.0102	0.0756 ^{a} ±0.0005	1.6852 ^{a} ±0.0018	14.4502 ^{a} ±0.1057
9 years old stem	0.0044 ^{b,c,d} ±0.0005	0.1878 ^{d} ±0.0018	0.1308 ^{d} ±0.0029	0.9078 ^{d} ±0.0100	0.0678 ^{b} ±0.0011	0.3354 ^{b} ±0.0009	14.3222 ^{b} ±0.1020

The heating value detected for the poplar ranged between 17.55 and 17.85 MJ·kg^{-1} with non-significant statistical differences (Kruskal Wallis test, $p > 0.05$).

In our study, an ash content of less than 1.9% for material of 6 years of age (with and without branches) and 9-year-old stems was detected; values between 2.27% and 2.78% were detected for the whole trees of 9 and 3 years of age, respectively. With reference to the latter parameter, the Welch F test and the Tuckey post hoc test revealed significant differences between the various cultivation cycles with the same fraction used, excepted for the material of 6 and 9 years of age without branches. At the same coppicing intervals, however, it was possible to see a statistically significant difference only in the 9-year-old trees, obtaining a reduction of 17.62% in ash content by using stems without branches instead of the whole trees.

By analyzing the data related to the heavy metals content we identified the types of pellets characterized by their lower concentration: As, Ni and Cr in the pellet obtained from the poplar of 6 years without branches, Pb and Cu in the pellet of 6 years with branches, Zn in the pellet of 3 years without branches, Cd in the poplar of 9 years without branches. The wood material with the highest concentration of heavy metals in absolute was the whole 9-year-old poplar, relative to the values of As, Cr, Pb, Ni, and Zn ($p < 0.05$). Ash melting point was always higher than 1400 °C. About this parameter, there were no significant differences between the whole 3-year-old poplar and the 6-year-old poplar without branches, and between 3 and 9-year-old poplar without branches. In all other cases, however, the differences were statistically significant at a level $p < 0.01$.

Other parameters examined, and not showed in Table 3, concern the concentration of sulfur (S), equal to 0.0%, nitrogen (N), with values between 0.04% and 0.26% and the mechanical durability, with values between 97.1% for the 9-year-old poplar without branches and 98.6% for the whole 6-year-old poplar. Regarding this last parameter, the Kruskal Wallis test did not reveal a significant difference ($p > 0.05$). Tables 5 and 6, unlike the previous ones, highlight the qualitative parameters of the refined material immediately before of the passage into the pellet machine. The heating value ranged between 16.35 and 17.74 MJ·kg^{-1}, while the ash content was always higher than 2% (1.95%), with peak of 2.97%. As showed in Table 7, with high statistical significance, the refined material would tend to be characterized by a lower heating and ash fusibility values compared to the pelletized material. The other percentage differences showed in Table 7, which refer to ash content and heavy metals content, were not statistically confirmed.

Table 5. Heating value, ash content and ash melting point of the different types of refined biomass (mean ± St. dev.).

Biomass Source before Pelletization	LHV (MJ·kg^{-1})	Ash (%)	Ash Melting Point (°C)
3 years old whole tree	16.93 ±0.07	2.14 ±0.21	1438.2 ±3.11
3 years old stem	17.07 ±0.15	2.97 ±0.22	1378.6 ±23.6
6 years old whole tree	17.47 ±0.09	1.95 ±0.34	1377.0 ±6.8
6 years old stem	16.72 ±0.34	2.39 ±0.21	1433.0 ±4.39
9 years old whole tree	17.74 ±0.31	2.57 ±0.20	1406.0 ±4.42
9 years old stem	16.35 ±0.32	2.42 ±0.13	1431.8 ±1.1

Table 6. Heavy metals of the different types of refined biomass (mean ± St. dev.).

Biomass Source before Pelletization	As (mg·kg^{-1})	Cd (mg·kg^{-1})	Cr (mg·kg^{-1})	Cu (mg·kg^{-1})	Pb (mg·kg^{-1})	Ni (mg·kg^{-1})	Zn (mg·kg^{-1})
3 years old whole tree	0.0018 ±0.0004	0.2222 ±0.0013	0.1522 ±0.0051	0.9126 ±0.0017	0.0256 ±0.0005	0.2226 ±0.0005	10.7319 ±0.1648
3 years old stem	0.0056 ±0.0009	0.2328 ±0.0062	0.2292 ±0.0059	0.9193 ±0.0091	0.1008 ±0.0008	0.3434 ±0.0013	11.9100 ±0.0624
6 years old whole tree	0.0058 ±0.0008	0.1912 ±0.0006	0.2072 ±0.0062	0.8707 ±0.0018	0.1338 ±0.1935	0.4136 ±0.0009	13.4410 ±0.1845
6 years old stem	0.0120 ±0.0010	0.2050 ±0.0009	0.1448 ±0.0033	1.0230 ±0.0024	0.0654 ±0.0005	0.3366 ±0.0011	12.6780 ±0.1146
9 years old whole tree	0.0070 ±0.0012	0.2064 ±0.0038	0.1242 ±0.0016	0.9782 ±0.0045	0.0750 ±0.0029	1.3270 ±0.0060	15.3179 ±0.3089
9 years old stem	0.0046 ±0.0009	0.1866 ±0.0009	0.2026 ±0.0032	0.9099 ±0.0056	0.0688 ±0.0004	1.4142 ±0.0008	14.8296 ±0.2872

Table 7. Average values of heating value, ash content, ash melting point and heavy metals relative to biomass (mean ± St. dev.) before and after pelleting, in addition the t-test result referred to independent samples (* statistically significant).

Parameter	Refined	Pellet	Δ (%)	*t*-Value	DF	*p*-Value
LHV (MJ·kg^{-1})	17.05 ± 0.52	17.72 ± 0.38	+3.95 *	−5.959	58	0.001
Ash content (%)	2.41 ± 0.38	2.22 ± 0.43	−7.88	1.819	58	>0.05
As (mg·kg^{-1})	0.006 ± 0.003	0.005 ± 0.002	−20.62	1.950	58	>0.05
Cd (mg·kg^{-1})	0.207 ± 0.020	0.207 ± 0.018	0	0.088	58	>0.05
Cr (mg·kg^{-1})	0.177 ± 0.039	0.156 ± 0.044	−11.45	1.872	58	>0.05
Cu (mg·kg^{-1})	0.936 ± 0.085	0.917 ± 0.053	−2.03	1.404	58	>0.05
Pb (mg·kg^{-1})	0.078 ± 0.079	0.049 ± 0.017	−36.34	1.919	58	>0.05
Ni (mg·kg^{-1})	0.676 ± 0.503	0.476 ± 0.552	−29.52	1.463	58	>0.05
Zn (mg·kg^{-1})	13.15 ± 2.023	12.90 ± 1.145	−1.90	0.656	58	>0.05
Ash melting point (°C)	1404.2 ± 38.2	1444.1 ± 25.2	+2.84 *	−4.773	58	0.001

Another aspect to consider was the improvement of the quality parameters that could be obtained by transforming the poplar wood chips into pellets (Table 7).

Comparing the surveyed data, before and after pelletization, there was an average increase in the heating value of 3.95% and an increase in the ash melting point of 2.84%, with statistical significance, while the possible reduction in heavy metals content was not statistically confirmed. Another important aspect was the noticeable increase in bulk density, which was not less than 80%.

4. Discussion

Populus spp. is considered an excellent source to produce wood, for technological [40] and energy purposes [2]. It is also one of the best choices for the SRWC establishment in Italy [41] resulting in the crop with the highest productive response, as showed in some recent studies [2,42]. In general, the Short Rotation Coppice guarantees a good biomass production, varying between 3 and 20 $Mg \cdot ha^{-1} \cdot year^{-1}$ of dry matter [23,43,44], an acceptable combustion quality [45–47] and a functional and well-structured logistics.

The raw materials utilized in this study had different starting characteristics. Taking as reference the diameters of insertion of the branches, the basal diameters of the stem and the heights, it is possible to highlight, respectively, average sizes higher than 12%, 47% and 48% in the comparison between shoots of 6 and 3 years and of 150%, 32% and 34% in the comparison between shoots of 9 and 6-years-old. The qualitative characteristics of the pellets produced, therefore, may differ as a function of coppicing cycle and wood fraction considered. As a result, this can make a general improvement of the fuel possible, as it offers longer coppicing and/or fractions without branches.

In fact, the pellet obtained from the stems without branches ensures a greater aggregation of the particles and it is characterized by a greater bulk density (between +3.6% and +5.5%) and by a greater average length (between +5% and +32%). The lengthening of the cycle would seem, instead, not to make direct improvements in these terms. Undoubtedly, the fraction of the branches mainly influences the densification phase of the material, creating discontinuity between the particles and favoring a more evident fragmentation of the cylinders which, consequently, turns out to be shorter.

The heating value is between 17.55 and 17.85 $MJ \cdot kg^{-1}$ confirming what reported in literature by other authors [48]. Concerning the ash content, the worst result was found for the 3-year-old material. In this case, the lengthening of the rotation leads to a significant reduction in average values. The presence of the branches, however, would seem to influence this parameter only for the long cycle.

The values of heavy metal content and the temperature of ash fusibility fully comply with the values indicated by the current regulations.

Every biomass contains a certain amount of metal compounds, but in the last decade, the metal-contaminated biomass achieved increasing attention [49]. There are various toxicological effects of heavy metals emission during combustions for human health or for the environment, and the pollutants generated depends on their amount in the biomass [50]. The heavy metals content in biomass fuels should be limited, especially considering their utilisation in small-scale systems, which are usually not equipped with dust precipitation devices [51].

Although none of the materials analyzed exceeded the limits set by the regulation, it is clear that the quality characteristics of the 6-year old poplar guarantee the attainment of a preferable product. The management of the SRWC could be oriented in this sense to avoid an excessive lengthening of the crop interval, benefiting from a general reduction of the metals content.An average value of 10% in moisture content of the raw material is considered to be optimal for the subsequent process of pelletization, as well as in terms of durability of the final product, as reported by Samuelsson et al. [52], Whittaker and Shield [9], Lehtikangas [53], and Filbakk et al. [54]. However, it is necessary to consider the loss of moisture content that is only directly attributable to the pelletization process, which has been quantified to range between 8.50% and 9.76%, as verified in all the tested theses, with statistically significant differences. This factor must be considered during the transformation phase, as it will affect the stability and final energy yield of the product.

Referring to EN ISO 17225-2 [55], we can identify the quality parameters that are respected by our poplar pellets (Table 8).

Table 8. Quality parameters of poplar pellets compared to [55]: ✓✓✓(A1); ✓✓(A2); ✓(B); □ (not classified).

Years	3		6		9	
Type	**Tree**	**Stem**	**Tree**	**Stem**	**Tree**	**Stem**
Diameter	✓✓✓	✓✓✓	✓✓✓	✓✓✓	✓✓✓	✓✓✓
Length	✓✓✓	✓✓✓	✓✓✓	✓✓✓	✓✓✓	✓✓✓
Moisture content	✓✓✓	✓✓✓	✓✓✓	✓✓✓	✓✓✓	✓✓✓
Ash content	□	□	✓	✓	□	✓
LHV	✓✓✓	✓✓✓	✓✓✓	✓✓✓	✓✓✓	✓✓✓
Bulk density	□	□	□	□	□	□
N	✓✓✓	✓✓✓	✓✓✓	✓✓✓	✓✓✓	✓✓✓
S	✓✓✓	✓✓✓	✓✓✓	✓✓✓	✓✓✓	✓✓✓
As	✓✓✓	✓✓✓	✓✓✓	✓✓✓	✓✓✓	✓✓✓
Cd	✓✓✓	✓✓✓	✓✓✓	✓✓✓	✓✓✓	✓✓✓
Cr	✓✓✓	✓✓✓	✓✓✓	✓✓✓	✓✓✓	✓✓✓
Cu	✓✓✓	✓✓✓	✓✓✓	✓✓✓	✓✓✓	✓✓✓
Pb	✓✓✓	✓✓✓	✓✓✓	✓✓✓	✓✓✓	✓✓✓
Ni	✓✓✓	✓✓✓	✓✓✓	✓✓✓	✓✓✓	✓✓✓
Zn	✓✓✓	✓✓✓	✓✓✓	✓✓✓	✓✓✓	✓✓✓
Ash melting point	✓✓✓	✓✓✓	✓✓✓	✓✓✓	✓✓✓	✓✓✓
Durability	✓✓✓	✓✓	✓✓✓	✓✓	□	□

As shown in Table 8, all the products meet the requirements of class A1 for diameter, length, moisture content, ash melting point, lower heating value, as well as N, S, and heavy metals.

In order to define the pellet's quality, it is also important to know the ash content. A high ash content can cause problems in the combustion of biomass, as it produces slag, incrustations and corrosion in the combustion device, with an inevitable reduction in performance of the plant itself [56]. The pellet of poplar can respect only the parameters of class B and not for all the types of material used, excluding the material of 3 years of age and the entire trees of 9 years of age.

As regards the mechanical durability, it is possible to respect the parameters of class A1 for whole poplar of 3 and 6 years and A2 for poplar of 3 and 6 years without branches.

Lastly, none of the theses analyzed can satisfy the bulk density in the three reference classes, according to Monedero et al. [57].

Other studies report the difficulty in producing good quality pellets from poplar or willow short rotation coppices [58], especially using solely this type of material [59].

There are different actions, however, that can be put in place to improve the quality of the material, such as preheating the feed material [60], binding additives [61], and increasing the pelleting pressure [62–64].

Monedero et al. [57] refers, for example, to the possibility of achieving quality classes A1 and A2 using poplar and pine mixtures, in variable percentages depending on the result to be obtained.

The increasing demand for wood pellets and the limited availability of the biomass traditionally used for its processing, requires the use of other sources of raw materials [65]. In this perspective, many resources and research activities are now focused on the exploitation of agricultural by-products or energy grasses. However, several studies carried out show problems of low durability of pellets produced from cereal residues [66], high ash content over 3.5%, 5.5% and 6.5% for miscanthus, rape straw and wheat straw respectively [58], 2.27% and 4.2% for apple pruning [67] and vineyard pruning [68].

The qualitative characteristics of these materials are therefore lower than those found in dedicated wood crops.

5. Conclusions

The performed research and the analysis of the results reveals that it is possible to produce a good quality pellet starting from wood chips obtained from Short Rotation Coppice of poplar and thereby converting a material with a low commercial value into superior merchandise material, in terms of quality, cost and energy. This represents an interesting development opportunity for potential small and medium scale agro-energy chains.

Among the various types of pelletized material, of the types palletized for six years produced the best results, especially in terms of ash content which, at the moment, is the most limiting factor. The qualitative improvement that can be achieved by cutting the branches before chipping appears to be convenient only in terms of bulk density. To enhance the value of the SRWCs through a densification process, it would therefore be advisable to direct the management of the crop cycles towards a 5–6 year interval, avoiding a too short rotation of 2 or 3 years, due to the excessive ash content, and cultivation cycles that are too long, which would lead to a greater accumulation in heavy metals and require, in any case, even more specialized harvest mechanization.

Despite the general qualitative improvement that can be achieved through the pelletizing process, some of the parameters required by the legislation are not respected using only this type of material, especially the ash content and the bulk density. The legislation that manages the market is very restrictive, compared to the traditional one of the wood chips, so the introduction of specific quality standards, otherwise, could facilitate and encourage the exploitation of wood biomass produced by dedicated plantations in a market of great interest, such as that of pellets.

Author Contributions: Conceptualization, V.C. and A.A. (Andrea Acampora); methodology, V.C., G.S. and A.A. (Andrea Acampora); validation, V.C. and A.A. (Andrea Acampora); investigation, A.A. (Andrea Acampora), R.P. and A.A. (Alberto Assirelli); data curation, V.C. and R.P.; writing-original draft preparation, G.S. and A.A. (Alberto Assirelli); writing-review and editing, R.P. and G.S; supervision, V.C., R.P. and A.A. (Alberto Assirelli).

Funding: This research received no external funding.

Acknowledgments: This work was supported by: (1) the Italian Ministry of Agriculture (MiPAAF) under the AGROENER project (D.D. n. 26329, 1 April 2016). http://agroener.crea.gov.it/ and (2) the Italian Ministry for education, University and Research (MIUR) for financial support (Law 232/2016, Italian University Departments of excellence)—UNITUS-DAFNE WP3.

Conflicts of Interest: The authors declare no conflict of interest.

References

1. Huang, Y.; Finell, M.; Larsson, S.; Wang, X.; Zhang, J.; Wei, R.; Liu, L. Biofuel pellets made at low moisture content e Influence of water in the binding mechanism of densified biomass. *Biomass Bioenergy* **2017**, *98*, 8–14. [CrossRef]
2. Picchio, R.; Spina, R.; Sirna, A.; Monaco, A.L.; Civitarese, V.; Del Giudice, A.; Suardi, A.; Pari, L. Characterization of Woodchips for Energy from Forestry and Agroforestry Production. *Energies* **2012**, *5*, 3803–3816. [CrossRef]

3. IPCC Climate Change. *Synthesis Report Contribution of Working Groups I, II and III to the Fourth Assessment Report of the Intergovernmental Panel on Climate Change*; Cambridge University Press: Cambridge, UK. Available online: https://www.researchgate.net/publication/262260453_IPCC_2007_Climate_Change_2007_Synthesis_Report_Contribution_of_Working_Groups_I_II_III_to_the_Fourth_Assessment_Report_of_the_Intergovernmental_Panel_on_Climate_Change_Geneva (accessed on 6 February 2019).
4. European Commission. 2020 Climate & Energy Package. Combating Climate Change the EU Leads the Way, European Move Service 24. Available online: https://ec.europa.eu/clima/policies/strategies/2020_en (accessed on 8 January 2019).
5. Tumuluru, J.S. Effect of process variables on the density and durability of the pellets made from high moisture corn stover. *Biosyst. Eng.* **2014**, *119*, 44–57. [CrossRef]
6. Tumuluru, J.S.; Wright, C.T.; Hess, J.R.; Kenney, K.L. A review of biomass densification systems to develop uniform feedstock commodities for bioenergy application. *Biofuels Bioprod. Biorefining* **2011**, *5*, 683–707. [CrossRef]
7. Sánchez, J.; Curt, M.D.; Sanz, M.; Fernández, J. A proposal for pellet production from residual woody biomass in the island of Majorca (Spain). *AIMS Energy* **2015**, *3*, 480–504. [CrossRef]
8. Associazione Italiana Energie Agroforestali (AIEL). Available online: http://www.aielenergia.it/public/rassegna_stampa/378_1%20febbraio%202019%20biomassapp.it.pdf (accessed on 10 July 2019).
9. Whittaker, C.; Shield, I. Factors affecting wood, energy grass and straw pellet durability—A review. *Renew. Sustain. Energy Rev.* **2017**, *71*, 1–11. [CrossRef]
10. Karkania, V.; Fanara, E.; Zabaniotou, A. Review of sustainable biomass pellets production—A study for agricultural residues pellets' market in Greece. *Renew. Sustain. Energy Rev.* **2012**, *16*, 1426–1436. [CrossRef]
11. Alberici, S.; Boeve, S.; van Breevoort, P. Subsidies and Costs of EU Energy. European Commission. Available online: https://ec.europa.eu/energy/en/content/inal-report-ecofys (accessed on 12 January 2019).
12. Peksa-Blanchard, M.; Dolzan, P.; Grassi, A.; Heinimoe, J.; Junginger, M.; Ranta, T.; Walter, A. *Global Wood Pellets Market and Industry: Policy, Drivers, Market Status and Raw Material Potential*; IEA bioenergy Task 40; International Energy Agency: Paris, France, 2007; Available online: https://www.researchgate.net/publication/27709106_Global_Wood_Pellets_Markets_and_Industry_Policy_Drivers_Market_Status_and_Raw_Material_Potential (accessed on 15 February 2019).
13. Gravelsins, A.; Muižniece, I.; Blumberga, A.; Blumberga, D. Economic sustainability of pellet production in Latvia. *Energy Procedia* **2017**, *142*, 531–537. [CrossRef]
14. GSE (Energy Services Manager). Rapporto Delle Attività. Energia in Movimento. Available online: https://www.gse.it/documenti_site/Documenti%20GSE/Rapporti%20delle%20attivit%C3%A0/GSE_RA2017.pdf (accessed on 5 March 2019).
15. Haruna, N.Y.; Afzalb, M.T. Effect of Particle Size on Mechanical Properties of Pellets Made from Biomass Blends. *Procedia Eng.* **2016**, *148*, 93–99. [CrossRef]
16. Supin, M. Wood Processing and Furniture Manufacturing Challenges on the World Market: Wood Pellet Global Market Development 2015. pp. 255–260. Available online: https://www.researchgate.net/publication/286443751_Wood_Pellet_Global_Market_Development (accessed on 7 February 2019).
17. Acda, M.N.; Jara, A.A.; Daracan, V.C.; Devera, E.E. Opportunities and Barriers to Wood Pellet Trade in the Philippines. *Ecosyst. Dev. J.* **2016**, *6*, 27–31. Available online: https://www.researchgate.net/publication/303680604_Opportunities_and_barriers_to_wood_pellet_trade_in_the_Philippines (accessed on 7 February 2019).
18. European Biomass Association (AEBIOM) Bioenergy. A Local and Renewable Solution for Energy Security. Available online: https://www.slideshare.net/AEBIOM2/biomass (accessed on 8 January 2019).
19. Verhoest, C.; Ryckmans, Y. Industrial Wood Pellets Report. Available online: http://www.bpa-intl.com/images/stories/present-1/PELLCERT%20-%20Industrial%20Wood%20Pellets%20Report%20(2012).pdf (accessed on 5 March 2019).
20. GSE (Energy Services Manager). Rapporto Statistico. Energia Da Fonti Rinnovabili. Available online: https://www.gse.it/documenti_site/Documenti%20GSE/Rapporti%20statistici/Rapporto%20statistico%20GSE%20-%202016.pdf (accessed on 5 March 2019).
21. Energia Dall'agricoltura: Innovazioni Sostenibili Per la Bioeconomia. Available online: http://agroener.crea.gov.it (accessed on 1 June 2019).

22. Di Candilo, M.; Facciotto, G. Colture da biomassa ad uso energetico Potenzialità e prospettive. In *Progetti Di Ricerca Suscace e Faesi. Recenti Acquisizioni Scientifiche per Le Colture Energetiche*; Supplemento n. 2 a Sherwood-Foreste ed Alberi Oggi n. 183; Compagnia delle Foreste Srl: Roma, Italy, 2012; pp. 10–19.
23. Paris, P.; Mareschi, L.; Sabatti, M.; Pisanelli, A.; Ecosse, A.; Nardin, F.; Scarascia-Mugnozza, G. Comparing hybrid Populus clones for SRF across northern Italy after two biennial rotations: Survival, growth and yield. *Biomass Bioenergy* **2011**, *35*, 1524–1532. [CrossRef]
24. Facciotto, G. Storia della SRF e suo sviluppo in Italia. In *Progetti Di Ricerca Suscace e Faesi. Recenti Acquisizioni Scientifiche per Le Colture Energetiche*; Supplemento n. 2 a Sherwood-Foreste ed Alberi Oggi n. 183; Compagnia delle Foreste Srl: Roma, Italy, 2012; pp. 26–30.
25. Verani, S.; Sperandio, G.; Picchio, R.; Marchi, E.; Costa, C. Sustainability Assessment of a Self-Consumption Wood-Energy Chain on Small Scale for Heat Generation in Central Italy. *Energies* **2015**, *8*, 5182–5197. [CrossRef]
26. Mitchell, C.P.; Kofman, P.D.; Angus-Hankin, C.M. *Guidelines for Conducting Harvesting Trials in Short Rotation Forestry*; Forestry Research Paper; Aberdeen University: Aberdeen, UK, 1997; pp. 1–50.
27. Bergström, D.; Israelsson, S.; Öhman, M.; Dahlqvist, S.-A.; Gref, R.; Boman, C.; Wästerlund, I. Effects of raw material particle size distribution on the characteristics of Scots pine sawdust fuel pellets. *Fuel Process. Technol.* **2008**, *89*, 1324–1329. [CrossRef]
28. Monaco, A.L.; Todaro, L.; Sarlatto, M.; Spina, R.; Calienno, L.; Picchio, R. Effect of moisture on physical parameters of timber from Turkey oak (Quercus cerris L.) coppice in Central Italy. *For. Stud. China* **2011**, *13*, 276–284. [CrossRef]
29. EN ISO 18122. *Solid Biofuels—Determination of Ash Content*; ISO: Geneva, Switzerland, 2015.
30. CEN/TS 15370-1. *Solid Biofuels—Method for the Determination of Ash Melting Behaviour. Part 1: Characteristic Temperatures Method*; CEN/TS: Geneva, Switzerland, 2006.
31. EN ISO 18125. *Solid Biofuels—Determination of Heating Value*; ISO: Geneva, Switzerland, 2015.
32. Ghafghazi, S.; Sowlati, T.; Sokhansanj, S.; Bi, X.; Melin, S. Particulate matter emissions from combustion of wood in district heating applications. *Renew. Sustain. Energy Rev.* **2011**, *15*, 3019–3028. [CrossRef]
33. EN ISO 16968. *Solid Biofuels—Determination of Minor Elements*; ISO: Geneva, Switzerland, 2015.
34. EN ISO 16948. *Solid Biofuels—Determination of Total Content of Carbon, Hydrogen and Nitrogen*; ISO: Geneva, Switzerland, 2015.
35. EN ISO 16994. *Solid Biofuels—Determination of Total Content of Sulfur and Chlorine*; ISO: Geneva, Switzerland, 2016.
36. EN ISO 17828. *Solid Biofuels—Determination of Bulk Density*; ISO: Geneva, Switzerland, 2015.
37. EN ISO 17829. *Solid Biofuels—Determination of Length and Diameter of Pellets*; ISO: Geneva, Switzerland, 2015.
38. EN ISO 17831–1. *Solid Biofuels—Determination of Mechanical Durability of Pellets and Briquettes. Part 1: Pellets*; ISO: Geneva, Switzerland, 2015.
39. EN ISO 18134-1. *Solid Biofuels—Determination of Moisture Content—Oven Dry Method*; ISO: Geneva, Switzerland, 2015.
40. Pelosi, C.; Agresti, G.; Calienno, L.; Lo Monaco, A.; Picchio, R.; Santamaria, U.; Vinciguerra, V. Application of spectroscopic techniques for the study of the surface changes in poplar wood and possible implications in conservation of wooden artefacts. In Proceedings of the SPIE—The International Society for Optical Engineering, Munich, Germany, 30 May 2013; Volume 8790, p. 879014.
41. Picchio, R.; Verani, S.; Sperandio, G.; Spina, R.; Marchi, E. Stump grinding on a poplar plantation: Working time, productivity, and economic and energetic inputs. *Ecol. Eng.* **2012**, *40*, 117–120. [CrossRef]
42. Civitarese, V.; Faugno, S.; Picchio, R.; Assirelli, A.; Sperandio, G.; Saulino, L.; Crimaldi, M.; Sannino, M. Production of selected short-rotation wood crop species and quality of obtained biomass. *Eur. J. For. Res.* **2018**, *137*, 541–552. [CrossRef]
43. Bergante, S.; Facciotto, G.; Minotta, G. Identification of the main site factors and management intensity affecting the establishment of Short-Rotation-Coppices (SRC) in Northern Italy through stepwise regression analysis. *Open Life Sci.* **2010**, *5*, 522–530. [CrossRef]
44. Di Matteo, G.; Sperandio, G.; Verani, S. Field performance of poplar for bioenergy in southern Europe after two coppicing rotations: Effects of clone and planting density. *iForest—Biogeosci. For.* **2012**, *5*, 224–229. [CrossRef]

45. Bungart, R.; Hüttl, R.F. Growth dynamics and biomass accumulation of 8-year-old hybrid poplar clones in a short-rotation plantation on a clayey-sandy mining substrate with respect to plant nutrition and water budget. *Eur. J. For. Res.* **2004**, *123*, 105–115. [CrossRef]
46. Hauk, S.; Knoke, T.; Wittkopf, S. Economic evaluation of short rotation coppice systems for energy from biomass—A review. *Renew. Sustain. Energy Rev.* **2014**, *29*, 435–448. [CrossRef]
47. Huber, J.A.; May, K.; Hülsbergen, K.J. Allometric tree biomass models of various species grown in short-rotation agroforestry systems. *Eur. J. For. Res.* **2017**, *136*, 75–89. [CrossRef]
48. Telmo, C.; Lousada, J. Heating values of wood pellets from different species. *Biomass Bioenergy* **2011**, *35*, 2634–2639. [CrossRef]
49. Kovacs, H.; Szemmelveisz, K. Heavy metal contaminated biomass combustion as treatment after phytoremediation—A review. *Mater. Sci. Eng.* **2016**, *41*, 69–78.
50. Ružinská, E.; Štollmann, V.; Hagara, V.; Jabłoński, M. Analysis of selected heavy metals in biomass for preparation of biofuels—Part I. Toxicological effects of heavy metals. *Ann. Wars. Univ. Life Sci. SGGW For. Wood Technol.* **2015**, *92*, 383–389.
51. Obernbergera, I.; Theka, G. Physical characterisation and chemical composition of densified biomass fuels with regard to their Combustion behavior. *Biomass Bioenergy* **2004**, *27*, 653–669. [CrossRef]
52. Samuelsson, R.; Thyrel, M.; Sjöström, M.; Lestander, T.A. Effect of biomaterial characteristics on pelletizing properties and biofuel pellet quality. *Fuel Process. Technol.* **2009**, *90*, 1129–1134. [CrossRef]
53. Lehtikangas, P. Quality properties of pelletised sawdust, logging residues and bark. *Biomass Bioenergy* **2001**, *20*, 351–360. [CrossRef]
54. Filbakk, T.; Skjevrak, G.; Høibø, O.; Dibdiakova, J.; Jirjis, R. The influence of storage and drying methods for Scots pine raw material on mechanical pellet properties and production parameters. *Fuel Process. Technol.* **2011**, *92*, 871–878. [CrossRef]
55. EN ISO 17225. *Solid Biofuels—Fuel specifications and classes—Part 2: Graded wood pellets*; ISO: Geneva, Switzerland, 2014.
56. Toscano, G.; Riva, G.; Pedretti, E.F.; Corinaldesi, F.; Mengarelli, C.; Duca, D. Investigation on wood pellet quality and relationship between ash content and the most important chemical elements. *Biomass Bioenergy* **2013**, *56*, 317–322. [CrossRef]
57. Monedero, E.; Portero, H.; Lapuerta, M. Pellet blends of poplar and pine sawdust: Effects of material composition, additive, moisture content and compression die on pellet quality. *Fuel Process. Technol.* **2015**, *132*, 15–23. [CrossRef]
58. Carroll, J.P.; Finnan, J. Physical and chemical properties of pellets from energy crops and cereal straws. *Biosyst. Eng.* **2012**, *112*, 151–159. [CrossRef]
59. Porsö, C.; Hansson, P.-A. Time-dependent climate impact of heat production from Swedish willow and poplar pellets—In a life cycle perspective. *Biomass Bioenergy* **2014**, *70*, 287–301. [CrossRef]
60. Kaliyan, N.; Morey, R.V. Factors affecting strength and durability of densified biomass products. *Biomass Bioenergy* **2009**, *33*, 337–359. [CrossRef]
61. Ahn, B.J.; Chang, H.-S.; Lee, S.M.; Choi, D.H.; Cho, S.T.; Han, G.-S.; Yang, I. Effect of binders on the durability of wood pellets fabricated from *Larix kaemferi* C. and *Liriodendron tulipifera* L. sawdust Renew. *Energy* **2014**, *62*, 18–23.
62. Nielsen, N.P.K.; Gardner, D.J.; Poulsen, T.; Felby, C. Importance of temperature, moisture content, and species for the conversion process of wood residues into fuelpellets. *Wood Fibre Sci.* **2009**, *41*, 414–425.
63. Stelte, W.; Holm, J.K.; Sanadi, A.R.; Barsberg, S.; Ahrenfeldt, J.; Henriksen, U.B. A study of bonding and failure mechanisms in fuel pellets from different biomass resources. *Biomass Bioenergy* **2011**, *35*, 910–918. [CrossRef]
64. Lee, S.M.; Ahn, B.J.; Choi, D.H.; Han, G.-S.; Jeong, H.-S.; Ahn, S.H.; Yang, I. Effects of densification variables on the durability of wood pellets fabricated with *Larix kaempferi* C. and *Liriodendron tulipifera* L. sawdust. *Biomass Bioenergy* **2013**, *48*, 1–9. [CrossRef]
65. Stelte, W.; Sanadi, A.R.; Shang, L.; Holm, J.K.; Ahrenfeldt, J.; Henriksen, U.B. Recent developments in biomass Pelletisation: A review. *Bioresources* **2012**, *7*, 4451–4490.
66. Adapa, P.; Tabil, L.; Shoenau, G.; Opoku, A. Pelleting characteristics of selected biomass with and without steam explosion pretreatment. *Int. J. Agric. Biol. Eng.* **2010**, *3*, 62–79.

67. Brand, M.A.; Barnasky, R.R.D.S.; Carvalho, C.A.; Buss, R.; Waltrick, D.B.; Jacinto, R.C. Thermogravimetric analysis for characterization of the pellets produced with different forest and agricultural residues. *Ciênc. Rural* **2018**, *48*, 1–10. [CrossRef]
68. Toscano, G.; Alfano, V.; Scarfone, A.; Pari, L. Pelleting Vineyard Pruning at Low Cost with a Mobile Technology. *Energies* **2018**, *11*, 2477. [CrossRef]

Article

Optimizing the 3D Distributed Climate inside Greenhouses Using Multi-Objective Optimization Algorithms and Computer Fluid Dynamics

Kangji Li [1,*], Wenping Xue [1], Hanping Mao [2], Xu Chen [1], Hui Jiang [1] and Gang Tan [3]

1 School of Electricity Information Engineering, Jiangsu University, Zhenjiang 212013, China
2 Institute of Agricultural Engineering, Jiangsu University, Zhenjiang 212013, China
3 Department of Civil and Architectural Engineering, University of Wyoming, Laramie, WY 82071, USA
* Correspondence: likangji@ujs.edu.cn

Received: 29 May 2019; Accepted: 6 July 2019; Published: 26 July 2019

Abstract: As one of the major production facilities in agriculture, a greenhouse has many spatial distributed factors influencing crop growth and energy consumption, such as temperature field, air flow pattern, CO_2 concentration distribution, etc. By introducing a hybrid computational fluid dynamics–evolutionary algorithm (CFD-EA) method, this paper constructs a micro-climate model of greenhouse with main environmental parameters optimized. Considering environmental factors' spatial influences together with energy usage simultaneously, the optimal solutions of control variables for crop growth are calculated. A commercial greenhouse located in east China is chosen for the method validation. Field experiments using temperature/velocity sensor matrix are carried out for CFD accuracy investigation. On this basis, the proposed optimization method is employed to search for the optimal control variables and parameters corresponding to the environmental Pareto frontier. By the proposed multi-objective scheme, we believe the method can provide set point basis for the design and regulation of large/medium-sized greenhouse production with high spatial resolution.

Keywords: greenhouse; multiple environmental parameters; interactive optimization scheme; spatial distributed factors; online–offline strategy; CFD-EA

---

1. Introduction

Greenhouse production plays an important role in the development of modern agriculture, especially in densely-populated areas with tight land, such as eastern China. The environmental conditions in greenhouses are essential for crop growth, pest/disease prevention, energy saving, etc. For most active or passive greenhouses, appropriate environmental parameters for crop growth are assessed based on the analysis of environmental factors as lumped parameters [1]. Meanwhile, for large/medium-sized greenhouses, the environmental parameters at the crop growing area are not equal to the values at the sensors' location. For more accuracy, parameters such as temperature, ventilation rate and CO_2 concentration, should be considered based on a precise micro-climate model considering the spatial distribution inference.

Improvements in computing facilities together with theoretical and experimental studies increased our understanding of the biophysical process in a greenhouse system. At present, computational fluid dynamics (CFD) has been widely applied for greenhouse climate simulation [2–6]. Ould Khaous et al. [3] applied CFD to greenhouse ventilation efficiency evaluation. Results showed that air velocity at plant level varied from 0.1 to 0.5 ms^{-1} according to opening configurations and compartment positions, whereas air temperature differences varied from 2 to 6 °C. Similarly, Lee and Short [2] used CFD to evaluate natural ventilation rates and airflow distributions in a multi-span greenhouse; Santolini et al. [5] studied the effect of shading screens on airflow patterns within the greenhouse through CFD simulations. To relieve computational burden of CFD simulation, alternative models based on an artificial neural network (ANN) [7] or support vector machine (SVM) [8] are also reported for greenhouse system analysis. For these data-driven models, the modeling error issue could not be ignored.

High resolution modeling may facilitate precise regulation. To search optimal parameters of greenhouse environment, researchers employed manual CFD simulations for evaluation of candidate parameters. Tong et al. [9] used CFD method for span dimension selection of Chinese solar greenhouses. Three groups of span configurations (10 m, 12 m and 14 m) were investigated using CFD simulations and their characteristics including solar heat gains, heat losses and temperature distributions were analyzed. Wang et al. [10] used CFD method for thermal performance improvement of solar greenhouses. Three solar greenhouses with different north walls were simulated and analyzed; to achieve the best north wall thickness of the greenhouse, Zhang et al. [11] carried out 18 case studies based on CFD simulation. An evaluation model using weighted entropy and fuzzy optimization scheme was then employed for decision making; to optimize the vent dimension and position in greenhouse design, He et al. [12] investigated all effects of different back wall vent configurations by two groups of CFD models.

By contrast with solution searching through manual simulations, an interactive optimization scheme using a hybrid simulation–optimization method has superior performances in terms of computation efficiency, solution accuracy, and design convenience. In the field of building enclosure optimal design, commercial middleware such as GenOpt [13] and OpenFOAM [14] have been used in various scenarios. Asadi et al. [15] combined building energy simulation program with GenOpt to regulate energy usage and thermal comfort of a residential building. Futrell et al. [16] used similar hybrid method to search optimal solutions of daylighting and thermal performance of a campus building. Liu and Chen [17] combined CFD simulation with genetic algorithm (GA) through OpenFOAM for optimization of indoor environmental parameters.

Compared with residential buildings, the optimization task of greenhouse environment is still challenging. Firstly, there are multiple optimization goals within greenhouse systems, such as crop quality, economic benefits, energy usage, etc.; secondly, many environmental variables contribute to the goals of optimization, for instance, air temperature, illumination intensity, wind speed, soil fertility, etc.; most differently, the optimization goals of the greenhouse environment are not constant, but changing with space and time. In this study, we design an interactive optimization scheme for environmental performance regulation in greenhouse system. To facilitate the combination of existing CFD simulation and proper optimization algorithms, an efficient interactive module is realized, which links CFD and Matlab software through data exchange mechanism. On this basis, the paper aims to understand how the spatial distribution of environmental factors influence crop growth as well as to present a multi-objective optimization case study based on the hybrid computational fluid dynamics—evolutionary algorithm (CFD-EA) method. A summer's 720 m^2 Venlo type commercial greenhouse located in east China is chosen as the case scenario. The temperature fields and airflow patterns at essential sections are validated by field experiments. Three indexes, i.e., proper temperature distribution, proper CO_2 concentration, and electrical energy consumption are chosen as optimization objectives. Simulation study provides precise set points at one selected period for feedback regulation within the greenhouse system. Results will show the feasibility and high resolution of the hybrid optimization method.

The rest of the paper is organized as the following. Section 2 provides the details of the proposed optimization method. Section 3 depicts the construction of the CFD model for the real greenhouse. Accuracy validation with field experiment is presented in Section 4. Section 5 provides the optimization procedure and results. Some concluding remarks and future works are given in Section 6.

2. Method

Generally, the greenhouse climate management can be resolved into two dynamic processes: slow crop growth dynamics and relatively fast greenhouse climate dynamics [1,18]. For fast climate dynamics, the set-points of the greenhouse environmental variables such as air temperature, humidity and CO_2 concentration, should be decided in advance by the grower or computer systems. Once these set-points are decided, the desired climate in the greenhouse can be achieved by proportional–integral (PI) or other classic feed-back control methods. However, these set-points are usually time-varying and spatially distributed. In recent years, many optimal control schemes have been reported for such greenhouse climate control problems [18–20]. The environmental parameters' spatial feature is seldom mentioned, which is important for crop growth and energy saving in large and medium-sized greenhouse. Figure 1 describes the basic block diagram of the climate control procedure and the highlighted optimization part is the focus of this study.

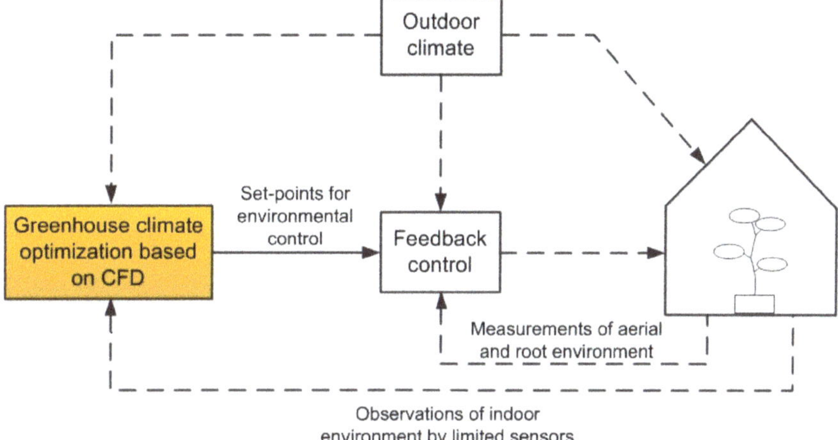

Figure 1. A block diagram of the climate control procedure.

2.1. Interactive Optimization Scheme

Considering the spatial influences of environmental parameters, CFD simulation is the most accurate way of modeling. Taking advantage of the current high-speed development of computing technology, hybrid simulation–optimization has been widely used in the field of building enclosure optimal design and operation. Combined with previous reported interactive optimization methods [17,21,22], we propose a set of greenhouse environment optimization solutions.

Using a validated CFD model, a global optimization scheme was combined for greenhouse environmental parameters optimization. At each iteration of optimization, distributed indoor micro-climate model was calculated by CFD (Airpak3.0.16 with Fluent engine), and the results including temperature field, CO_2 concentration, and energy consumption were format converted and transmitted to the optimization scheme through a middle module; in order to adapt to the features of the greenhouse environment optimization, we developed a middleware (C++ program) instead of directly using commercial softwares [21].

The spatial distributions of multiple environmental parameters are non-linearity, discontinuity and with high uncertainty. EAs working with a population of stochastic solutions can be used to find multiple Pareto-optimal solutions in one single simulation run. For this study, the non-dominant genetic algorithm with elite strategy (NSGA-II) [23] was chosen, which can find much better spread of solutions and better convergence near the true Pareto-optimal front compared to its previous version. This algorithm is realized in the Matlab environment. After finding out the Pareto frontier of main environmental parameters, the corresponding control variables were updated and exchanged back to CFD for next simulation. The interactive optimization loop continues until the stop criteria is reached.

Figure 2 describes the basic flow chart of the proposed interactive scheme.

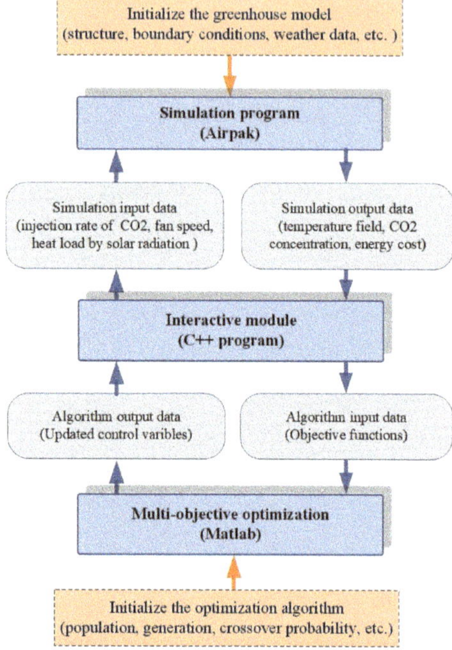

Figure 2. Interactive optimization scheme.

In the scheme, the C++ program is applied for data exchange between CFD and optimization algorithms. For each iteration of optimization algorithm, the data exchange module works as the follows.

```
Start
Read control variables from Matlab;
Delete old model file of Airpak (Airpak{i}.cas);
Create new model file (Airpak{i+1}.cas) according to template file's format
(Template.cas) and update the control variables;
Call Airpak to run the CFD simulation;
WHEN the CFD simulation meets the stop criteria
  Save the results and transfers them from Airpak to Matlab by txt files;
End
```

2.2. Multiple Objectives and Control Variables

For greenhouse production, the optimal indoor environmental parameters have characteristics of time-varying and spatial heterogeneity, especially for the uneven planting of crops. The indoor environmental parameters that this article focuses on include air temperature, CO_2 concentration and corresponding energy consumption.

2.2.1. Objectives

The suitability of environmental temperature is a primary assessment criterion for crop growth. In a greenhouse, solar radiation and other external factors affect indoor temperature dramatically.

How to reduce external disturbance and improve the indoor temperature distribution is an important issue in large greenhouse systems. For greenhouse production, the ideal temperature field is a function of time and crop growth. In this paper, we refer to the crop growth model based on temperature management technology [24,25], and choose one period in summer for the later case study.

Another important factor affecting crop growth is indoor CO_2 concentration. The optimum concentration in the greenhouse depends on several factors: crop photosynthetic rate, CO_2 loss rate, indoor temperature, and CO_2 cost. A benchmark curve of CO_2 concentration in one day was referred [24], and one segment was chosen for the later case study. Although relative humidity also has a great influence on crop growth, it was not controlled in this study because its value is usually greater than 80% in summer of east China, and the spatial difference can be ignored.

Three assessment indexes for the multi-objectives optimization were set, which were temperature distribution index, CO_2 distribution index, and energy consumption index. It should be noted that the essential economic cost for greenhouse crop production is heating energy. Considering the climate characteristics of eastern China in summer (air temperature is up to 40 °C), we only chose the economic costs of electricity and CO_2 injection representing the main energy consumption in this study. The three indexes are formulated below,

$$\begin{cases} J_T = \sqrt{\dfrac{\sum_{i=1}^{N_p}(T_{meas,i} - \hat{T})^2}{N_p}}, \\ J_{co_2} = \sqrt{\dfrac{\sum_{j=1}^{N_p}(C_{meas,j} - \hat{C})^2}{N_p}}, \\ J_{cost} = (\sum\limits_{k=1}^{N_{fan}} \dfrac{\Delta P \dot{V}_{air,k}}{\eta_{fan}} Prc_e + \sum\limits_{p=1}^{N_{co2}} \dot{V}_{co_2,p} \omega \rho Prc_{co_2}) T_{hour}, \end{cases} \tag{1}$$

where $T_{meas,i}$ and $C_{meas,j}$ represent values at i_{th}/j_{th} sampling points in interested crop areas. $\hat{T}$ and $\hat{C}$ are idea values of temperature and CO_2 concentration in crop areas. N_p is the total number of sampling points in crop areas and N_{fan} is the number of fans. $\dot{V}_{air}$ and $\dot{V}_{co_2}$ are the overall volumetric flow rate of supply air and CO_2 (m^3/s) respectively. ΔP is the pressure rise through the supply fan (Pa) and η_{fan} is the fan efficiency. Prc_e and Prc_{co_2} are prices of electricity and CO_2 at local market. The parameters ω and ρ are mass fraction and density of injected CO_2. T_{hour} is the duration time of the regulation.

2.2.2. Control Variables

To achieve these three objectives, three control variables (setting points) were set in the optimization scheme including fans' speeds, injection rate of CO_2, and heat load of solar radiation. According to previous research [18,19], these variables have obvious effects on indoor micro-climate. Since the optimization is simulation based, the heat load of solar radiation is assumed to be changed by the shading coefficient, and the practical implementation of these control variables are beyond the scope of the article.

2.3. Online–Offline Strategy

To derive the multiple environmental parameters' steady fields of the greenhouse, the full CFD procedure may be very time consuming. This further makes the computational cost of interactive optimization very expensive.

For efficient computation, an online–offline scheme was employed. In the offline phase, the full CFD simulations were executed to calculate the original greenhouse model with precise solutions. In the online phase, on the basis of ensuring convergence, the mesh was simplified and the converge conditions are relaxed to acceptable extents. The purpose of this setting was to relieve the calculation burden and improve the optimization efficiency.

3. Model Construction Using CFD

Since the summer weather in east China is very hot, crop growth and energy usage are significantly affected by spacial distributed micro-climate factors of the greenhouse such as temperature field and air flow distribution. In this study, we choose a medium-large greenhouse in east China for model construction and method validation.

3.1. Structure of the Venlo Greenhouse

The greenhouse for investigation is located in Zhenjiang, Jiangsu Province (32.080248(N), 119.503427(E), 50 m(ASL)). The greenhouse has three spans with the length × width: 40 m × 18 m in total (720 m^2). The roof goes along the north-south direction. The eaves height is 3.175 m and the roof height is 5.0 m. The main surrounding shelter and roofing materials are polycarbonate (PC) sunshine plates (thickness 4 mm, light transmission rate 87%). Three ventilation fans are installed on the north wall of the greenhouse (one fan each span). To facilitate experiment validation of temperature and airflow distributions, the greenhouse is almost empty and no crops are growing inside (We carried out the experiment after weeding, but the weeds grown out fast within two weeks during the experiment. In the CFD simulation, the latent transfer and water vapour exchanges of weeds are not considered). The location and profile of the greenhouse are shown in Figure 3.

Figure 3. The location and structure of the target greenhouse.

3.2. CFD Model Construction

In the greenhouse thermal procedure, temperature distribution is thermally coupled with airflow field, which is mainly governed by the incompressible Navier–Stokes equations. Boussinesq approximation is applied for representing buoyancy effect in the CFD simulation. The nondimensional form of basic conservation equations are shown below:

$$\begin{cases} \nabla \cdot \mathbf{V} = 0, \\ \dfrac{\partial \mathbf{V}}{\partial t} = -(\mathbf{V} \cdot \nabla)\mathbf{V} + \dfrac{1}{Re}\nabla^2 \mathbf{V} - \nabla p + \dfrac{Gr}{Re^2} T\delta, \\ \dfrac{\partial T}{\partial t} = -\mathbf{V} \cdot \nabla T + \dfrac{1}{RePr}\nabla^2 T, \end{cases} \tag{2}$$

in the domain $\Omega \times [0, T]$, where Ω represents a spatial domain, δ is a unit vector in the direction of gravitational acceleration, the Prandtl number $Pr = \mu c_p/\kappa$, Grash of number $Gr = \beta \ell^3 g \Delta T/\nu^2$, and Reynolds number $Re = V_{max}\ell/\nu$.

Considering the turbulent characteristics of the airflow during the ventilation procedure, renormalization group $k - \varepsilon$ model (RNG) was used to predict indoor microclimate of greenhouse, which was reported more accurate than standard $k-\varepsilon$ model with weak air velocities [26]. The discrete ordinates (DO) radiation model was chosen for indoor radiation simulation, and the effect of solar radiation on the room temperature is simplified using Airpak's solar load model. The latent transfer and water vapour exchanges of weeds are ignored.

Airpak 6.3 is the simulation tool based on a finite volume method as a Fluent engine. The calculation domain consists of two parts: indoor area (40 m × 18 m × 5 m) and outdoor area (200 m × 100 m × 25 m). In early summer in eastern China, the main heat source of greenhouse comes from outdoor heat radiation. The ventilation condition was set as negative pressure ventilation (the velocity inlet type is applied for fans, and the zero gradient boundary condition (outlet type) was applied for windows). The roof is considered as semitransparent medium, and the surrounding shelter's absorption coefficient of radiation is set as a constant. The basic parameters setting and initial conditions are listed in Table 1. There were a total 348,134 unstructured hexahedral mesh elements generated for model calculation. The structure of the greenhouse model with mesh is shown in Figure 4.

Table 1. The basic parameters setting for computational fluid dynamics (CFD) simulation.

Parameter type	Value (Unit)
Operating pressure	101,325 N/m^2
Sunshine fraction	0.9
Ground reflectance	0.1
Default temperature	24 °C
Material of roof and wall	Float glass
- Density of glass	2400 kg/m^3
- Specific heat of glass	790 J/kgK
- Conductivity of glass	2.58 W/mK

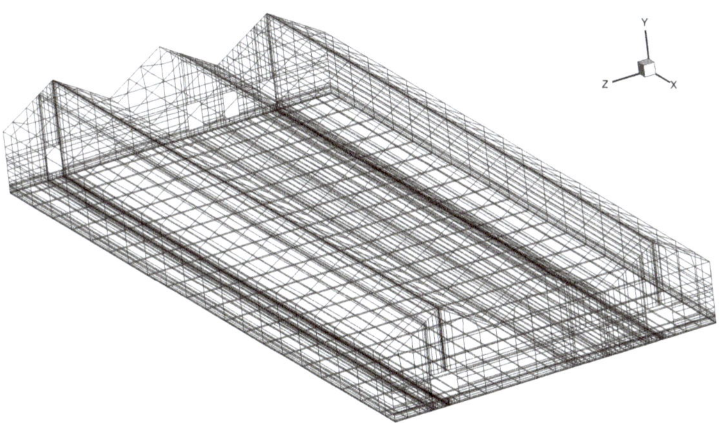

Figure 4. The structure of the greenhouse model with mesh.

3.3. Simulation Results

Figure 5 provides steady temperature contour and airflow vector field at one vertical section facing the window and fan. From the figure, it is seen that indoor temperature and airflow are not evenly distributed. The assumption of "well mixed" does not exactly fit into the real situation, which was also reported by many previous literature [2,3,5]. Concretely, at noon during a warm day (May in east China), the temperature in the upper part of the greenhouse was generally higher than the temperature in the lower area because of the outdoor heat radiation through the sunshine-plate roof. The temperature near the natural ventilation windows was lower than the temperature of middle area because the average temperature inside the greenhouse was higher than outside at noon. Because of the negative pressure ventilation by three fans, the lower air flows faster than the upper part.

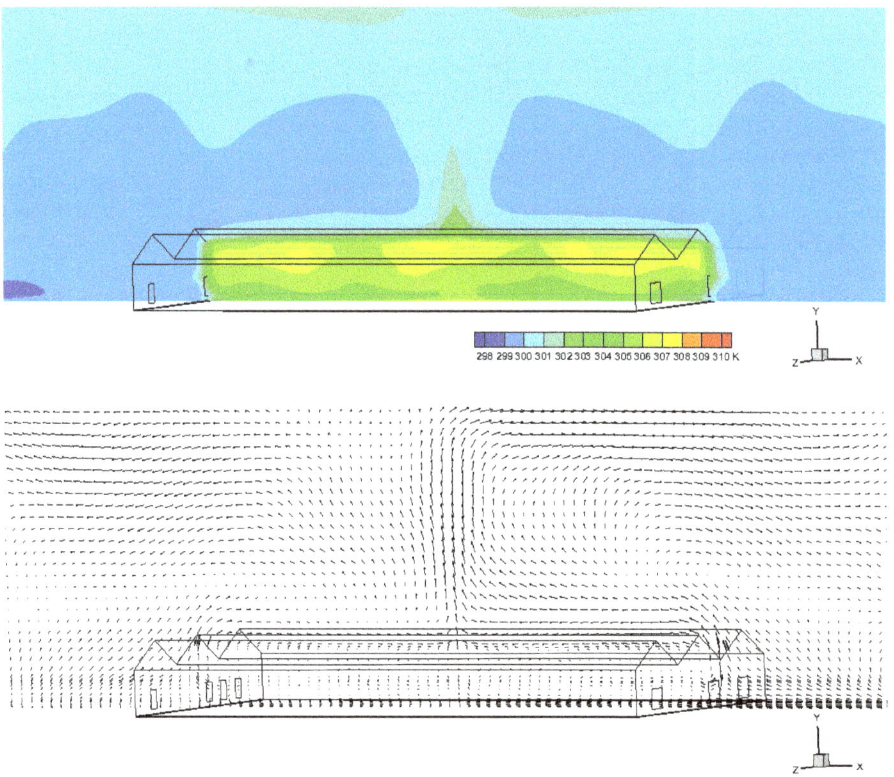

Figure 5. Temperature (**top**) and airflow (**bottom**) fields at the vertical section (Z = 15.5 m).

4. Model Validation with Field Experiment

4.1. Experiment Setting

To validate the CFD model's accuracy, field experiments are employed for this study. Both temperature and air speed measurements are executed in this case study.

There were 33 temperature sensors ($T_{1b} - T_{33b}$) forming a measurement matrix in single vertical section from south to north (length direction × height direction: 11 × 3). The section location was selected facing the middle fan (Z = 7.089 m). At each height, there were 11 temperature sensors uneven distributed: sparse in middle area, dense near boundaries. The side view of the 2D sensor matrix from east direction is provided in Figure 6a. Airflow patterns were also investigated by 20 observing points ($V_{1a/b} - V_{10a/b}$) at two vertical sections with two heights (V_a: 0.7 m, V_b: 1.6 m). The top view of the air velocity sensor matrix is provided in Figure 6b.

The type of temperature sensor is TP4029POS, ±0.2 °C; the type of ultrasonic three-dimensional wind speed sensor is WS-A2, ±0.1 m/s. The solar radiation intensity and outside temperament/airflow speed are also recorded by meteorological station. The experimental scene is shown in Figure 7.

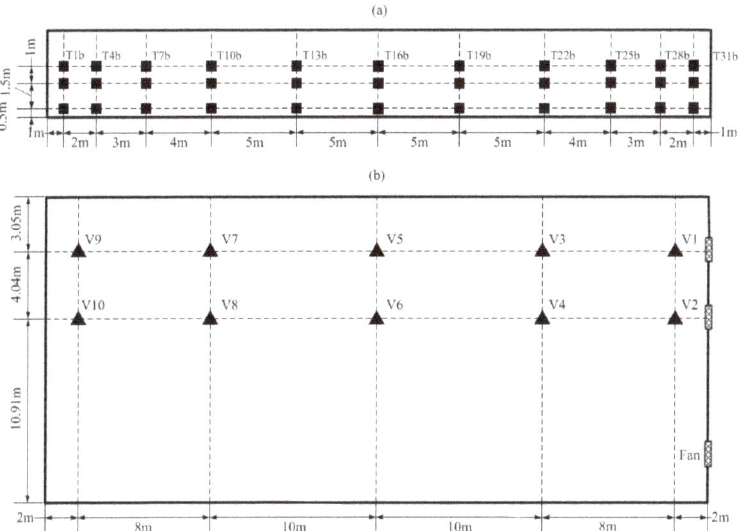

Figure 6. Locations of temperature and air speed sensors: (**a**) the side view of temperature sensor matrix, (**b**) the top view of the air velocity sensor matrix.

Figure 7. The experimental scene in the Venlo type commercial greenhouse.

4.2. Experiment Results

The experiment was carried out at 12 a.m.–15 p.m., 5 May 2017. At this time period, the outside temperature was relative high, which is helpful to observe obvious temperature gradient distribution. Each sensor sampled ten times at the same location (1 min interval), and the average results were recorded and saved by wireless data collector. Under natural and mechanical ventilation conditions, temperature measurements are employed using above experiment settings. Detailed data is provided in Appendix section in Tables A1 and A2. Under mechanical ventilation condition, the air velocity field is also sampled by wind speed sensors. Results are shown in Table A3.

Figure 8 shows the 2D temperature distributions under natural and mechanical ventilation conditions. In the figure, the obvious temperature gradient distribution in vertical section is observed like simulation results. Moreover, compared with natural ventilation results (Figure 8a), mechanical ventilation (Figure 8b) made the indoor temperature more even, especially at the lower area. This homogeneous distribution of temperature was better for crops production in greenhouses. Using absolute error percentage formula ($Percentage = \frac{|T_{sim} - T_{exp}|}{T_{exp}}$), Figure 8c shows the percentage of the 2D temperature difference between experiment and simulation results in mechanical ventilation condition.

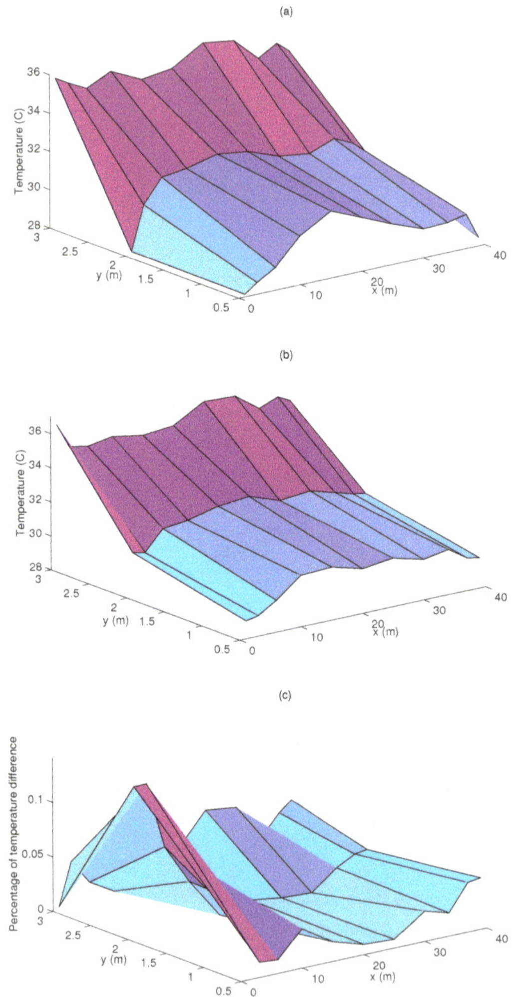

Figure 8. Results of 2D temperature measurements: (**a**) natural ventilation, (**b**) mechanical ventilation, (**c**) percentage of temperature difference between experiment and simulation.

For model validation, simulation results are compared with experiment results in both natural and mechanical ventilation conditions. The average errors of temperature simulations in different conditions are less than 15%. Due to the sensor accuracy limitations, the air velocity field is hard to validated accurately. The average errors are about 20%. It should be noted that the errors are also partly due to the ignoring of the latent transfer and water vapour exchanges of weeds cover.

5. Multi-Objectives Optimization of the Greenhouse Environment

5.1. Optimization Setting

On the basis of the experiment validation, the optimization case study using the proposed interactive scheme was employed in this section. The optimization focuses on solving two problems: (1) how to find out multiple variables' optimum setting points according to multiple environmental requirements? (2) For local planting areas, how to adjust environmental variables with high spatial resolution to find the balance of energy saving and environment suitability?

During summer in east China, the indoor temperature at noon is higher than 45 °C, which is harmful to crop growth. To find out the optimal heat flux of the roofs and the ventilation speeds of fans in the greenhouse, we choose an optimization scenario of summer noon (25 July, 10:00–14:00), and an ideal temperature is set (306 K) according to Section 2.2's discussion.

The control variables include: (1) the speeds of three ventilation fans on the north wall (m/s), (2) the injection rate of the simulated CO_2 device at the center of the greenhouse (m/s), and (3) the simulated heat loads of solar radiation on the east roofs (W/m^2). The simulated crop growing area ($X \times Z \times Y : 30$ m $\times$ 4 m $\times$ 1.5 m) and the three kinds of control devices are shown in Figure 9.

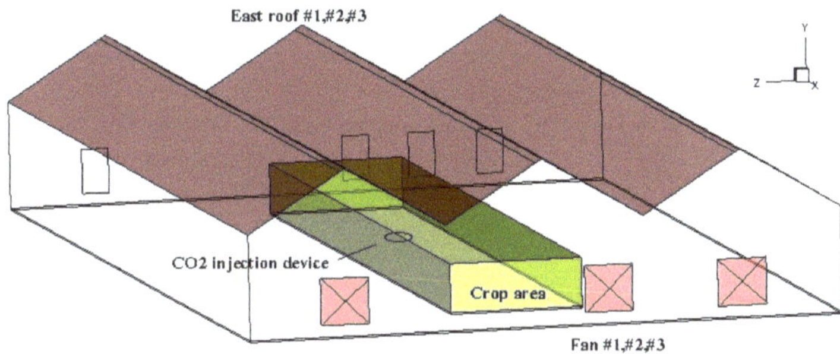

Figure 9. Profile of the control variables and simulated crop area in the greenhouse.

In addition to Equation (2), The CFD solver uses the multi-species transport model to simulate CO_2 concentration variation, and the detailed equations and panel setting can be found in Airpak manual [27]. For simplicity purposes, we only recorded the value of spacial CO_2 concentration in simulated crop area, but the photosynthetic/respiration of crops are not considered in this case. The multiple objective indexes including temperature, CO_2 concentration, and energy consumption are already described in Section 2, and the related parameters in the formulas are listed in Table 2. In this table, the variation ranges of the control variables are also specified.

We realized the multi-objective based genetic algorithm (NSGA-II) on the basis of Kalyanmoy Deb's report [23]. According to the literature analysis and to meet a trade-off with available computational capacity, the population size and maximum number of generations are chosen as 25 and 10, respectively. The online–offline strategy described in Section 2 was applied for computational efficiency improvement. Table 2 specifies the main parameters of NSGA-II for this study.

The computer used to run the simulation is an Intel Core E3 (4 cores, 3.2 GHz) with 16 GB of RAM and the whole optimization procedure requires about 84 h to obtain the Pareto frontier.

Table 2. Main parameters of the optimization case.

Category	Parameter	Design Range
Control variables	Speed of fan #1, #2, #3	0.5–5.0 m/s
	Heat loads of the east roof #1, #2, #3	100–300 W/m^2
	Velocity of CO_2 emission	0.01–0.10 m/s
Objectives & related parameters	Objective I: ideal temperature $\hat{T}$	306 K
	Objective II: ideal CO_2 concentration $\hat{C}$	1000 ppm(mass)
	Objective III: energy consumption	minimum
	Mass fraction of CO_2 injection	95%
	Density of CO_2	1.977 kg/m^3
	Local prices of electricity & CO_2	0.56 CNY/kWh & 1.614 CNY/KG
NSGA-II	Number of generations	25
	Population size	10
	Crossover probability	0.8
	Number of elites produced	2

5.2. Results and Analysis

5.2.1. Objective Space Analysis

The results of the multi-objective optimization are shown in Figure 10. From a total of 243 chromosomes, 25 pairs of chromosomes of the last generation identified by the NSGA-II algorithm belong to the Pareto frontier. It is noted that seven chromosomes of total 250 are abandoned because of the CFD convergence issue. From the figure, we can find that the target sets represented by particles spread toward the optimal directions. Each particle performs differently with respect to the individual objective indexes. Concretely, among the best 25 solutions, the variation of the index of J_T was [5.08, 20.90] (K), the variation of the index of J_{CO_2} was [0.0, 0.0055] (ppm), and the variation of the index of J_{COST} was [17.4, 59.9] (CNY).

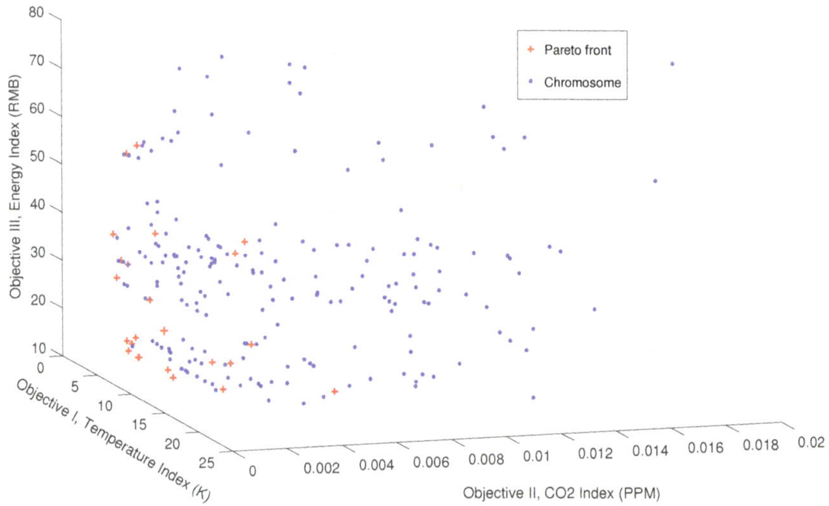

Figure 10. Tri-dimensional projection of the multiple objectives represented by particles.

By narrowing down the scope appropriately, we also fit an optimal surface using nearest neighbor method [28], which is shown in Figure 11. In this figure, we can roughly observe the distribution characteristics of the optimal solutions more directly. Basically, as the requirements of environmental indexes (distributions of temperature and CO_2 in this study) became more stringent, the energy consumption rose sharply, which was consistent with the results of previous studies. If we separately extract temperature and energy cost indexes and plot, they show a significant negative correlation that is shown in Figure 12.

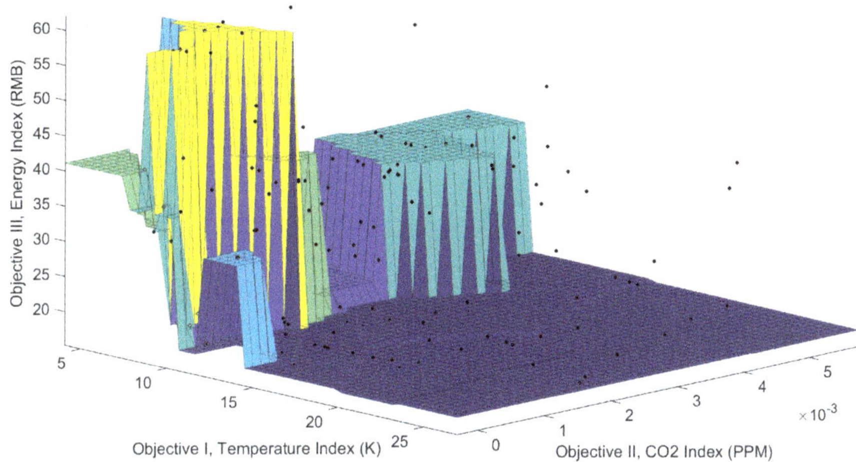

Figure 11. Fitting surface of optimal solutions using nearest neighbor method.

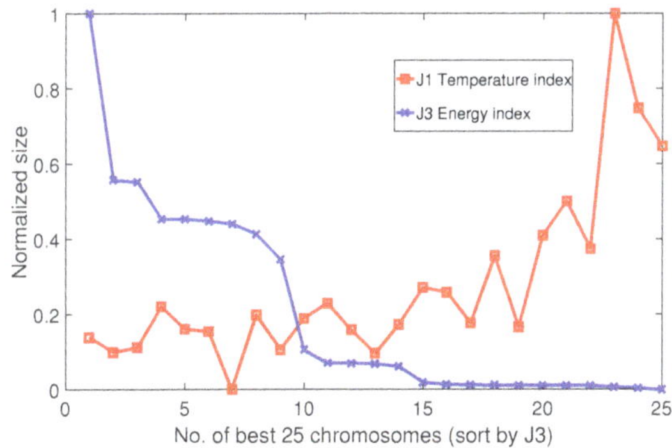

Figure 12. Negative correlation between temperature and energy cost indexes (25 pairs of chromosomes belonging to the Pareto front).

5.2.2. Control Variable Space Analysis

The multiple objective optimization scheme also provide the optimal control variables that are recorded in Figure 13. In total, five different control quantities were divided into two tri-dimensional projections. In the top projection, three main control variables, i.e., heat load of roofs, speed of three fans, and velocity of CO_2 emission are selected. The variables were clustered in several groups representing different searching orientations of the objective functions. Quantitatively, corresponding to the best 25 solutions, the variation of roofs' heat load was [154.56, 291.86] (W/m^2), the variation of fans' speed is [1.61, 4.63] (m/s), and the variation of CO_2 emission's velocity was [0.010, 0.077] (m/s). Note that the maximum speed values of three fans (Fan #1, #2, #3) are extracted for this subgraph, and each heat load of the east roofs (Roof #1, #2, #3) remains the same in this case study.

In the bottom subgraph of Figure 13, the individual speeds of three fans are chosen and shown in the form of tri-dimensional projection. From the subgraph, it is found that three fans' speeds are different because of the relative positions to the crop area. Due to the constraints of energy cost index, the fans' speeds do not tend to be maximum even at noon in summer. For heat-resistant crops, this was an optimized choice. But for heat-sensitive crops, the weights of energy cost and temperature indexes must be further improved, otherwise the results cannot be adopted in real situation.

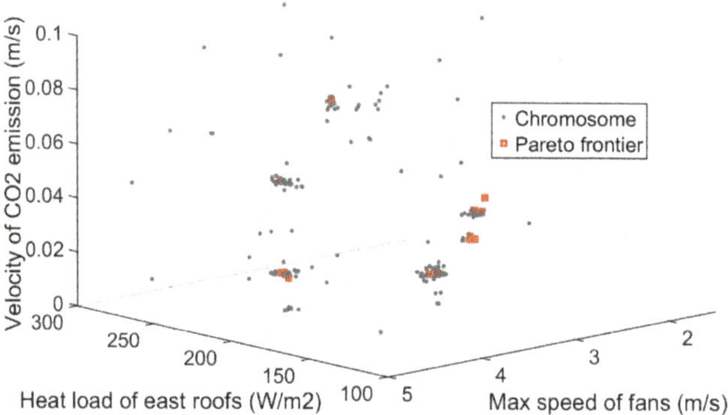

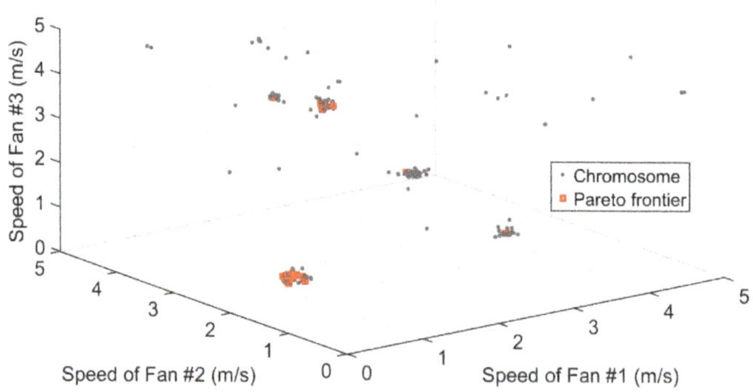

Figure 13. Tri-dimensional projections of the multiple control variables represented by particles (top projection: CO_2 emission speed - heat load of roofs - max speed of fans; bottom projection: speed of fan #1, #2, #3).

5.2.3. Full Simulation of Optimized Greenhouse Model

After the optimization, a steady CFD simulation is carried out using one of the best 25 solutions from the 3D Pareto frontier. Considering the temperature, CO_2, and the energy cost in a balanced way, we choose the following pair of control variables: The heat load of east roofs is 154.56 W/m^2; the speeds of three fans are 3.56 m/s (Fan #1), 3.448 m/s (Fan #2), and 1.228 m/s (Fan #3); the velocity of CO_2 emission is 0.021 m/s.

To roughly describe the resulted temperature distribution of this solution, we record the main temperature contour at center plane in Figure 14 (top). Although the ideal value in the temperature index was set at 306 K, a iso-surface of 311 K (38 °C) is drawn in this figure considering the heat resistance of crops. From the iso-surface, we can find that most of controlled area's temperature is below 38 °C, and the temperature near the roof (above three meters height) rise sharply because of the solar radiant heat. In the bottom sub-figure, we also record the resulted CO_2 distribution at 0.3 m height. Quantitatively, the average error from standard value (1000 ppm) within crop area is less than 0.005 ppm.

To describe the temperature distribution's feature of crop area, we also extract three temperature fields at different heights (1.5 m, 1.0 m, 0.3 m) and record them in Figure 15. From the figure, it was found that the temperature values within the most crop area (red rectangle) are between 35 °C and 37 °C. This was a result of compromise between environmental parameters and energy usage by multi-objective optimization. Note that the results were based on the premise that keep the ventilation design of the greenhouse unchanged, and some boundary conditions were simplified for simulation iterations.

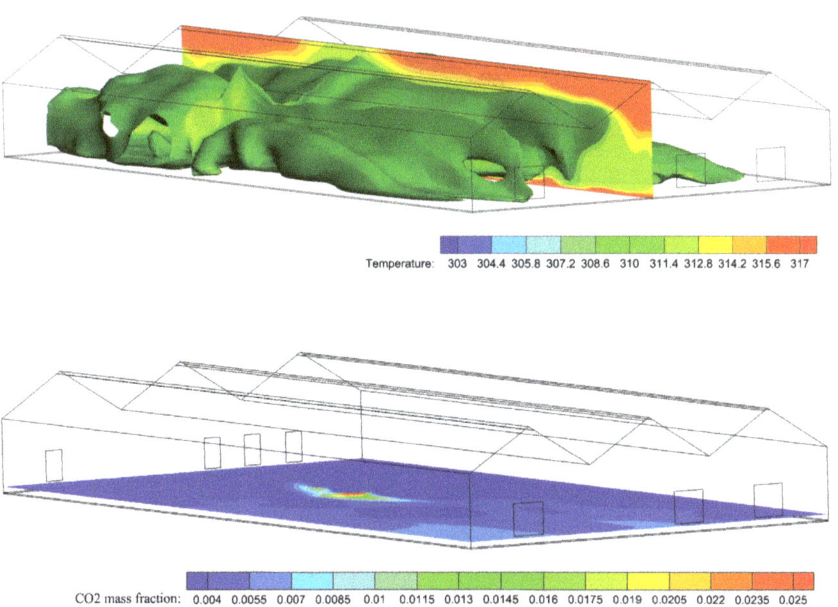

Figure 14. Resulted temperature contour at center Z plane combined with iso-surface of 311 K (**top**), and CO_2 fraction distribution at 0.3 m height (Y plane, **bottom**).

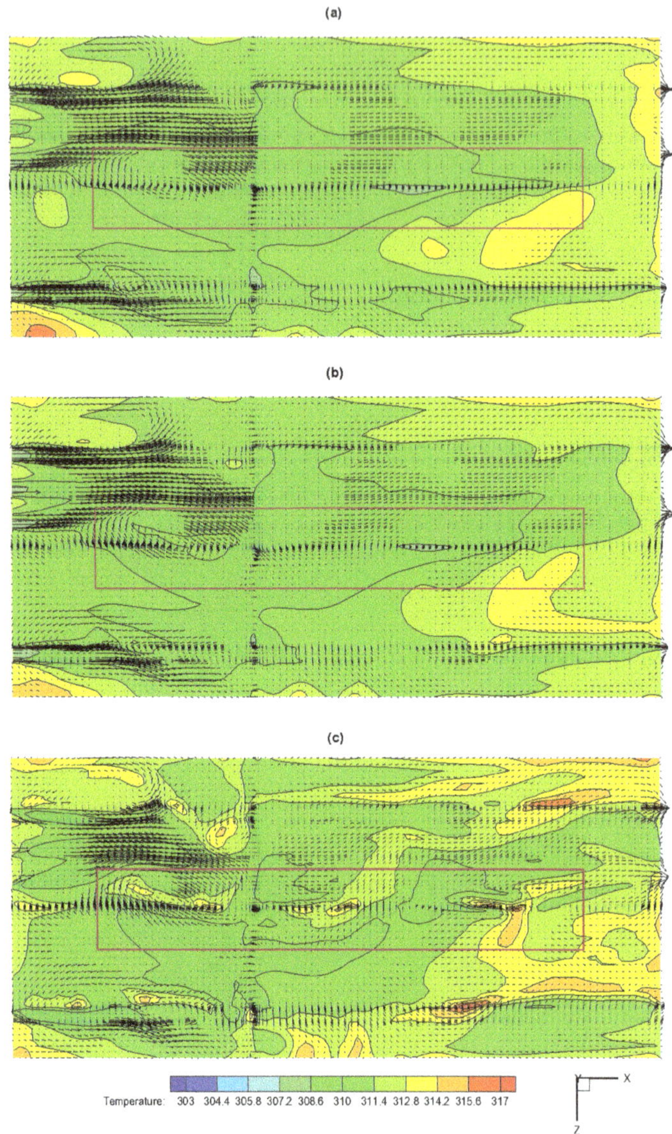

Figure 15. Temperature contours and air velocity vectors at three heights (Y plane): (**a**) 1.5 m, (**b**) 1.0 m, (**c**) 0.3 m.

To our knowledge, the multi-objective algorithm NSGA-II had a better sorting algorithm and better incorporated elitism than its previous version, NSGA [23]. The proposed CFD-EA optimization platform also may adopt other latest stochastic algorithms, which is suitable for discontinuous, complicated, and distributed parameter systems. However, the time-consuming issue is a big holdback to the proposed method's applications. Next step, a supercomputer with 260 computing cores will be hired to do such calculations.

One of the purposes of the CFD-EA platform is to serve as a tool for searching the set-points of the greenhouse environmental variables with high spatial resolution. The Pareto frontier of solutions provides useful information for decision-making. However, the set points vary due to different plants, regions, and seasons. The above results cannot be directly used as a control basis without concrete analysis for specific issues.

6. Conclusions

Considering the environmental factors' spatial influences in greenhouses, this paper presents a CFD-EA optimization scheme that combines CFD simulations with multi-objective evolutionary algorithms. The NSGA-II is stochastic in nature and able to extract the inter features between environmental parameters and energy costs, providing information on optimal control variables and performances of greenhouse systems with high spatial resolution. A field greenhouse located in Jiangsu Province, east China, is used for the CFD model construction and validation. A simulated crop growing area (180 m^3) in the greenhouse in a summer noon scenario is chosen for the optimization case study. The used multiple objectives include: indoor temperature field, CO_2 distribution and energy costs. The heat load of roofs, the speeds of ventilation fans, and the simulated CO_2 emission are involved as control variables. As a result, 25 pairs of control variables from a total of 250 chromosomes were identified belonging to the optimum set point basis. Using this method, we can adjust optimal environmental variables with high spatial resolution to find the balance of energy saving and environment suitability. A detailed analysis may be provided that helps find the potential of the crop yield and energy conservation.

In the future works, a supercomputer with high computing speeds will be hired to solve the time-consuming problem; the interactive optimization scheme will be applied for optimal design of size, materials, and layout of greenhouse models.

Author Contributions: Conceptualization, K.L. and W.X.; methodology, K.L. and H.M.; software, X.C.; validation, H.J. and G.T.; resources, H.M.; writing—original draft preparation, K.L.; writing—review and editing, K.L. and G.T.; project administration, K.L.; funding acquisition, K.L. and W.X.

Funding: This research was funded by the National Natural Science Foundation of China (Grant No. 61873114, 51705206), China Postdoctoral Science Foundation (Grant No. 2018T110457, 2016M601741), and Project Foundation for Priority Academic Program Development of Jiangsu Higher Education Institutions.

Conflicts of Interest: The authors declare no conflict of interest. The funders had no role in the design of the study; in the collection, analyses, or interpretation of data; in the writing of the manuscript, or in the decision to publish the results.

Appendix A

Table A1. Results of temperature measurements, 2D, natural ventilation.

Sensor No.	T (°C)	Sensor No.	T (°C)	Sensor No.	T (°C)
T_{1b}	35.75	T_{12b}	30.55	T_{23b}	31.24
T_{2b}	28.17	T_{13b}	34.71	T_{24b}	29.53
T_{3b}	28.12	T_{14b}	32.26	T_{25b}	34.49
T_{4b}	35.38	T_{15b}	31.51	T_{26b}	31.76
T_{5b}	30.47	T_{16b}	34.93	T_{27b}	29.58
T_{6b}	28.60	T_{17b}	32.05	T_{28b}	35.06
T_{7b}	34.88	T_{18b}	30.82	T_{29b}	31.31
T_{8b}	31.69	T_{19b}	35.93	T_{30b}	29.80
T_{9b}	29.18	T_{20b}	31.46	T_{31b}	34.53
T_{10b}	35.48	T_{21b}	30.14	T_{32b}	30.76
T_{11b}	31.86	T_{22b}	35.68	T_{33b}	28.45

Table A2. Results of temperature measurements, 2D, mechanical ventilation.

Sensor No.	T (°C)	Sensor No.	T (°C)	Sensor No.	T (°C)
T_{1b}	36.44	T_{12b}	31.02	T_{23b}	31.93
T_{2b}	30.57	T_{13b}	34.72	T_{24b}	30.37
T_{3b}	29.05	T_{14b}	32.21	T_{25b}	34.79
T_{4b}	35.00	T_{15b}	31.10	T_{26b}	31.51
T_{5b}	30.40	T_{16b}	34.89	T_{27b}	30.63
T_{6b}	29.19	T_{17b}	32.40	T_{28b}	35.23
T_{7b}	34.79	T_{18b}	30.61	T_{29b}	31.21
T_{8b}	31.61	T_{19b}	35.90	T_{30b}	29.92
T_{9b}	29.86	T_{20b}	31.75	T_{31b}	34.88
T_{10b}	35.04	T_{21b}	30.84	T_{32b}	31.08
T_{11b}	31.77	T_{22b}	35.82	T_{33b}	29.77

Table A3. Results of air velocity measurements, mechanical ventilation.

Sensor No.	V (m/s)	Sensor No.	V (m/s)
V_{1a}	2.20	V_{1b}	1.80
V_{2a}	2.20	V_{2b}	2.40
V_{3a}	0.65	V_{3b}	0.60
V_{4a}	0.71	V_{4b}	0.75
V_{5a}	0.60	V_{5b}	0.30
V_{6a}	0.80	V_{6b}	0.70
V_{7a}	0.20	V_{7b}	0.11
V_{8a}	0.10	V_{8b}	0.20
V_{9a}	0.38	V_{9b}	0.57
V_{10a}	0.40	V_{10b}	0.50

References

1. Van Henten, E. *Greenhouse Climate Management: An Optimal Control Approach*; Agricultural University: Kerala, India, 1994.
2. Lee, I.B.; Short, T.H. Two-dimensional numerical simulation of natural ventilation in a multi-span greenhouse. *Trans. ASAE* **2000**, *43*, 745–753. [CrossRef]
3. Khaoua, S.O.; Bournet, P.E.; Migeon, C.; Boulard, T.; Chasseriaux, G. Analysis of Greenhouse Ventilation Efficiency based on Computational Fluid Dynamics. *Biosyst. Eng.* **2006**, *95*, 83–98. [CrossRef]
4. Piscia, D.; Muñoz, P.; Panadès, C.; Montero, J. A method of coupling CFD and energy balance simulations to study humidity control in unheated greenhouses. *Comput. Electron. Agric.* **2015**, *115*, 129–141. [CrossRef]
5. Santolini, E.; Pulvirenti, B.; Benni, S.; Barbaresi, L.; Torreggiani, D.; Tassinari, P. Numerical study of wind-driven natural ventilation in a greenhouse with screens. *Comput. Electron. Agric.* **2017**, *149*, 41–53. [CrossRef]
6. Tang, P.; Li, H.; Issaka, Z.; Chen, C. Impact forces on the drive spoon of a large cannon irrigation sprinkler: Simple theory, CFD numerical simulation and validation. *Biosyst. Eng.* **2017**, *159*, 1–9. [CrossRef]
7. Nabavi-Pelesaraei, A.; Abdi, R.; Rafiee, S. Neural network modeling of energy use and greenhouse gas emissions of watermelon production systems. *J. Saudi Soc. Agric. Sci.* **2016**, *15*, 38–47. [CrossRef]
8. Yu, H.; Chen, Y.; Hassan, S.G.; Li, D. Prediction of the temperature in a Chinese solar greenhouse based on LSSVM optimized by improved PSO. *Comput. Electron. Agric.* **2016**, *122*, 94–102. [CrossRef]
9. Tong, G.; Christopher, D.M.; Zhang, G. New insights on span selection for Chinese solar greenhouses using CFD analyses. *Comput. Electron. Agric.* **2017**, *149*, 3–15. [CrossRef]
10. Wang, J.; Li, S.; Guo, S.; Ma, C.; Wang, J.; Jin, S. Simulation and optimization of solar greenhouses in Northern Jiangsu Province of China. *Energy Build.* **2014**, *78*, 143–152. [CrossRef]
11. Zhang, X.; Wang, H.; Zou, Z.; Wang, S. CFD and weighted entropy based simulation and optimisation of Chinese Solar Greenhouse temperature distribution. *Biosyst. Eng.* **2016**, *142*, 12–26. [CrossRef]

12. He, X.; Wang, J.; Guo, S.; Zhang, J.; Wei, B.; Sun, J.; Shu, S. Ventilation optimization of solar greenhouse with removable back walls based on CFD. *Comput. Electron. Agric.* **2017**, *149*, 16–25. [CrossRef]
13. Coffey, B.; Haghighat, F.; Morofsky, E.; Kutrowski, E. A software framework for model predictive control with GenOpt. *Energy Build.* **2010**, *42*, 1084–1092. [CrossRef]
14. Jasak, H.; Jemcov, A.; Tukovic, Z. A C++ library for complex physics simulations. In *International Workshop on Coupled Methods in Numerical Dynamics*; IUC Dubrovnik Croatia: Dubrovnik, Croatia, 2007; Volume 1000, pp. 1–20.
15. Asadi, E.; Silva, M.; Antunes, C.; Dias, L. A multi-objective optimization model for building retrofit strategies using TRNSYS simulations, GenOpt and MATLAB. *Build. Environ.* **2012**, *56*, 370–378. [CrossRef]
16. Futrell, B.J.; Ozelkan, E.C.; Brentrup, D. Bi-objective optimization of building enclosure design for thermal and lighting performance. *Build. Environ.* **2015**, *92*, 591–602. [CrossRef]
17. Liu, W.; Chen, Q. Optimal air distribution design in enclosed spaces using an adjoint method. *Inverse Probl. Sci. Eng.* **2015**, *23*, 760–779. [CrossRef]
18. van Henten, E.J.; Bontsema, J. Time-scale decomposition of an optimal control problem in greenhouse climate management. *Control Eng. Pract.* **2009**, *17*, 88–96. [CrossRef]
19. Hasni, A.; Taibi, R.; Draoui, B.; Boulard, T. Optimization of Greenhouse Climate Model Parameters Using Particle Swarm Optimization and Genetic Algorithms. *Energy Procedia* **2011**, *6*, 371–380. [CrossRef]
20. van Beveren, P.; Bontsema, J.; van Straten, G.; van Henten, E. Optimal control of greenhouse climate using minimal energy and grower defined bounds. *Appl. Energy* **2015**, *159*, 509–519. [CrossRef]
21. Li, K.; Xue, W.; Liu, G. Exploring the Environment/Energy Pareto Optimal Front of an Office Room Using Computational Fluid Dynamics-Based Interactive Optimization Method. *Energies* **2017**, *10*, 231. [CrossRef]
22. Li, K.; Pan, L.; Xue, W.; Jiang, H.; Mao, H. Multi-Objective Optimization for Energy Performance Improvement of Residential Buildings: A Comparative Study. *Energies* **2017**, *10*, 245. [CrossRef]
23. Deb, K.; Pratap, A.; Agarwal, S.; Meyarivan, T. A fast and elitist multiobjective genetic algorithm: NSGA-II. *IEEE Trans. Evol. Comput.* **2002**, *6*, 182–197. [CrossRef]
24. Kroner, O. Crop Based Climate Regimes for Energy Saving in Greenhouse Cultivation. Ph.D. Thesis, Wageningen University, Wageningen, The Netherlands, 2003.
25. Raviv, M.; Medina, S.; Wendin, C.; Lieth, J.H. Development of alternate cut-flower rose greenhouse temperature set-points based on calorimetric plant tissue evaluation. *Sci. Hortic.* **2010**, *126*, 454–461. [CrossRef]
26. Coussirat, M.; Guardo, A.; Jou, E.; Egusquiza, E.; Cuerva, E.; Alavedra, P. Performance and influence of numerical sub-models on the CFD simulation of free and forced convection in double-glazed ventilated facades. *Energy Build.* **2008**, *40*, 1781–1789. [CrossRef]
27. Fluent Inc. *Airpak 3.0 User Guide*; Fluent Inc.: Lebanon, NH, USA, 2007.
28. Lehmann, T.M.; Gonner, C.; Spitzer, K. Survey: Interpolation methods in medical image processing. *IEEE Trans. Med. Imaging* **1999**, *18*, 1049–1075. [CrossRef] [PubMed]

Article

Oscillations Analysis of Front-Mounted Beet Topper Machine for Biomass Harvesting

Volodymyr Bulgakov [1], Simone Pascuzzi [2,*], Alexandros Sotirios Anifantis [2] and Francesco Santoro [2]

1 Department of Mechanics, Faculty of Construction and Design, National University of Life and Environmental Sciences of Ukraine, 03041 Kyiv, Ukraine
2 Department of Agricultural and Environmental Science, University of Bari Aldo Moro, Via Amendola, 165/A, 70126 Bari, Italy
* Correspondence: simone.pascuzzi@uniba.it; Tel./Fax: +39-080-544-2214

Received: 14 June 2019; Accepted: 17 July 2019; Published: 19 July 2019

Abstract: The beet leaves and tops, which currently are excluded from the production process of sugar, could be an interesting opportunity for the production of renewable energy. Usually, the defoliators are joined with root collar remover machines, which are installed in front of the tractor. In working conditions on soils having natural roughness these front-mounted beet topper machines carried by tractors are affected by angular oscillations in a longitudinal-vertical plane that strongly affect the cutting uniformity. A theoretical study of these oscillations was carried out in this paper using Lagrange II kind equations, with the aim to assess the design and kinematic parameters of a front-mounted beet topper, corresponding to more stable and suitable movements in the longitudinal-vertical plane. A numerical simulation was then performed adopting the developed mathematical model. In order to improve the efficiency of this harvesting machine, a significant role is assumed by the soil preparation. In this work the stiffness and damping parameters of the feeler wheels pneumatic tires have been considered constant but further studies are in progress to assess their effective importance and influence for reducing the vibration of the front-mounted beet topper machine with the final aim to achieve a better machine design.

Keywords: beet tops; rotary cutting device; tractor; oscillations; differential equations; optimal parameters; biomass production

1. Introduction

Sugar beet is an extensive crop of great agronomic value with important production and economic results if properly managed [1]. The cultivation of sugar beet requires the knowledge of agronomic, nutritional, and physiological elements that characterize the soil-plant-atmosphere ecosystem [2]. In particular, the most important aspect to keep in mind is that from this plant the sugar is obtained, which is produced in the leaves (photosynthetic process) and accumulated in the root. The process of accumulation occurs obviously when the foliar apparatus reaches enough development to guarantee a production of sucrose higher than the daily consumption of the plant itself, considering that the solar radiation intercepted by the leaf surface affects the growth processes of the plant, and the production of dry matter [3,4].

The cultivation of sugar beet and the by-products deriving from their industrial processing represent also an important renewable energy resource within the agricultural sector [5]. As known, electricity can be produced from the biogas derived from the fermentation of the pulp, which is a by-product of sugar refineries [6]. Furthermore, agricultural residues represent another important biomass resource to be used as a substrate in anaerobic digestion for the production of renewable energy. The use of beet leaves and tops, currently excluded from the production process of sugar, could be an interesting incoming opportunity [7].

For all the above-mentioned reasons is a rather complicated and urgent task for beet-growing sector to develop a high-performance and high-quality harvesting of sugar beet tops [8,9].

Sugar-beet harvesting requires a set of specific operations: (i) defoliation; (ii) root collar remover; (iii) uprooting; (iv) cleaning. The defoliators usually are joined with root collar remover machines that can be installed in front of the tractor, thus allowing the use of a harvester machine mounted at the back of the tractor itself [10].

Experimental tests carried out with a front-mounted beet topper machine highlight that the cutting uniformity over the machine operative width as well as the full harvesting and transportation without loss, is strongly affected by machine oscillating movements which are, in turn, produced by soil roughness, tractor forward speed, location of feeler wheels related to machine suspension system, etc. In particular, the oscillations in the vertical plane of front-mounted beet topper machines could be related with pneumatic feeler wheels usage but, despite the widespread use of such kind of machines, there are no detailed analytical studies concerning these movements [11–13]. In this regard, a theoretical study of the oscillations of front-mounted beet topped machine was carried out using Lagrange II kind equations even if not all acting forces on the dynamic system were considered [14,15]. Conversely, a more accurate mathematical model should analyze also the effect of the design parameters of the above-mentioned machine, on the oscillatory movements along sugar beet roots rows due to soil roughness [16]. Taking in mind the aforesaid, the aim of this work is to assess the design and kinematic parameters of a front-mounted beet topper, corresponding to more stable and suitable movements in the longitudinal-vertical plane and reduction of the indicated oscillations. Therefore, a mathematical model of the movement of a front-mounted beet topper machine, mounted on a wheeled tractor, has been developed that describe oscillations of the corresponding rotary cutting apparatus in a longitudinally vertical plane when it moves on irregularity ground using pneumatic feeler-wheels.

2. Materials and Methods

2.1. The Developed and Built Front-Mounted Beet Topper Machine

Based on the principle of a mower-shredder, a new universal front-mounted beet topper machine has been developed, which has been mounted on a wheeled tractor (Figure 1a). This rotary beet topper machine cuts in a continuous way both the rosette of the leaves and the weeds, transporting the cut material into a vehicle that follows sideways [8].

The machine performs its technological process as follows. During the machine movement in the forward direction, the feeler wheels (2 in Figure 1b) located in the front part of the movable frame (1 in Figure 1b), adjusts the rotor (3 in Figure 1b) putting the knives at the required cutting height. The knives themselves are arc-shaped, hinged mounted on a cylindrical surface along the length of the rotor (3 in Figure 1b) assures an effective cut of the top of the beets [9–11]. Knives have high absolute speed (from 20–25 m s^{-1} to 40–50 m s^{-1} for thicker-stemmed crops), ensuring the effective cut of the entire array of tops [14].

The beet tops already cut by the knives move upper-ward inside the case and then they fall onto the screw conveyor so they can be moved to the rear part of the machine where the loading mechanism (4 in Figure 1b), through the chute (5 in Figure 1b), discharges them into a sideway running vehicle. The front-mounted beet topper machine receives motion and power through a cardan universal joint (6 in Figure 1b) connected to the aggregating tractor front power take-off. The final part of the beet tops harvesting is completed by a rear mounted machine that cut the beet roots collar.

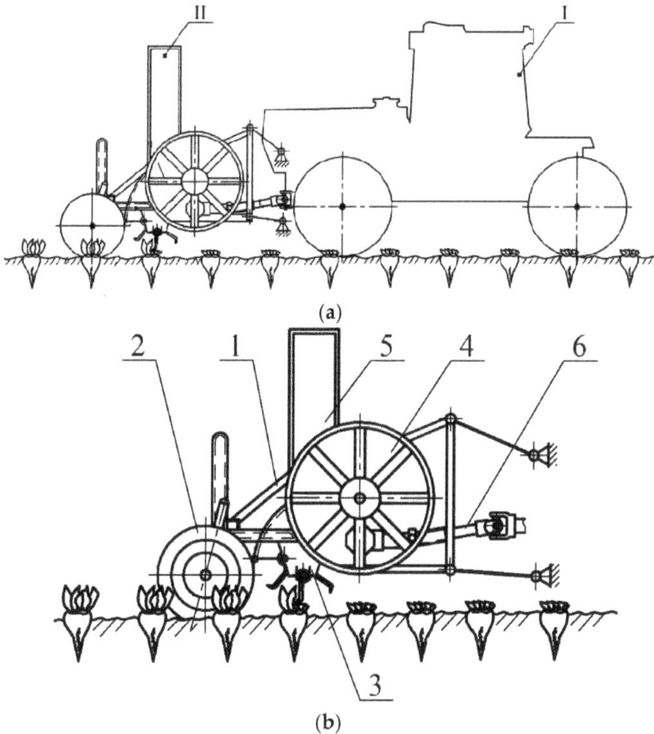

Figure 1. Front-mounted beet topper machine: (**a**) general view: I—carrying tractor; II beet topper machine; (**b**) constructive-technological scheme: 1—frame; 2—pneumatic feeler wheel; 3—rotary beet top cutting device; 4—loading mechanism of cutting beet top; 5—discharge chute; 6—drive working bodies.

From the mechanisms described above, it is shown that the quantity of harvested beet tops and their qualitative characteristics using such beet top harvesting machine, are strictly affected by: (i) the stability of the movement of the rotary beet harvesting device in the longitudinal-vertical plane; (ii) the efficiency in following soil surface roughness over all the working width; (iii) overall design parameters, above all, those concerning the tractor-machine interface frame.

2.2. Theoretical Study of the Oscillations of the Front-Mounted Beet Topper Machine

A mathematical model of the rotary beet harvesting device was set up in order to evaluate its oscillation amplitude in the longitudinal-vertical plane, which depends on soil roughness and, in turn, is affected by the structural and kinematic parameters of the front-mounted beet topper machine itself [17].

An equivalent scheme of movement only in the longitudinal-vertical plane of the front-mounted beet topper machine has been initially set up (Figure 2) [8]. In the equivalent scheme two different conditions have been considered: (i) the feeler wheel is reaching the top of soil irregularity; (ii) the feeler wheel is on the top of this irregularity.

As represented in Figure 2, the front-mounted beet topper machine is connected to the tractor by means of a frame based on three beams (two lower—OK and one upper—DM), having hinges in the points O, D, M, and K. The radii of the feeler wheels and the beet top cutting device are respectively r and r_1.

M is the mass of the front-mounted beet topper machine and the related weight $\overline{G}$ is considered to be applied in the center of mass C of the machine itself; $m = m_1 + m_2$ is the mass of the two feeler wheels, which is supposed to be concentrated in point B.

A dynamic system to the fixed cartesian coordinates $xOyz$ is considered in such a way in which the xOz plane is the longitudinal plane of the front-mounted beet topper machine and it is vertical and perpendicular to the soil surface.

Pneumatic feeler wheels are considered as elastic-damping models having a total stiffness coefficient $2c$ and total damping coefficient 2μ.

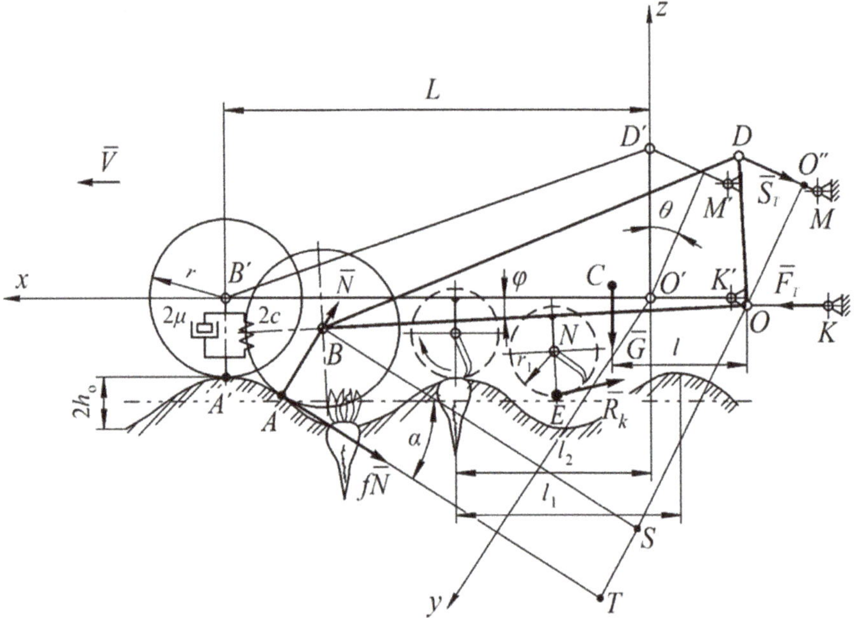

Figure 2. Equivalent scheme of the front-mounted beet topper machine.

It has been considered that both feeler wheels move almost in the same conditions when moving between the rows of sugar beet crops and it is possible to assume that while moving and crushing the upper and most loose soil layer they are in contact in point A with the soil surface. At the same time, the profile representing the soil roughness can be represented as a harmonic function like the following [1,2]:

$$h = h_o\left(1 - cos\frac{2\pi x}{l_1}\right) \tag{1}$$

where, referring to Figure 2:

- h [m]: Soil surface height irregularity;
- h_o [m]: Half of the maximum soil roughness;
- l_1 [m]: Horizontal distance between two consecutives soil points having the same characteristics;
- $x = V{\cdot}t$ [m]: Is the current coordinate with V [m s^{-1}] forward speed of the front-mounted beet topper machine.

In the mechanical behavior of the front-mounted beet topper machine jointed with the carrying tractor, it needs to be considered that during their common forward movement, the vertical oscillations of the tractor center of mass, even if not completely cancelled, are significantly smoothed due to its high inertia and the flatting effect produced on the soil by the big tractor weight.

Therefore, the hinges located at the points K and M (Figure 2), as belonging to the tractor, also perform vertical oscillations, even if reduced in amplitude. Furthermore, these weak oscillations can be considered to not affect the oscillation of the front-mounted beet topper machine (hinged at the points O and D) due its considerable mass, so it can be assumed that the suspension points of the front-mounted beet topper machine move in a first approximation on a straight line.

The front-mounted beet topper machine, due to the soil roughness received through the feeler wheels, only has angular oscillations in the vertical plane around the point O by an angle of rotation φ affected by the value h of soil roughness in accordance to where the point A of the feeler wheels, time by time, touch the soil surface.

The rotation of the frame of the machine around point O are due to the torque of all the forces whose action line do not pass through the point O that are:

(a) the normal $\overline{N}$ soil reactions applied at point A of contact between the feeler wheels and the soil;
(b) the tangential $f\overline{N}$ soil reactions applied at point A of contact between the feeler wheels and the soil (where f is the rolling friction coefficient related to the relative movement between the feeler wheels and the soil surface);
(c) the weight $\overline{G}$ of the front-mounted beet topper machine applied in its center of gravity;
(d) the cutting resistance reaction $\overline{R}_k$ of beet tops applied at point E;
(e) the traction tension $\overline{S}_T$ in the upper beam DM of the mounting frame between the front-mounted beet topper machine and the tractor in the direction from point D to point M.

Furthermore, along the lower traction OK of the mounting frame should be considered the tractor traction force $\overline{F}_T$, directed from point K to point O, but it does not contribute to moments around point O.

Finally, the elastic damping properties of the feeler wheels pneumatic tires also play a significant role in the vertical oscillation of the front-mounted beet topper machine. In this regard it is necessary to assess the potential energy P and dissipative function R of this dynamic system to consider the elastic viscosity properties of the tires.

3. Results and Discussion

3.1. Mathematical Model

The use of Lagrange equations of the second kind are suitable to describe the motion of the front-mounted beet topper machine, which incorporate the constraints directly by judicious choice of generalized coordinates that; therefore, have to be defined [9].

The position of the frame of the front-mounted beet topper machine, including its center of mass (point C in Figure 2), in the longitudinal-vertical plane is completely determined by the independent coordinate φ (Figure 2) of rotation of the frame of the machine which is, in turn, affected by the irregularity of the soil surface and by the feeler wheels elastic-damping properties.

Considering that the feeler wheels axis is rigidly connected to the main supporting frame, the vertical displacement [m] of their center of mass (point B in Figure 2) can be defined as follows:

$$z = L \cdot \varphi \tag{2}$$

where, referring to Figure 2, L [m] is BO distance (the length of the frame of the machine).

This dynamic system in the longitudinal-vertical plane has only one degree of freedom having one generalized coordinate φ and, so, the Lagrange equations of the second kind is [18–20]:

$$\frac{d}{dt}\left(\frac{\partial T}{\partial \dot{\varphi}}\right) - \frac{\partial T}{\partial \varphi} = Q_{\varphi} - \frac{\partial P}{\partial \varphi} - \frac{\partial R}{\partial \dot{\varphi}}, \tag{3}$$

where:

- T: kinetic energy;
- Q_φ: generalized force;
- P: potential energy;
- R: dissipative function (Rayleigh function);
- φ: generalized coordinate;
- $\dot{\varphi}$: generalized speed.

After a series of transformation (Appendix A), from Equation (3) is possible to obtain the nonlinear differential Equation (4), which models the angular oscillations of the frame of the front-mounted beet topper machine in the longitudinal-vertical plane.

$$\ddot{\varphi} + \frac{2L^2}{I_{oy}+mL^2}\left(c\varphi + \mu\dot{\varphi}\right) = \frac{1}{I_{oy}+mL^2}\left\{ -NLcos(\alpha+\varphi) + fN[r + Lsin(\alpha+\varphi)] + R_k r_1 cos\varphi + Glcos\varphi - S_T ODcos(\theta+\varphi) \right\} \quad (4)$$

In Equation (4), as well as constructive parameters (L, OD, l, r, r_1, m, G, c, μ), appear the modulus of two forces: The modulus of the cutting resistance reaction $\overline{R}_k$ of beet tops applied at point E and the modulus of the normal $\overline{N}$ soil reactions applied at point A of contact between the feeler wheels and the soil.

Experimentally evaluating the modulus of the cutting resistance reaction of beet tops (R_k), the modulus of the normal soil reaction between the feeler wheels and the soil is given by (Appendix B):

$$N = \frac{R_k r_1 + Gl - S_T ODcos\theta}{L - fr} \quad (5)$$

With the initial condition at $t = 0 :\ \varphi = 0; \varphi^{\cdot} = 0$ is possible to solve Equation (4) using the Runge-Kutta method in order to obtain $\varphi(t)$ and then the time dependent vertical movement of both the knife belonging to beet top cutting device (point E, Figure 2) and the center of mass (point C, Figure 2) of the whole machine using, respectively, the following equations:

$$z_E(t) = l_2 \cdot \varphi(t) \quad (6)$$

$$z_C(t) = l \cdot \varphi(t) \quad (7)$$

where:

- l_2: horizontal distance between the front-mounted beet topper machine rotation axis and cutting device rotation axis;
- l: horizontal distance between the front-mounted beet topper machine rotation axis and its center of mass.

3.2. Numerical Simulation

The developed mathematical model has been adopted as a base for numerical simulations using the parameters reported in the following Table 1.

Table 1. Numerical simulations parameters.

Parameter	Symbol	Unit	Value
Machine weight	G	N	9480.0
Feeler wheels weight ($G_k = G_{k1} + G_{k2}$)	G_k	N	48.9
Machine moment of inertia relative to its rotation axis	I_{oy}	Kg m^2	60.0
Machine rotation axis-feeler wheels axis distance $\left(\overline{OB}\right)$ Machine frame length	L	m	1.800
Machine frame height	OD	m	0.580

Table 1. *Cont.*

Parameter	Symbol	Unit	Value
Angle between the vertical frame beam (OD) and the perpendicular to the upper suspension beam (DM)	θ	rad	0.087
Machine rotation axis-cutting device axis horizontal distance	l_2	m	1.100
Machine rotation axis-center of mass horizontal distance	l	m	0.800
Forward speed of the tractor	V	m s^{-1}	1.5–3.0
Cutting device radius	r_1	m	0.365
Feeler wheels radius	r	m	0.300
Feeler wheels pneumatic tires stiffness coefficient	2C	N m^{-1}	4000
Feeler wheels pneumatic tires damping coefficient	2μ	N s m^{-1}	150
Half of the maximum soil roughness	h_o	m	0.040
Horizontal distance between two consecutives soil irregularities	l_1	m	0.700
Soil-feeler wheels friction coefficient	f	-	0.30
Cutting resistance reaction of three beet tops	R_k	N	300
Normal reaction component between feeler wheels and soil	N	N	4117
Tangential reaction component between feeler wheels and soil	fN	N	1235
Traction tension in the upper beam of the mounting frame	S_T	N	209
Traction tension in the lower beam of the mounting frame	F_T	N	1750

Several numerical simulations of the oscillation behavior of the front-mounted beet topper machine have been carried out, considering the agrotechnical requirements concerning the restriction of forward velocity of a front-mounted beet topper machine, which should be in the range $1.5 \leq V \leq 3.0$ [7] and the results are reported in the following Figures 3–5.

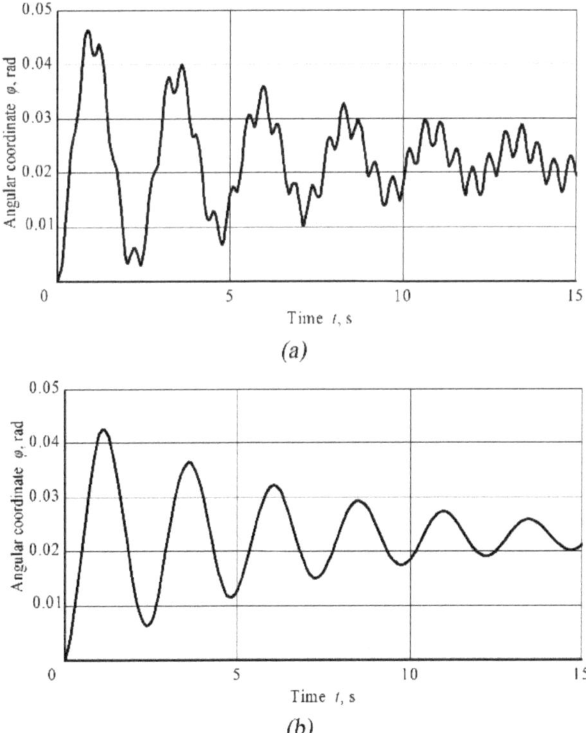

Figure 3. Angular oscillation φ of the front-mounted beet topper machine in the first period of its motion for a forward speed of 1.5 (**a**) and 3.0 m s^{-1} (**b**).

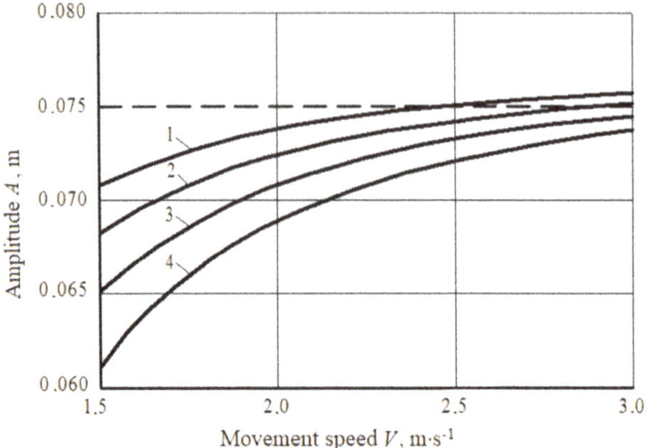

Figure 4. Relation between the amplitude of oscillations of the cutting device of the front-mounted beet topper machine A and the forward velocity V for different soil roughness step l_1 (1: 0.6 m; 2: 0.7 m; 3: 0.8 m; 4: 0.9 m).

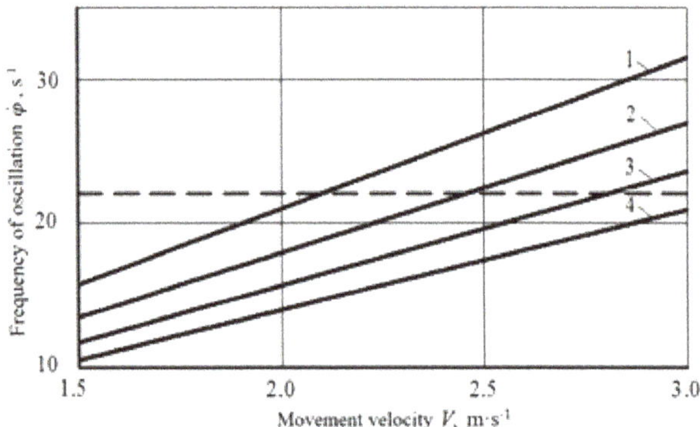

Figure 5. Relation between the frequency of oscillations of the cutting device of the front-mounted beet topper machine $\dot{\varphi}$ and the forward velocity V for different soil roughness step l_1 (1: 0.6 m; 2: 0.7 m; 3: 0.8 m; 4: 0.9 m).

In particular in Figure 3a,b are reported, for two different tractor velocities (1.5 m s^{-1} and 3.0 m s^{-1}, respectively), the angular oscillation trend of the front-mounted beet topper machine center of mass at the beginning of its movement in the case of a distance of between two consecutive soil irregularities (l_1) equal to 0.7 m (Table 1). In both cases a damped behavior is shown with the main difference that at a low value of forward speed the corresponding trend curve has a jagged appearance as the feeler wheels have enough time to follow all the soil irregularities, while at higher forward speed the trend curve has a smoother shape as the feeler wheels have both not enough time to follow all the soil irregularities and these are crushed by the feeler wheels themselves. Furthermore, in both cases, the angular oscillation extends asymptotically to a common value equal to $\phi = 0.02$ rad which, in turn, strictly depends upon the stiffness and damping coefficients of the feeler wheels 2C and 2μ, respectively (Table 1).

The improvement of soil preparation, aimed at reducing its roughness, significantly reduces the amplitude of oscillations of the cutting device of front-mounted beet topper machine as shown in Figure 4.

In Figure 5 it is also shown that, regardless of soil roughness step, increasing the forward velocity within the considered range, the amplitude of oscillation increases and extends asymptotically to a common value close to A = 0.075 m. Furthermore, a better soil preparation (Line 4—Figure 4) produces a variation equal to 19.7% of the amplitude of oscillation while a worst soil preparation (Line 1—Figure 4) produces a variation equal to 7.0% increasing the velocity from 1.5 to 3.0 m s^{-1}.

The frequency of forced oscillations of the cutting device of front-mounted beet topper machine have a linear trend in the entire range of forward velocity acceptable by an agrotechnical point of view, as shown in Figure 5, with an average percentual increase close to 100% for each of the considered soil roughness step. Furthermore, in case of better soil preparation, corresponding to step of irregularities equal to 0.9 m, it is less than 22 s^{-1} (Figure 5).

The numerical simulation shows that the developed mathematical model can be further used to study the influence of different constructive and kinematic parameters of the front-mounted beet topper machine in order be optimized with the final aim to assess and reduce the oscillations arising during its operation.

4. Conclusions

1. In working conditions on soils having natural roughness, front-mounted beet topper machine carried by a suitable tractor is affected by angular oscillations in a longitudinal-vertical plane that can be considered kinematic disturbances.

2. A motion equivalent scheme in a longitudinal-vertical plane has been developed and used, which considered all the external forces, all the dimensional and constructive parameters, and the soil roughness characteristics of soil roughness in the form of a harmonic function. The front-mounted beet topper machine frame rotation φ was used as a generalized coordinate in the Lagrange II kind nonlinear differential equation adopted as the mathematical model of the machine motion.

3. The numerical simulation of the oscillation characteristics in a longitudinal-vertical plane of the front-mounted beet topper machine has shown that, with the used constructive parameters, this oscillatory system is able to damp disturbing influences arising from soil roughness.

4. In order to improve the efficiency of this kind of machine, a significant role is assumed by the soil preparation, which should be carried out in order to flatten as much as possible the soil itself. As shown in the carried out numerical simulation, an increase in soil preparation leads to a reduction of the amplitude of oscillation close to 14.1% and the frequency oscillation close to 50% at the lowest forward velocity (1.5 m s^{-1}). Furthermore, low forward velocities allow to reduce all oscillatory effects considering that, in the carried out numerical simulation, the amplitude and the frequency of oscillation are the lowest and, respectively equal to 0.061 m and 11.3 s^{-1} in the case of better soil preparation. It has to be considered, furthermore, that both soil preparation and forward velocity have to be balanced, considering the productivity aspect of the harvesting process as, in general cases, it is carried out by third-part companies that are required to optimize time efforts.

5. In this work the stiffness and damping parameters of the feeler wheels pneumatic tires have been considered to be constant. Further studies are in progress in which different values of these dynamic parameters will be considered variable in order to assess their effective importance and influence for reducing the vibration of the front-mounted beet topper machine with the final aim to achieve a better machine design.

Author Contributions: Conceptualization, V.B. and S.P.; methodology, V.B., S.P. and F.S.; software, validation, formal analysis and data curation, V.B., F.S. and A.S.A.; writing—original draft preparation, V.B., F.S., S.P.; writing—review and editing, F.S., S.P.; supervision, V.B. and S.P.

Funding: This research received no external funding.

Acknowledgments: Italian Authors wish to thank Giulia De Candia (Bari, Italy) for her help in translating and understanding Russian language references.

Conflicts of Interest: The authors declare no conflict of interest.

Appendix A

$$\frac{d}{dt}\left(\frac{\partial T}{\partial \dot{\varphi}}\right) - \frac{\partial T}{\partial \varphi} = Q_\varphi - \frac{\partial P}{\partial \varphi} - \frac{\partial R}{\partial \dot{\varphi}}, \tag{A1}$$

where:

- T: kinetic energy;
- Q_φ: generalized force;
- P: potential energy;
- R: dissipative function (Rayleigh function);
- φ: generalized coordinate;
- $\dot{\varphi}$: generalized speed.

The total kinetic energy T is, so, given by:

$$T = \sum_{i=1}^{4} T_i = T_1 + T_2 + T_3 + T_4 \tag{A2}$$

where:

- T_1: kinetic energy of the translational motion;
- T_2: kinetic energy of the rotational motion of machine frame around a point O;
- T_3: kinetic energy of the vertical oscillations of the feeler wheels;
- T_4: kinetic energy of rotational motion feeler wheels around their axes.

In particular:

$$T_1 = \frac{MV^2}{2} \tag{A3}$$

where:

- M [kg]: mass of the front-mounted beet topper machine;
- V [m s^{-1}]: forward speed of the center of mass C of the machine;

$$T_2 = \frac{I_{oy}\dot{\varphi}^2}{2} \tag{A4}$$

where:

- I_{oy} [kg m^2]: moment of inertia of the frame the front-mounted beet topper machine relative to the Oy axis (perpendicular to the longitudinal-vertical plane and passing through the point O);
- $\dot{\varphi}$ [s^{-1}]: angular speed of the frame of the machine;

$$T_3 = \frac{m \cdot \dot{z}^2}{2} \overset{z=L\cdot\varphi}{\Rightarrow} \frac{mL^2 \cdot \dot{\varphi}^2}{2} \tag{A5}$$

where:

- m [kg]: mass of the feeler wheels;
- $\dot{z}$ [m s^{-1}]: speed of vertical oscillations of the feeler wheels;
- $\dot{\varphi}$ [s^{-1}]: angular speed of the rotation of the frame of the machine;

$$T_4 = \frac{I_k \omega^2}{2} \tag{A6}$$

where:

- I_k [kg m^2]: moment of inertia of the two feelers wheels relative to their axis of rotation;
- ω [s^{-1}]: angular speed of the feeler wheels.

It is possible to evaluate the angular speed ω of the feeler wheels by mean of the length of the circular arc using the following:

$$\omega = \frac{dS}{dt} \cdot \frac{1}{r} \tag{A7}$$

where:

- r [m]: radius of the feeler wheels.

Considering that the feeler wheels move along the soil roughness profile defined by (1)

$$h = h_o\left(1 - cos\frac{2\pi x}{l_1}\right),$$

$$\begin{gathered} dS = \sqrt{dx^2 + dh^2} = \sqrt{dx^2 + \left(\frac{h_o 2\pi}{l_1} sin\frac{2\pi x}{l_1} dx\right)^2} = \\ = dx \cdot \sqrt{1 + \frac{h_o^2 4\pi^2}{l_1^2} sin^2\frac{2\pi x}{l_1}} \overset{dx = V \cdot dt}{\Rightarrow} V \cdot dt \cdot \sqrt{1 + \frac{h_o^2 4\pi^2}{l_1^2} sin^2\frac{2\pi x}{l_1}} \end{gathered} \tag{A8}$$

Substituting (A8) in (A7):

$$\omega = \frac{V}{r}\sqrt{1 + \frac{h_o^2 4\pi^2}{l_1^2} sin^2\frac{2\pi x}{l_1}} \tag{A9}$$

and, so:

$$T_4 = \frac{I_k V^2}{2r^2}\left(1 + \frac{h_o^2 4\pi^2}{l_1^2} sin^2\frac{2\pi x}{l_1}\right) \tag{A10}$$

Considering Equations (A3)–(A5), and (A10) and substituting them into (A2):

$$\begin{gathered} T = \frac{MV^2}{2} + \frac{I_{oy}\dot{\varphi}^2}{2} + \frac{mL^2 \cdot \dot{\varphi}^2}{2} + \frac{I_k V^2}{2r^2}\left(1 + \frac{h_o^2 4\pi^2}{l_1^2} sin^2\frac{2\pi x}{l_1}\right) = \\ \frac{1}{2}\left(MV^2 + I_{oy}\dot{\varphi}^2 + mL^2\dot{\varphi}^2\right) + \frac{1}{2r^2} I_k V^2\left(1 + \frac{h_o^2 4\pi^2}{l_1^2} sin^2\frac{2\pi x}{l_1}\right) \end{gathered} \tag{A11}$$

The potential energy P of the dynamic system is equal to the work of the elastic deformation forces of the pneumatic tires of both feeler wheels and is given by the following expression:

$$P = c \cdot L^2 \cdot \varphi^2 \tag{A12}$$

where:

- c [N m^{-1}]: stiffness coefficient of the pneumatic tires of the feeler wheels;
- L [m]: distance between the axis of suspension of the front-mounted beet topper machine (point O) and the axis of the feeler wheels (point B).

The dissipative function R of the dynamic system is due to the viscous resistance forces of the pneumatic tires of both feeler wheels which are proportional to speed:

$$R = \mu \cdot L^2 \cdot \dot{\varphi}^2 \tag{A13}$$

where:

- μ [N s m^{-1}]: damping coefficient of the pneumatic tires of the feeler wheels;
- L [m]: distance between the axis of suspension of the front-mounted beet topper machine (point O) and the axis of the feeler wheels (point B).

In order to evaluate the last remaining term Q_φ to be used in the (A1), let us consider the elementary work of all active forces related to an infinitesimal rotation $\delta\varphi$:

$$\delta W_\varphi = -N \cdot BS\delta\varphi + fN \cdot TO\delta\varphi + R_k \cdot EN\, cos\, \varphi\, \delta\varphi + G \cdot l\, cos\, \varphi\, \delta\varphi - S_T \cdot OO''\, \delta\varphi \tag{A14}$$

where:

- N [N]: modulus of normal soil reactions of feeler wheels with the soil;
- BS [m]: arm of $\overline{N}$;
- fN [N]: modulus of tangential soil reactions of feeler wheels with the soil;
- TO [m]: arm of $f\overline{N}$;
- R_k [N]: modulus of cutting resistance reaction of beet tops;
- $EN\, cos\, \varphi$ [m]: arm of $\overline{R}_k$;
- G [N]: modulus of weight of the front-mounted beet topper machine;
- $l\, cos\, \varphi$ [m]: arm of $\overline{G}$;
- S_T [N]: modulus of traction tension in the upper connection beam DM;
- OO'' [m]: arm of $\overline{S}_T$;

The generalized force Q_φ acting on the dynamic system is, thus, the algebraic sum of the moments of all active forces relative to point O:

$$Q_\varphi = \frac{\delta W_\varphi}{\delta\varphi} = -N \cdot BS + fN \cdot TO + R_k \cdot EN\, cos\, \varphi + G \cdot l\, cos\, \varphi - S_T \cdot OO'' \tag{A15}$$

- Calling α the slope of the tangent to the profile $h = h_o\left(1 - cos\frac{2\pi x}{l_1}\right)$ representing the soil roughness: $\alpha = \arctan\left(\frac{dh}{dx}\right) = \arctan\left(\frac{2\pi h_o}{l_1} sin\frac{2\pi x}{l_1}\right)$;
- calling θ the angle between the vertical part of the front-mounted beet topper machine frame and the normal to the upper connection beam DM
- remembering that L [m] is the length of the frame of the machine;
- remembering that OD [m] is the height of the frame of the machine;
- remembering that $EN = r_1$ [m] is the radius of the beet top cutting device in (A13), considering Figure 2;
- $BS = L \cdot cos(\alpha + \varphi)$;
- $TO = r + L \cdot sin(\alpha + \varphi)$;
- $OO'' = OD \cdot cos(\theta + \varphi)$

and, so, Equation (A15) can be written as:

$$\begin{gathered} Q_\varphi = -N \cdot L \cdot cos(\alpha + \varphi) + fN \cdot [r + L \cdot sin(\alpha + \varphi)] + \\ + R_k \cdot r_1\, cos\, \varphi + G \cdot l\, cos\, \varphi - S_T \cdot [OD \cdot cos(\theta + \varphi)] \end{gathered} \tag{A16}$$

Is it now possible, using (A11), (A12) and (A15) to write four of the five terms that appear in (A1).

$$\frac{\partial T}{\partial \dot{\varphi}} = \left(I_{oy} + m \cdot L^2\right)\dot{\varphi} \Rightarrow \frac{d}{dt}\left(\frac{\partial T}{\partial \dot{\varphi}}\right) = \left(I_{oy} + m \cdot L^2\right)\ddot{\varphi} \tag{A17}$$

$$\frac{\partial T}{\partial \varphi} = 0 \tag{A18}$$

$$\frac{\partial P}{\partial \varphi} = 2c \cdot L^2 \cdot \varphi \tag{A19}$$

$$\frac{\partial R}{\partial \dot{\varphi}} = 2\mu \cdot L^2 \cdot \dot{\varphi} \tag{A20}$$

Finally, using (A17), (A18), (A16), (A19), and (A20) it is possible to write the final expression of (A1):

$$\begin{gathered} \left(I_{oy} + mL^2\right)\ddot{\varphi} + 2cL^2\varphi + 2\mu L^2\dot{\varphi} = \\ = -NL\,cos(\alpha + \varphi) + fN[r + L\,sin(\alpha + \varphi)] + R_k r_1\,cos\,\varphi + Gl\,cos\,\varphi - S_T[OD\,cos(\theta + \varphi)] \end{gathered} \tag{A21}$$

that can be written in the following form:

$$\begin{gathered} \ddot{\varphi} + \frac{2L^2}{I_{oy} + mL^2}\left(c\varphi + \mu\dot{\varphi}\right) = \\ \frac{1}{I_{oy} + mL^2}\left\{ -NL\,cos(\alpha + \varphi) + fN[r + L\,sin(\alpha + \varphi)] + R_k r_1\,cos\,\varphi + Gl\,cos\,\varphi - S_T OD\,cos(\theta + \varphi) \right\} \end{gathered} \tag{A22}$$

Appendix B

The modulus of the normal reaction $\overline{N}$ between the feeler wheels and the soil can be determined from the equilibrium condition of the system at a fixed time, setting equal to zero the algebraic sum of the moments of all forces acting on the system relative to point O.

The above-mentioned condition is raised, when (A16):

$$\begin{gathered} Q_\varphi = -NL\,cos(\alpha + \varphi) + fN[r + L\,sin(\alpha + \varphi)] + \\ + R_k r_1\,cos\,\varphi + Gl\,cos\,\varphi - S_T[OD\,cos(\theta + \varphi)] = 0 \end{gathered} \tag{A23}$$

As the angles α and φ in (A23) are time-dependent, it is convenient to fix the time at which the point of contact between the feeler wheels and the profile of soil roughness is in the upper most part of the profile itself: In this case (Figure 2) $\alpha = \varphi = 0$ so (A23) simplifies in:

$$Q_\varphi = -NL + fNr + R_k r_1 + Gl - S_T OD cos\theta = 0 \tag{A24}$$

and, finally:

$$N = \frac{R_k r_1 + Gl - S_T OD cos\theta}{L - fr} \tag{A25}$$

References

1. Bulgakov, V.M. *Beet Harvesting Machines*; Agrarian Science: Kiev, Ukraine, 2011; 351p.
2. Khvostov, V.A.; Reingart, E.S. *Machines for Harvesting Root Crops and Onions (Theory, Design, Calculation)*; VISHOM: Moscow, Russia, 1995; 391p.
3. Cerruto, E.; Manetto, G.; Santoro, F.; Pascuzzi, S. Operator Dermal Exposure to Pesticides in Tomato and Strawberry Greenhouses from Hand-Held Sprayers. *Sustainability* **2018**, *10*, 2273. [CrossRef]
4. Pascuzzi, S.; Cerruto, E.; Manetto, G. Foliar spray deposition in a "tendone" vineyard as affected by airflow rate, volume rate and vegetative development. *Crop Prot.* **2017**, *91*, 34–48. [CrossRef]
5. Manetto, G.; Cerruto, E.; Pascuzzi, S.; Santoro, F. Improvements in citrus packing lines to reduce the mechanical damage to fruit. *Chem. Eng. Trans.* **2017**, *58*, 391–396.
6. Pascuzzi, S.; Cerruto, E. An innovative pneumatic electrostatic sprayer useful for tendone vineyards. *J. Agric. Eng.* **2015**, *XLVI:458*, 123–127. [CrossRef]
7. Bulgakov, V.; Adamchuk, V.; Ivanovs, S.; Ihnatiev, Y. Theoretical investigation of aggregation of top removal machine frontally mounted on wheeled tractor. *Eng. Rural Dev. Jelgava* **2017**, *16*, 273–280.

8. Bulgakov, V.; Adamchuk, V.; Nozdrovický, L.; Ihnatiev, Y. Theory of Vibrations of Sugar Beet Leaf Harvester Front-Mounted on Universal Tractor. In *Acta Technologica Agriculturae*; Slovaca Universitas Agriculturae Nitriae: Nitra, Slovakia, 2017; Volume 20, pp. 96–103.
9. Vasilenko, P.M. *Introduction to Agricultural Mechanics*; Agriculture: Kiev, Ukraine, 1996; 252p.
10. Vasilenko, I.F. *Theory of Cutting Machines Reaping Machines*; Works of the VISHOM: Moscow, Russia, 1937; Volume 5, pp. 7–14.
11. Reznik, N.E. *The Theory of Cutting with a Blade and the Basics of Calculating the Cutting Apparatus*; Mechanical Engineering: Moscow, Russia, 1975; 311p.
12. Bosoy, E.S. *Cutting Apparatus of the Harvesting Machines. Theory and Calculation*; Mechanical Engineering: Moscow, Russia, 1967; 167p.
13. Tatyanko, N.V. The calculation of the working bodies for cutting machines of sugar beet tops. *Tract. Agric. Mach.* **1962**, *11*, 18–21.
14. Kolychev, E.I. On the choice of the design impact case in the study of the smoothness of the movement of tractors and agricultural machinery. *Tract. Agric. Mach.* **1976**, *3*, 9–11.
15. Anifantis, A.S.; Pascuzzi, S.; Scarascia Mugnozza, G. Geothermal source heat pump performance for a greenhouse heating system. An experimental study. *J. Agric. Eng.* **2016**, *47*, 167–173. [CrossRef]
16. Bulgakov, V.; Pascuzzi, S.; Nadykto, V.; Ivanovs, S. A mathematical model of the plane-parallel movement of an asymmetric machine-and-tractor aggregate. *Agriculture* **2018**, *8*, 151. [CrossRef]
17. Bulgakov, V.; Pascuzzi, S.; Santoro, F.; Anifantis, A.S. Mathematical Model of the Plane-Parallel Movement of the Self-Propelled Root-Harvesting Machine. *Sustainability* **2018**, *10*, 3614. [CrossRef]
18. Tucki, K.; Mruk, R.; Orynycz, O.; Gola, A. The Effects of Pressure and Temperature on the Process of Auto-Ignition and Combustion of Rape Oil and Its Mixtures. *Sustinability* **2019**, *11*, 3451. [CrossRef]
19. Bulgakov, V.; Pascuzzi, S.; Ivanovs, S.; Kaletnik, G.; Yanovich, V. Angular oscillation model to predict the performance of a vibratory ball mill for the fine grinding of grain. *Biosyst. Eng.* **2018**, *171*, 155–164. [CrossRef]
20. Anifantis, A.S.; Colantoni, A.; Pascuzzi, S.; Santoro, F. Photovoltaic and Hydrogen Plant Integrated with a Gas Heat Pump for Greenhouse Heating: A Mathematical Study. *Sustainability* **2018**, *10*, 378. [CrossRef]

Article

Performance Analysis of a Small-Scale ORC Trigeneration System Powered by the Combustion of Olive Pomace

Andrea Colantoni [1], Mauro Villarini [1,*], Vera Marcantonio [1], Francesco Gallucci [2] and Massimo Cecchini [1]

1 Department of Agricultural and Forestry Sciences (DAFNE), University of Tuscia, Via S. Camillo de Lellis, SNC, 01100 Viterbo, Italy; colantoni@unitus.it (A.C.); vera.marcantonio@unitus.it (V.M.); cecchini@unitus.it (M.C.)

2 CREA Centro di Ricerca Ingegneria e Trasformazioni agroalimentari, Via della Pascolare 16, 00015 Monterotondo, Rome, Italy; francesco.gallucci@crea.gov.it

* Correspondence: mauro.villarini@unitus.it

Received: 24 April 2019; Accepted: 7 June 2019; Published: 14 June 2019

Abstract: The utilisation of low- and medium-temperature energy allows to reduce the energy shortage and environmental pollution problems because low-grade energy is plentiful in nature and renewable as well. In the past two decades, thanks to its feasibility and reliability, the organic Rankine cycle (ORC) has received great attention. The present work is focused on a small-scale (7.5 kW nominal electric power) combined cooling, heating and power ORC system powered by the combustion of olive pomace obtained as a by-product in the olive oil production process from an olive farm situated in the central part of Italy. The analysis of the employment of this energy system is based on experimental data and Aspen Plus simulation, including biomass and combustion tests, biomass availability and energy production analysis, Combined Cooling Heat and Power (CCHP) system sizing and assessment. Different low environmental impact working fluids and various operative process parameters were investigated. Olive pomace has been demonstrated to be suitable for the energy application and, in this case, to be able to satisfy the energy consumption of the same olive farm with the option of responding to further energy users. Global electrical efficiency varied from 12.7% to 19.4%, depending on the organic fluid used and the working pressure at the steam generator.

Keywords: renewable energy; biomass; olive pomace; combustion; ORC; working fluid

---

1. Introduction

Using fossil fuels has negative environmental impacts due to greenhouse gas emissions and air pollution problems. For this reason, in order to satisfy the continuous increase in energy demands and to reduce the environmental impact, the replacement of fossil fuels with renewable and sustainable resources is necessary. Biomass is considered an ideal energy source, thanks to its availability and its clean relationship with the environment [1,2]. Environmental analysis demonstrated that the biomass conversion processes can give a good performance [3,4]. A wide variety of biomass exists, e.g., food crops, energy crops, municipal solid wastes, green wastes and agricultural residues [5,6]. The use of biomass wastes avoids the "biomass vs. food" contrast [7] and solves the problem of waste disposal. In the Mediterranean areas of Europe, agro-industrial activities are very important, and a lot of residue is produced.

Among the major activities is the olive oil industry. Olive oil production generates a significant number of by-products, solids and liquids. It has been estimated [8] that one hectare of olives produces about 5000 kg of olives and, from these, about 2250 kg of olive pomace can been obtained.

The production process of olive oil typically brings an oily component, a solid residue and an aqueous component given by the water content of the olive pulp [9].

Anaerobic fermentation of the aqueous component, which has a high biochemical oxygen demand (BOD), has been suggested as a mean of solving this problem [10]. Although the solid residue does not present such a serious environmental problem, the option of producing a clean gaseous fuel composed of a residue that can be used as a fertiliser rich in nitrogen should be well thought out [11]. The global annual olive pomace production has been estimated at around 400 million tons on a dry basis [8].

For the management of mill solid wastes, some solutions have already been explored in the literature [10,12]: animal feed, biogas production, extraction of useful materials and fertilisers. Pelletising residue olive oil to increase the density and energy was also investigated. However, this solution is affected by the high oil content of the pomace, which reduces the quality of the pellets [13].

Biomass combustion is one of the most promising ways to reuse it, however it is necessary to consider the limits due to the low thermal efficiency of olive pomace [13]. In the literature, several studies have been reported on the combustion of olive mill wastes for energy production, alone or in combination with other fuels. Miranda et al. [14] investigated the combustion characteristics of solid mill waste (kernels, pulp and olive pomace) with different proportions of semi-solid waste (such as mill wastewater). Their results showed that the combustion of olive stones and olive-pomace gives a good efficiency and a reduced presence of non-combusted components, while lower combustion efficiencies were obtained in the case of pulp. Atimtay and Topal [15] have estimated co-combustion of various blends of olive pomace with lignite coal using various excess air ratios. Their study showed that with an air ratio of 40%–50%, considerable amounts of CO and non-fuelled hydrocarbons are formed, and the combustion efficiency drops to 84%–87%. According to the results, the combustion efficiency increased with an increasing excess air ratio. The authors suggested the addition of secondary air within the freeboard to improve the efficiency of the combustion process. This solution was also considered by Varol and Atimtay [16]. Combustion efficiencies in the range of 83.6%–90.1% were obtained from olive pomace.

The potential utilisation of solid olive mill wastes in combined heat and power (CHP) plants has been investigated by several authors [17–20] who highlighted the economic viability of such plants. The organic Rankine cycle (ORC) is one of the preferred and most studied CHP systems because of its reliability, versatility and low maintenance needs [21]. The ORC system can be considered a valuable application, also because the combustion is external with respect to the power generation system and unrelated to the features of the fuel.

The working fluid plays a key role in the ORC process. Organic fluids have higher pressures and lower boiling points compared to steam and since most of them are dry or isentropic fluids they do not require superheating before expansion [22]. Many studies have been carried out to select the best working fluids. Liu et al. [23] discovered that some working fluids at specific evaporations and condensation temperatures showed a similar thermal efficiency and the thermal efficiencies were found to increase with the critical temperature of the working fluid. Wang et al. [24] analysed the performance of the ORC system with different working fluids and showed that R245fa and R245ca are the most environment-friendly working fluids. Cataldo et al. [25] proposed to choose the fluid which has a low value of critical temperature and a high value of the latent heat of vaporisation. However, no single pure fluid has been found as optimal for the ORC due to the strong interdependence between the optimal working fluid, the working conditions and the cycle architecture [21].

Another important aspect to take into account is the heat transfer efficiency of the evaporator [26]. The pinch point temperature difference is often used to analyse the coupling heat transfer in the evaporator. Chen et al. [27] suggested a method to optimise the operating parameters of an ORC with a constrained inlet temperature of the heat source and the pinch point temperature difference in the evaporator.

The growing energy consumption of the agricultural field requests a fast evolution of the technologies aimed to biomass waste energy conversion because the application of more sophisticated processes and technologies within the chain of the agricultural and food industries requests more and more resources [28,29].

The aim of the present work is to evaluate the potential of olive pomace produced in a mill located in Central Italy as an energy resource, which represents the considered case study, whose combustion feeds an ORC unit combined with an absorption chiller. After the preliminary analysis of the biomass waste, the energy performance achieved by a small size CCHP ORC system powered by olive pomace combustion was investigated, considering various working fluids and different operating conditions.

2. Materials and Methods

2.1. Raw Materials

The olive pomace considered for the study derives from an Italian mill, located in Province of Rieti northeast of Rome. Composition and lower calorific value of the olive pomace has been evaluated, to determine a possible use as biofuel. Olive pomace (OP) samples at different stages (S1, S2, ... , S5) were taken, and for each stage, the analysis was carried out both on the surface of the olive pomace and at 25 cm depth inside the olive pomace. The first stage was the period between the olive pomace production and two weeks afterwards. The second stage began at the end of the first stage and lasted for two weeks. The third stage began at the end of the second stage and lasted two weeks as well. In Table 1, the results of the samples analysis during the first and the third stage are reported because only the conditions at the beginning and at the end of the ripening process were of interest for our study.

Table 1. Biomass proximate and ultimate analysis.

Biomass		Proximate Analysis (%wt, Dry Basis)			Ultimate Analysis (%wt, Dry Basis)				LHV (MJ/kg)	HHV (MJ/kg)	Moisture (%wet)
		Ash	Volatile Matter	Fixed Carbon	C	H	N	O			
S1	OP surface 1st stage	2.9	78.1	19	50.86	8.22	1.20	36.82	21.86	23.55	18.00
S2	25 cm OP depth 1st stage	3.3	77.9	18.8	51.96	8.49	1.46	34.79	22.71	24.45	19.00
S3	OP surface, 3rd stage	3.8	77.6	18.6	52.90	8.94	2.54	31.82	23.75	25.59	14.6
S4	OP surface, 3rd stage	3.6	77.7	18.7	58.28	8.92	2.35	26.85	26.13	27.97	17.4
S5	25 cm OP depth, 3rd stage	3.5	77.6	18.9	56.73	8.13	2.92	28.72	24.62	26.30	16.8

Biomass samples were analysed according to the principles of the standard ISO 18134-1:2015 [30], and moisture content values of the samples were determined. Ash content values were determined according to the standard of ISO 18122:2016 [31].

Hydrogen, nitrogen and carbon concentrations on a dry basis were determined according to ISO 16948:2015 which describes the method for the determination of total carbon, hydrogen and nitrogen contents in solid biofuels [32]. The Lower Heating Value (LHV) was calculated for each sample of olive residue analysed, starting from the Higher Heating Value (HHV) and hydrogen content, and, in particular, the HHV was determined using the following formula [33]:

$$HHV = 0.3941\,C + 1.1783\,H + 0.1005\,S - 0.1034\,O - 0.0151\,N - 0.0211\,Ash \quad (1)$$

The abovementioned analysis results include both higher and lower heating values of the samples of biomass and results of the proximate and ultimate analyses of the olive pomace under investigation.

After the days elapsed between the first stage and the third stage, the contact with air (sample 3 and sample 4) eased the fermentation process which degraded the chemical and physical properties of biomasses. This is the reason why the ash content in sample 2 increased compared to sample 1 and then decreased in the following ones. Sample 4 had the highest C% concentration because of the presence of fungus and microorganisms from fermentation despite the increment of the ash content.

2.2. Combustion Process Model

For the model developed in the present study, the Aspen Plus process model simulator was used. The following assumptions were considered for the simulation:

- the process is steady-state and isothermal [34];
- drying and pyrolysis take place instantaneously and volatile products mainly consist of H_2, CO, CO_2, CH_4 and H_2O [35,36];
- char is 100% carbon [37];
- all gases behave ideally.

The flowsheet of the simulation, developed in Aspen Plus, is shown in Figure 1.

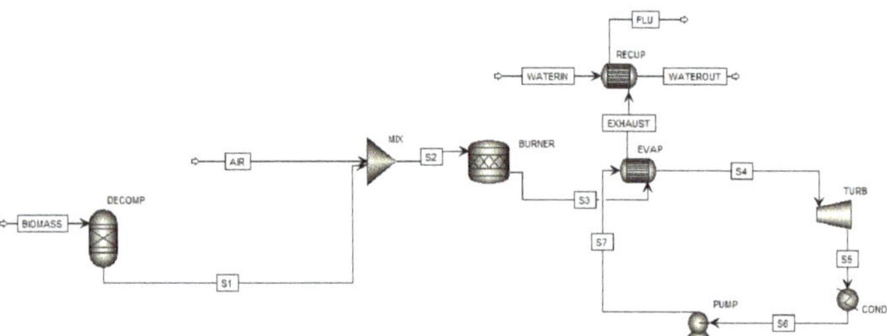

Figure 1. Flowchart of the plant evaluated in this study.

The stream BIOMASS was considered a non-conventional stream, as defined by its proximate and ultimate analysis. The BIOMASS stream goes to the RYIELD reactor used to simulate the decomposition of the unconventional feed into its conventional components (carbon, hydrogen, oxygen, sulphur, nitrogen, and ash by specifying the yield distribution according to the biomass ultimate analysis in Table 1 [1]. Off-products from DECOMP move into a mixer MIX in order to add the oxidising fluid, composed of pure air, to the combustible, and then the resulting mixed stream S2 goes into the BURNER, which simulates the combustion process. The fumes out of the combustor move to the ORC, composed of an evaporator called EVAP, a turbine called TURB, a condenser called COND and a pump called PUMP. The exhausted fumes out of the EVAP move into an exchanger called RECUP that generates the cogeneration effect: the heat from the fumes heats up a mass of water until a fixed value is reached.

Equations (2)–(7) are the chemical reactions considered in this work for the combustion process [38].

$$C + 2H \rightarrow CH \quad (2)$$

$$C + O \rightarrow CO \quad (3)$$

$$C + 0.5O \rightarrow CO \quad (4)$$

$$H + 0.5O \rightarrow H\ O \quad (5)$$

$$N + 3H \rightarrow 2NH \quad (6)$$

$$CH + 2O \rightarrow CO + 2H\ O \quad (7)$$

The following Table 2 showed the Aspen units of the flowchart presented in Figure 1.

Table 2. Description of Aspen Plus flowsheet unit operation presented in Figure 1.

Aspen Plus Name	Block ID	Description
RYIELD	DECOMP	Yield reactor—converts the non-conventional stream "BIOMASS" into its conventional components
MIXER	MIX	Mixer—mixes oxidising fluid with S1 stream, which represents combustible fluid
RSTOIC	BURNER	Rstoic reactor—simulates the combustion process
HEATX	EVAP	HeatX—represents the evaporator of the ORC
-	RECUP	HeatX—heats up the temperature of the water until the utilisation level is reached
TURBINE	TURB	Turbine—represents the turbine of the ORC
HEATER	COND	Heater—represents the condenser of the ORC
PUMP	PUMP	Pump—represents the pump of the ORC

2.3. ORC-Based CCHP System Model

The CCHP is composed of an ORC system coupled with an absorption chiller aimed at cooling. In particular, the ORC is composed of the evaporator (EVAP), condenser (COND), turbine (TURB) and pump (PUMP). Low-boiling working fluid is pressurised in the pump and then flows into the evaporator. Within the evaporator, the working fluid receives heat from the combustion process and evaporates. After the vaporisation of working fluid, it moves into the turbine and a mechanical power is then produced by means of expansion. Finally, the working fluid is cooled within the condenser at a low pressure of the cycle, and the pump restarts the cycle again. The assumptions of the present evaluation are reported in Table 3 [24], and the pressure drops within the thermodynamic cycle are shown in Table 4 and vary depending on the working fluid.

Table 3. Organic Rankine cycle (ORC) system conditions.

Heat source temperature (°C)	500
Heat source mass flow rate (kg/h)	200
Isentropic efficiency of turbine	0.85
Mechanical efficiency of turbine	0.75
Isentropic efficiency of pump	0.80

Table 4. Selected working fluids for organic Rankine cycle.

Working Fluid	Fluid Type	Condensing Pressure at 30 °C (bar)	Critical Temperature (°C)	Critical Pressure (bar)	ODP
R245fa	Isentropic	1.77	154.29	36.50	0
R245ca	Isentropic/dry	1.21	174.58	39.30	0
Cyclobutane	Isentropic	1.83	187.05	49.88	0
Cyclopentene	Isentropic/dry	0.61	234.11	48.05	0

The Aspen Plus model simulated the operations of the CHP system and the available and residual downstream exhaust thermal power was converted into cooling power by means of a Yazaki absorption chiller and determined by means of its COP (coefficient of performance) declared by the manufacturer.

The performance of the system is strictly dependent on the properties of the working fluid which can be typically categorised as dry, isentropic and wet according to the slope of the saturation curve in an S-T diagram. It can be respectively positive, infinite or negative. The dry or isentropic working fluids allow to avoid the superheating, which is usually needed to prevent the impingement of liquid droplets on the turbine blades, increasing the economic efficiency of the ORC systems [24]. For this simulation, dry and isentropic organic fluids are chosen. The adopted criteria for the fluid selection were: trifling ozone depletion potential (ODP), higher critical temperature and higher critical pressure. In this way, fluorinated alkanes such as R245fa and R245ca were selected for their application in many operating plants nowadays. Then, cyclobutene and cyclopentene were selected for the critical parameters fit to the range of the heat source temperatures [38]. The characteristics of the working fluids used in the simulation are reported in Table 4 [39].

Finally, the considered absorption chiller is a 35 kWc and 0.7 COP Yazaki WFC SC 10 which receives part of the exhaust thermal power. The available exhaust thermal power is overabundant with respect to the 50 kWth thermal power required by the Yazaki unit. The functional scheme of the overall system has been shown in Figure 2.

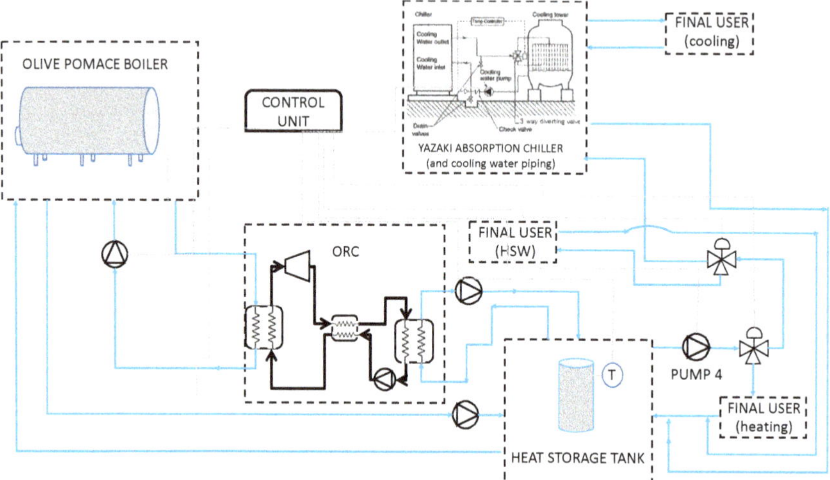

Figure 2. Functional scheme of the ORC-based trigeneration system.

3. Results and Discussion

3.1. Combustion Products

The examined mill produces 200,000 kg of olive pomace per year. As a consequence, 5000 h of operation per year have been considered with a biomass flow rate consumption of 40 kg/h. The process parameters of the burner are 500 °C and 1 bar.

3.2. Mechanical, Thermal and Electrical Efficiencies

The first results of the simulation are reported in Table 5, showing the comparison of the working fluids analysed in terms of 5 system performance. The thermal duty Q of the combustion product (stream S3) is equal to 75 kW, and for each working fluid, the thermal duty Q of the exhausted fumes out of evaporator (stream EXHAUST), the power P produced from the expansion in turbine and the mechanical and thermal efficiency of the cycle η and η, respectively, were investigated, varying the pressure of the pump. Then η is calculated according to Equation (8), where p_{pump}(bar) is the pressure of the pump, ΔP_{pump} (kW) is the power required by the pump, P (kWe) is the electrical power produced and Q_{evap} (kW) is the thermal power required by the evaporator. Then η is calculated according to Equation (9). η is calculated considering an alternator efficiency equal to 0.97 [40].

$$\eta_{mecc} = \frac{P_{expansion} - W_{pump}}{Q_{evap}} \tag{8}$$

$$\eta_{therm} = \frac{Q}{Q_{source}} \tag{9}$$

Table 5. Comparison of the working conditions of the selected working fluids.

p_{pump} (bar)	R245fa		R245ca		Cyclobutene			Cyclopentene		
	20	30	20	30	20	30	40	20	30	40
$Q_{exhaust}$ (kW_{th})	65.41	65.22	61.91	61.20	64.00	63.20	63.00	61.00	59.30	59.00
ΔP_{pump} (kW)	0.21	0.31	0.19	0.28	0.37	0.57	0.78	0.36	0.56	0.75
$P_{expansion}$ (kW)	3.62	4.01	4.52	5.01	4.33	5.00	5.30	5.92	6.51	6.93
P_{el} (kW_e)	3.41	3.80	4.30	4.85	4.17	4.85	5.12	5.72	6.30	6.68
η_{mecc}	13.10%	14.51%	14.34%	15.60%	14.91%	16.72%	17.80%	17.80%	19.00%	20.00%
η_{therm}	70.30%	70.00%	66.60%	65.81%	68.81%	68.00%	68.00%	65.00%	63.70%	63.52%
η_{el}	12.77%	14.10%	13.93%	15.12%	14.41%	16.20%	17.20%	17.20%	18.40%	19.40%

From the comparison of the working fluid conditions, shown in Table 5, it is possible to ascertain that the mechanical power release by the turbine is comprehended between 3.62 kW and 6.9 kW, depending on the organic fluid used and the pressure of the pump. The lower value is 3.62 kW for the R245fa at 20 bar of pump pressure, and the highest value is 6.9 kW for the cyclopentene at a pump pressure of 40 bar. The mechanical efficiency varies from a minimum of 13.1% for the R245fa at a pump pressure of 20 bar to a maximum of 20% for the cyclopentene at a pump pressure of 40 bar. The thermal efficiency ranges between 63.5% to 70.3%, where the minimum value is for the R245fa at 20 bar of pump pressure, and the maximum one is for the cyclopentene at 40 bar pump pressure. Furthermore, Q is enough to allow a small size cogeneration.

3.3. Sensitivity Analysis

A sensitivity analysis is carried out in order to evaluate the impact of the pressure pump on the trends of the mechanical and thermal efficiency for each working fluid considered. The trends resulted from the simulation are showed in Figures 3–6.

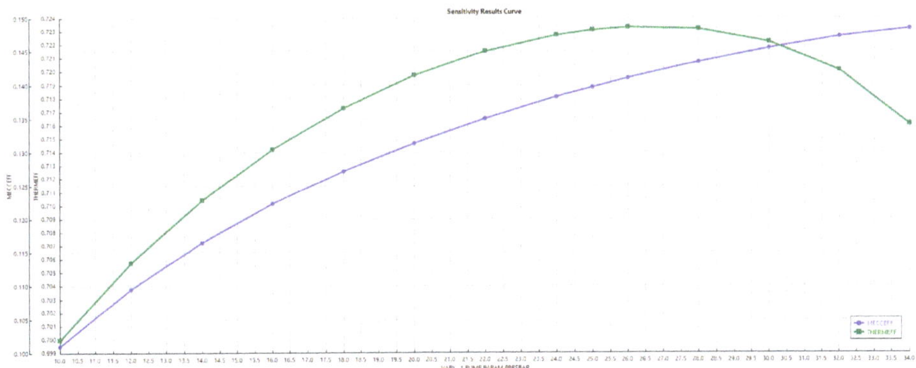

Figure 3. Mechanical and thermal efficiency vs. operating pressure. R245FA.

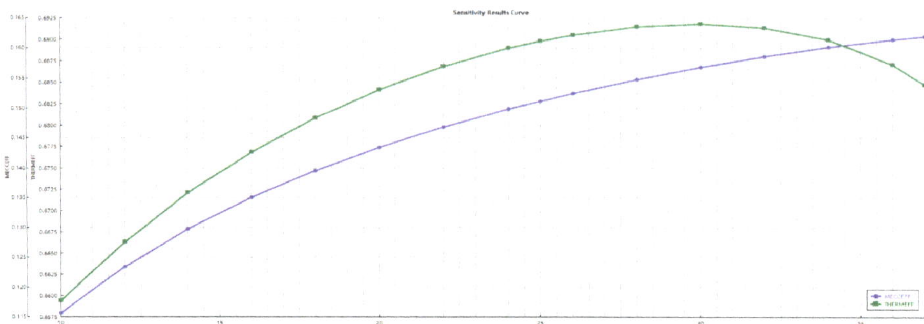

Figure 4. Mechanical and thermal efficiency vs. operating pressure. R245CA.

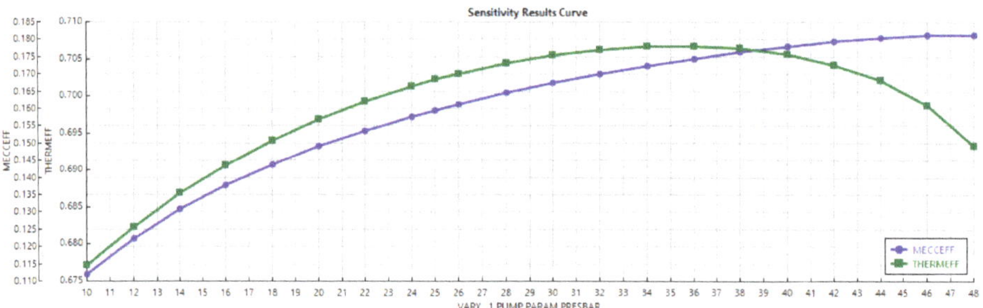

Figure 5. Mechanical and thermal efficiency vs. operating pressure. Cyclobutene.

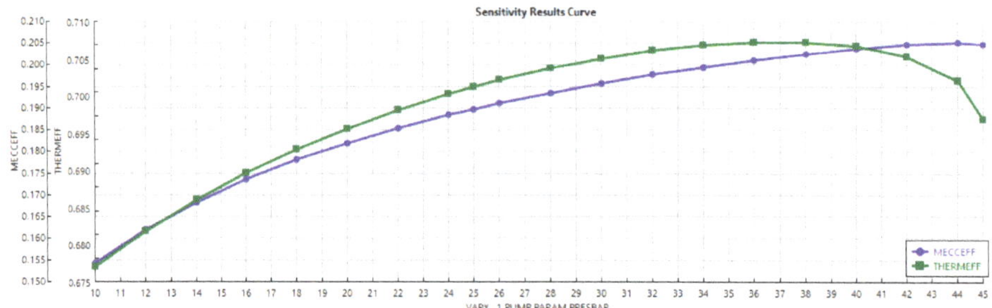

Figure 6. Mechanical and thermal efficiency vs. operating pressure. Cyclopentene.

Figures 3–6 shows a crescent trend for the mechanical efficiency with the rise of the cycle operating pressure included between 10 bar and the critical pressure for each considered working fluid. Instead, the thermal efficiency shows a crescent trend until a maximum is reached and after that a decreasing trend. In the cases of R245fa and R245ca, the mechanical efficiency increases from 10 bar to the critical pressure. In case of Cyclobutene, the efficiency is stable for a pressure higher than 46 bar and, in the case of Cyclopentene, the maximum of mechanical efficiency is reached at 44 bar and decreases between 44 and 45 bar.

3.4. Proposed Scenario

Considering operations of 5000 h/year, the working period could be split between hot season (from June to September), cold season (from 15th of November to 15th of April according to the relevant Italian decree [41]) and middle season (all other months). During the cold season, the thermal power Q from the exhaust stream could be used for heat uses as industrial processes, hot sanitary water, and space heating. During the hot season, it has been mostly considered as the thermal input to a Yazaki absorption chiller unit aimed to address the cooling needs. The remaining thermal power can be used for residual heat uses. The two scenarios proposed are reported in Tables 6 and 7, where Q (kWth) is the thermal power required for the cooler, P (kW) is the electrical power produced from the cooler, Q (kW) is the thermal power available for heating water and W (kg/h) is the flow rate of heated water.

Table 6. Hot season proposed scenario.

Working Fluid	R245fa		R245ca		Cyclobutene			Cyclopentene		
P_{pump} (bar)	20	30	20	30	20	30	40	20	30	40
$Q_{exhaust}$ (kW_{th})	65.41	65.22	61.91	61.2	64	63.2	63	61	59.3	59
$Q_{chiller\_feeding}$ (kW_{th})	50	50	50	50	50	50	50	50	50	50
$P_{cooling}$ (kW_e)	35	35	35	35	35	35	35	35	35	35
Q_{heat_use} (kW_{th})	15.41	15.22	11.91	11.2	14	13.20	13	11	9.30	9
W_{water} (kg/h)	205	202	159	149	186	176	173	146	124	120

Table 7. Cold season proposed scenario.

Working Fluid	R245fa		R245ca		Cyclobutene			Cyclopentene		
P_{pump} (bar)	20	30	20	30	20	30	40	20	30	40
$Q_{exhaust}$ (kW_{th})	65.41	65.22	61.91	61.20	64	63.20	63	61	59.30	59
W_{water} (kg/h)	870	868	824	815	852	841	838	812	789	785

The results of the present case study indicated that the proposed approach is suitable for the energy supply of a farm, considering the yearly production shown in Table 8. As discussed, the use of different working fluids changes the distribution between thermal and electric energy. The cooling energy produced is the same because it was obtained by the combination of operating hours and the cooling power supplied by the absorption chiller. The latter is equal to the nominal cooling power of the absorption chiller as the thermal power feeding the absorber is always major than the required thermal power. The COP considered is also the nominal one. This representation of energy use can be different according to the energy consumption profiles depending on the kind of activity. The cooling energy production has been determined only for the hot season as for a residential user even if it could be differently distributed along the year according to possible industrial processes requiring cooling energy during the entire year. Table 8, regarding the yearly energy productivity of the CCHP system, shows that the Cyclopentene differs from the others in terms of the electric energy production. As a consequence, the thermal energy available for heating uses is smaller.

Table 8. Yearly energy production.

P_{pump} (bar)	R245fa		R245ca		Cyclobutene			Cyclopentene		
	20	30	20	30	20	30	40	20	30	40
Electric energy (MWh_e)	17	19	22	24	21	24	26	29	32	33
Thermal Energy (MWh_{th})	195	194	180	177	189	185	184	176	169	167
Cooling Energy (MWh_e)	58	58	58	58	58	58	58	58	58	58

4. Conclusions

In this work, the chemical composition and energy performance of the olive pomace has been analysed, showing a good value of LHV (around 22 MJ/Kg) and good values of the proximate and ultimate analysis. A trigeneration system composed of an ORC unit, powered by a biomass boiler, developed in Aspen Plus and coupled with an absorption chiller was investigated with different working conditions. Four dry/isentropic organic fluids were considered for their negligible ozone depletion potential, higher critical temperature and critical pressure, and a sensitivity study was carried out in order to determine the mechanical, thermal and electrical efficiency of the plant at varying operative conditions for each working fluid. The minimum and maximum level obtained whilst varying the organic fluid and the pressure of the pump were 13.1%–20.0%, 63.5%–70.3% and 12.7%–19.4% for mechanical, thermal and electrical efficiency, respectively. Furthermore, a heat recovery of the exhaust gas out of the evaporator was managed in order to heat up an amount of water (120 kg/h) to 85 °C temperature and feeding a 35 kWc absorption chiller. The present analysis deepened the coupling between an ORC CCHP system and olive pomace for the first time in the literature and lead to the conclusion that this biomass waste can be an effective and available by-product of the olive oil production process suitable for an energy-from-biomass-waste trigeneration system. In particular, the most important findings of the present work were:

- At a national level, considering an olive pomace production ratio of 2250 kg/ha and about 1052000 ha of Italy being occupied with olive trees aimed at oil production in 2017 [42], the energy potential of this energy system would be of 5.9 GWh, 31.1 GWh and 10.7 GWh, corresponding to a 0.286 Mtoe (Million Tonnes of Oil Equivalent) in terms of primary energy, which would give an important contribution to the overall primary energy national production.
- The triple energy could be used within the agricultural chain, representing a virtuous case of distributed generation.
- At a local level, the developed model shows that Cyclopentene is the most highly performing fluid in terms of electricity production, while R245fa is the least.
- The increasing pressure entails an energy benefit in terms of power production and electric efficiency, but the increasing trend is slower at the highest pressures of the considered range.

Author Contributions: Conceptualization, M.V. and A.C.; Methodology, M.V.; Software, V.M.; Validation, M.V., M.C. and A.C.; Formal Analysis, F.G.; Investigation, V.M. and M.V.; Resources, M.C. and F.G.; Data Curation, V.M. and M.V.; Writing-Original Draft Preparation, V.M.; Writing-Review & Editing, M.V. and A.C.; Visualization, V.M. and M.V.; Supervision, A.C., M.C. and F.G.; Project Administration, M.C.; Funding Acquisition, A.C. and M.C.

Funding: The research was partially supported by the Italian Ministry for Education, University and Research (MIUR) according to the Italian Law 232/2016 within the fund for the financing of universities "Departments of Excellence". This study was also partially supported by the HBF 2.0 Project, funded in the framework of the RDS Ricerca di Sistema Programme of the Italian Ministry of Economic Development and by the Ekogroup srl company within the SANSA 4.0 project.

Conflicts of Interest: The authors declare no conflict of interest.

References

1. Marcantonio, V.; De Falco, M.; Capocelli, M.; Bocci, E.; Colantoni, A.; Villarini, M. Process analysis of hydrogen production from biomass gasification in fluidized bed reactor with different separation systems. *Int. J. Hydrogen Energy* **2019**, *44*, 10350–10360. [CrossRef]
2. Villarini, M.; Marcantonio, V.; Colantoni, A.; Bocci, E. Sensitivity Analysis of Different Parameters on the Performance of a CHP Internal Combustion Engine System Fed by a Biomass Waste Gasifier. *Energies* **2019**, *12*, 688. [CrossRef]
3. Rajabi Hamedani, S.; Villarini, M.; Colantoni, A.; Moretti, M.; Bocci, E. Life Cycle Performance of Hydrogen Production via Agro-Industrial Residue Gasification—A Small Scale Power Plant Study. *Energies* **2018**, *11*, 675. [CrossRef]
4. Hamedani Rajabi, S.; Bocci, E.; Villarini, M.; Di Carlo, A.; Naso, V. Techno-economic Analysis of Hydrogen Production Using Biomass Gasification -A Small Scale Power Plant Study. *Energy Procedia* **2016**, *101*, 806–813. [CrossRef]
5. Begum, S.; Rasul, M.G.; Akbar, D.; Ramzan, N. Performance analysis of an integrated fixed bed gasifier model for different biomass feedstocks. *Energies* **2013**, *6*, 6508–6524. [CrossRef]
6. Anifantis, A.S.; Colantoni, A.; Pascuzzi, S. Thermal energy assessment of a small scale photovoltaic, hydrogen and geothermal stand-alone system for greenhouse heating. *Renew. Energy* **2017**, *103*, 115–127. [CrossRef]
7. Büyüktahtakın, E.; Cobuloglu, H.I. Food vs. biofuel: An optimization approach to the spatio-temporal analysis of land-use competition and environmental impacts. *Appl. Energy* **2015**, *140*, 418–434. [CrossRef]

8. Barbanera, M.; Lascaro, E.; Stanzione, V.; Esposito, A.; Altieri, R.; Bufacchi, M. Characterization of pellets from mixing olive pomace and olive tree pruning. *Renew. Energy* **2016**, *88*, 185–191. [CrossRef]
9. Tekin, A.R.; Dalgıç, A.C. Biogas production from olive pomace. *Resour. Conserv. Recycl.* **2000**, *30*, 301–313. [CrossRef]
10. Marco IDe Riemma, S.; Iannone, R. Global Warming Potential Analysis of Olive Pomace Processing. *Chem. Eng. Trans.* **2017**, *57*, 601–606. [CrossRef]
11. Shen, L.; Zhang, D.-K. An experimental study of oil recovery from sewage sludge by low-temperature pyrolysis in a fluidised-bed. *Fuel* **2003**, *82*, 465–472. [CrossRef]
12. Borello, D.; De Caprariis, B.; De Filippis, P.; Di Carlo, A.; Marchegiani, A.; Pantaleo, A.M.; Shah, N.; Venturini, P. Thermo-Economic Assessment of a Olive Pomace Gasifier for Cogeneration Applications. *Energy Procedia* **2015**, *75*, 252–258. [CrossRef]
13. Christoforou, E.; Fokaides, P.A. A review of olive mill solid wastes to energy utilization techniques. *Waste Manag.* **2016**, *49*, 346–363. [CrossRef] [PubMed]
14. Miranda, M.T. Combined combustion of various phases of olive wastes in a conventional combustor. *Fuel* **2007**, *86*, 367–372. [CrossRef]
15. Atimtay, A.T.; Topal, H. Co-combustion of olive cake with lignite coal in a circulating fluidized bed. *Fuel* **2004**, *83*, 859–867. [CrossRef]
16. Varol, M.; Atimtay, A.T. Combustion of olive cake and coal in a bubbling fluidized bed with secondary air injection. *Fuel* **2007**, *86*, 1430–1438. [CrossRef]
17. Caputo, A.C.; Palumbo, M.; Pelagagge, P.M.; Scacchia, F. Economics of biomass energy utilization in combustion and gasification plants: Effects of logistic variables. *Biomass Bioenergy* **2005**, *28*, 35–51. [CrossRef]
18. Celma, A.R.; Rojas, S.; López-Rodríguez, F. Waste-to-energy possibilities for industrial olive and grape by-products in Extremadura. *Biomass Bioenergy* **2007**, *31*, 522–534. [CrossRef]
19. Celma, A.R.; Rojas, S.; López, F.; Montero, I.; Miranda, T. Thin-layer drying behaviour of sludge of olive oil extraction. *J. Food Eng.* **2007**, *80*, 1261–1271. [CrossRef]
20. Monarca, D.; Cecchini, M.; Colantoni, A.; Marucci, A. Feasibility of the Electric Energy Production through Gasification Processes of Biomass: Technical and Economic Aspects. In Proceedings of the International Conference on Computational Science and Its Applications, Santander, Spain, 20–23 June 2011.
21. Guo, C.; Du, X.; Yang, L.; Yang, Y. Performance analysis of organic Rankine cycle based on location of heat transfer pinch point in evaporator. *Appl. Therm. Eng.* **2014**, *62*, 176–186. [CrossRef]
22. Saadatfar, B.; Fakhrai, R.; Fransson, T. Conceptual modeling of nano fluid ORC for solar thermal polygeneration. *Energy Procedia* **2014**, *57*, 2696–2705. [CrossRef]
23. Liu, H.; Shao, Y.; Li, J. A biomass-fired micro-scale CHP system with organic Rankine cycle (ORC)-Thermodynamic modelling studies. *Biomass Bioenergy* **2011**, *35*, 3985–3994. [CrossRef]
24. Wang, J.; Diao, M.; Yue, K. Optimization on pinch point temperature difference of ORC system based on AHP-Entropy method. *Energy* **2017**, *141*, 97–107. [CrossRef]
25. Cataldo, F.; Mastrullo, R.; Mauro, A.W.; Vanoli, G.P. Fluid selection of Organic Rankine Cycle for low-temperature waste heat recovery based on thermal optimization. *Energy* **2014**, *72*, 159–167. [CrossRef]
26. Collings, P.; Yu, Z.; Wang, E. A dynamic organic Rankine cycle using a zeotropic mixture as the working fluid with composition tuning to match changing ambient conditions. *Appl. Energy* **2016**, *171*, 581–591. [CrossRef]
27. Chen, T.; Zhuge, W.; Zhang, Y.; Zhang, L. A novel cascade organic Rankine cycle (ORC) system for waste heat recovery of truck diesel engines. *Energy Convers. Manag.* **2017**, *138*, 210–223. [CrossRef]
28. TERNA. Consumi Energia Elettrica Per Settore Merceologico Italia. Available online: https://www.terna.it/it-it/sistemaelettrico/statisticheeprevisioni/consumienergiaelettricapersettoremerceologico.aspx (accessed on 22 April 2019).
29. Colonna N (ENEA). Valorizzazione energetica dei residui agroindustriali per produrre energia a servizio delle filiere agroalimentari, tra tecnologie ed efficienza energetica. *Agriregionieuropa*. 2014. Available online: https://www.eimaenergy.it/pdf/Colonna-15novembre.pdf (accessed on 22 April 2019).
30. International Organization for Standardization. ISO 18134-1:2015-Solid Biofuels-Determination of Moisture Content -Oven Dry Method-Part 1: Total Moisture–Reference Method. Available online: https://www.iso.org/standard/61538.html (accessed on 22 May 2019).
31. International Organization for Standardization. ISO 18122:2015-Solid Biofuels-Determination of Ash Content. Available online: https://www.iso.org/standard/61515.html (accessed on 22 May 2019).

32. ISO 16948:2015-Olid Biofuels-Determination of Total Content of Carbon, Hydrogen and Nitrogen. Available online: https://www.iso.org/standard/58004.html (accessed on 22 May 2019).
33. Channiwala, S.A.; Parikh, P.P. A unified correlation for estimating HHV of solid, liquid and gaseous fuels. *Fuel* **2002**, *81*, 1051–1063. [CrossRef]
34. Ye, G.; Xie, D.; Qiao, W.; Grace, J.R.; Lim, C.J. Modeling of fluidized bed membrane reactors for hydrogen production from steam methane reforming with Aspen Plus. *Int. J. Hydrogen Energy* **2009**, *34*, 4755–4762. [CrossRef]
35. Lv, P.; Xiong, Z.; Chang, J.; Wu, C.; Chen, Y.; Zhu, J. An experimental study on biomass air–steam gasification in a fluidized bed. *Bioresour. Technol.* **2004**, *95*, 95–101. [CrossRef]
36. Sadaka, S.S.; Ghaly, A.E.; Sabbah, M.A. Two phase biomass air-steam gasification model for fluidized bed reactors: Part I—Model Development. *Biomass Bioenergy* **2002**, *22*, 439–462. [CrossRef]
37. Demirbaş, A. Carbonization ranking of selected biomass for charcoal, liquid and gaseous products. *Energy Convers. Manag.* **2001**, *42*, 1229–1238. [CrossRef]
38. Yang, X.; Liu, B.; Song, W.; Lin, W. Process simulation of emission and control for NO and N2O during coal combustion in a circulating fluidized bed combustor based on Aspen Plus. *Energy Fuels* **2011**, *25*, 3718–3730. [CrossRef]
39. Rajabloo, T.; Iora, P.; Invernizzi, C. Mixture of working fluids in ORC plants with pool boiler evaporator. *Appl. Therm. Eng.* **2016**, *98*, 1–9. [CrossRef]
40. Wieland, C.; Meinel, D.; Eyerer, S.; Spliethoff, H. Innovative CHP concept for ORC and its benefit compared to conventional concepts. *Appl. Energy* **2016**, *183*, 478–490. [CrossRef]
41. Decree 412/93. 1993. Available online: http://www.gazzettaufficiale.it/eli/id/1993/10/14/093G0451/sg (accessed on 29 September 2017).
42. ISTAT. Tavola C27: Superficie (ettari) e produzione (quintali): Olivo, olive da tavola, olive da olio, olio di pressione. Dettaglio per Ripartizione Geografica –Anno. 2017. Available online: http://agri.istat.it/jsp/dawinci.jsp?q=plC270000010000011000&an=2017&ig=1&ct=311&id=15A%7C21A%7C30A%7C32A (accessed on 26 May 2019).

Article

Sensitivity Analysis of Different Parameters on the Performance of a CHP Internal Combustion Engine System Fed by a Biomass Waste Gasifier

Mauro Villarini [1,*], Vera Marcantonio [1], Andrea Colantoni [1] and Enrico Bocci [2]

[1] Department of Agricultural and Forestry Sciences (DAFNE), Tuscia University, Via San Camillo de Lellis, 01100 Viterbo, Italy; vera.marcantonio@unitus.it (V.M.); colantoni@unitus.it (A.C.)

[2] Department of Nuclear, Subnuclear and Radiation Physics, Marconi University, 00193 Rome, Italy; e.bocci@unimarconi.it

* Correspondence: mauro.villarini@unitus.it; Tel.: +39-340-2266196

Received: 10 January 2019; Accepted: 13 February 2019; Published: 20 February 2019

Abstract: The present paper presents a study of biomass waste to energy conversion using gasification and internal combustion engine for power generation. The biomass waste analyzed is the most produced on Italian soil, chosen for suitable properties in the gasification process. Good quality syngas with up to 16.1% CO–4.3% CH_4–23.1% H_2 can be produced. The syngas lower heating value may vary from 1.86 MJ/ Nm^3 to 4.5 MJ/Nm^3 in the gasification with air and from 5.2 MJ/ Nm^3 to 7.5 MJ/Nm^3 in the gasification with steam. The cold gas efficiency may vary from 16% to 41% in the gasification with air and from 37% to 60% in the gasification with steam, depending on the different biomass waste utilized in the process and the different operating conditions. Based on the sensitivity studies carried out in the paper and paying attention to the cold gas efficiency and to the LHV, we have selected the best configuration process for the best syngas composition to feed the internal combustion engine. The influence of syngas fuel properties on the engine is studied through the electrical efficiency and the cogeneration efficiency.

Keywords: biomass waste; gasification; power generation; internal combustion engine; CHP; Aspen Plus

---

1. Introduction

Sustainability and environmental issues regarding energy are becoming of more and more concern in this present age, and proper policies can determine the future low-carbon profile of the global system [1,2]. In Europe, every year, a large quantity of biomass waste is produced. This biomass is mostly vegetable waste from the agri-food chain (pruning of vines, olives, fruit trees, shells, etc.) and from wood [3,4].

The reuse of biomass waste is essential for a circular economy and sustainability [5,6]; in fact, biomass is considered one of the most important renewable energy sources as it can increase global energy sustainability and reduce greenhouse gas emissions [7,8].

There are many technologies by which to convert biomass into energy [9–11]. The two most important methods of conversion are conversion to power and to biofuel [12]. For the former, one of the most feasible and productive ways is thermochemical conversion [13]. Among the thermochemical processes, gasification is one of the most effective and studied methods to produce energy and fuels from biomass due to its capacity to handle different biomass feedstock [13–16]. Gasification, through the partial oxidation of the biomass at high temperature in the range of 800–1000 °C [17,18], with air, oxygen, and/or steam as a gasifying agent, allows for the production of a final product called syngas [19]. The application of biomass gasification to power generation has shown many important environmental benefits [20]. Syngas is mainly made up of CO, H_2, and CH_4; the remaining part consists of the non-combustible gases N_2 and CO_2 [21,22]. Obviously, depending on the quality of the biomass, particle size, gasifier type, operating conditions, and gasification agents, there are different compositions of the resulting syngas [23,24]. High-quality syngas is often characterized by low N_2 content, high H_2 content, low tar levels [25], and a high heating value [26].

Many studies have focused on the use of syngas produced from biomass gasification as an alternative fuel in engines in order to substitute fossil fuels with clean energy [27,28] and on the response of rapid compression machines to the composition of different fuels [29,30]. A gasification process combined with the cogeneration of heat and power has been considered more and more important, especially as a consequence of the growing interest toward small sizes plant [31]. In recent years, a considerable number of syngas-powered engines [32,33] have been developed, but the majority of them are based on a spark-ignition (SI) combustion system and studies have demonstrated that this engine is not suitable for this kind of fuel because the fluctuation of the syngas components makes it difficult to achieve stable combustion [21,34,35]. Therefore, the best approach seems to be the high pressure ratio [27,36].

Roy et al. [37] studied the effect of hydrogen content in syngas produced from biomass on the performance of a fuel engine and demonstrated that the engine power with a high H_2 content was greater than that obtained with low H_2. Akansu et al. [25] experimentally investigated the combustion and emissions characteristics of internal combustion engines fueled by natural gas/hydrogen blends and concluded that NO_x emissions generally increased with increased hydrogen content. Pilatau et al. [23] studied the ICE behavior with syngas from different biomass sources in a system where the exhausted ICE gases fed the gasifier and provided a method for selecting the type of main fuel used for the engine based on the chemical composition of the syngas and taking into account the engine operating parameters.

The first aim of this paper was to investigate the most available biomass waste on Italian soil in order to choose those with the best features to be gasified. Next, we proposed a simulation plant using Aspen Plus software that considered both the gasification system and the ICE. The type of gasifier analyzed was a bubbling fluidized bed gasifier, which has a lot of benefits for biomass conversion due to the good heat and mass transfer between the gas and solid phase, the high fuel flexibility and uniformity, and the easier to control temperature [38]. The gasification model was based on the restricted chemical equilibrium. The ICE was simulated with a gas turbine [39,40], fixing the pressure drop corresponding to the chosen engine. Gasification is very sensitive to some operation parameters often considered in the performance analysis such as the steam to biomass ratio, air equivalent ratio, or stoichiometric ratio [41,42]. The sensitivity analysis conducted in the simulation varied the gasification temperature, the air equivalent ratio (ER), and the steam to biomass ratio (S/B), thus allowing us to determine the best syngas composition to feed the ICE. Then, the ICE behavior was investigated through the electrical efficiency and the cogeneration efficiency.

2. Materials and Methods

2.1. Biomass Waste

The biomass waste can be classified in four categories: shells (i.e., hazelnuts, walnuts, almonds); pruning (i.e., of olives, vines, hazelnuts), straws (i.e., of wheat, corn, barley); and agro-industrial residues (i.e., exhausted olives). The chemical and physical characteristics of these biomass wastes are shown in Table 1.

Table 1. Preliminary and definitive biomass waste properties [13].

Biomass Waste	Moisture (wt %)	Bulk Density (kg/m^3)	Ash (wt %)	VM (wt %)	FC (wt %)	C (%)	H (%)	N (%)	O (%)	Cl (%)	S (%)	LHV (MJ/kg$_{dry}$)
Shells	11–14	300–500	1–2	74–78	20–25	48–51	6	0.2–05	41–44	0.02–0.03	0.01–0.03	18–20
Pruning	7–25	200–300	0.5–4	70–85	12–20	45–49	5–6	0.1–08	36–44	0.01–0.08	0.01–0.08	16–18
Straw	7–12	20–140	5–15	67–76	16–18	41–47	5–6	0.3–6	36–44	0.03–0.4	0.04–0.2	15–18
Exhausted olive	9	350	4	77	19	51	6	0.3	38	0.02	0.02	20

To select the biomass waste to be used in the gasification process, the first criterion to be considered was the feedstock availability on a significant scale (t/year). Then, the second criterion was the LHV, which has to be high, so biomass waste with a lower humidity is preferable. Another aspect to pay attention to is the density of the biomass waste as it significantly affects the storage and, in a fluidized bed gasifier, should be comparable with that of the bed. The size and shape of the biomass waste are also important, in fact, the waste must be processed to a uniform size or shape to feed into the gasifier to ensure homogeneous and efficient gasification. Chemical composition is another important characteristic that must be considered, especially the content of sulfur, chlorine, and ash [14].

To sum up, the following characteristics have to be taken into account during the choice of biomass waste for the gasification process:

1. Availability;
2. LHV;
3. Bulk density;
4. Cutting and shape;
5. Elemental composition, volatile substances, and ash.

From the research published by Enama [43], which referred to Italy, we selected the most suitable biomass with the characteristics listed above and report them in Table 2 [18,44,45].

Table 2. Biomass waste most available on Italian soil.

Biomass	Availability (t_{dry}/year)	LHV (MJ/kg$_{dry}$)	HHV (MJ/kg$_{dry}$)	Bulk Density (kg/m^3)	Moisture (wt %)
Olive pomace	3,246,000	20	23.50	350	9
Wheat	3,050,556	18.90	20.10	42	9.5
Olive pruning	1,548,711	19.90	21.20	200	20
Corn	1,269,980	17.60	18.60	58	8.50
Vine pruning	1,123,372	18.60	19.90	260	17.60
Barley	687,733	18.60	19.70	80	8
Hazelnut pruning	67,904	17.90	19	230	15
Hazelnut shells	58,000	18.85	20.20	319.14	12.45

Table 3 [44,46,47] shows the proximate and ultimate analysis of the various biomass waste types used in this study.

Table 3. Biomass proximate and ultimate analysis.

Biomass	Proximate Analysis (wt %, Dry Basis)			Ultimate Analysis (wt %, Dry Basis)					
	Ash	Volatile Matter	Fixed Carbon	C	H	N	O	Cl	S
Hazelnut shells	0.77	62.70	24.08	46.76	5.76	0.22	45.83	0.76	0.67
Olive pruning	3.67	82.35	13.98	47.50	6.00	1.06	43.66	1.74	0.04
Vine pruning	2.62	80.84	16.54	50.84	5.82	0.88	40.08	1.87	0.05
Hazelnut pruning	3.20	79.60	17.20	47.40	5.23	0.70	43.50	3.14	0.03
Olive pomace	4	77	19	51	6	0.30	38	0.02	0.02
Corn	7	69.5	15	47.30	5.50	0.90	45.3	0.5	0.5
Wheat	11	66.3	21.4	48.86	6.80	0.59	43.4	0.15	0.2
Barley	7	65	19	46.88	7	0.60	44.70	0.70	0.12

For the analysis presented in this paper, we selected four biomass waste sources from Table 2, each of them belonging to one of the four categories of biomass waste reported in Table 1. The biomass sources chosen, with consideration of the greatest availability and LHV within the categories, were:

1. Hazelnut shells, belonging to the shell category;
2. Olive pruning, belonging to the pruning category;
3. Olive pomace, belonging to the exhausted oil; and
4. Wheat, belonging to the straw category.

The diffusion of the biomass waste under examination on Italian soil is reported in Figure 1.

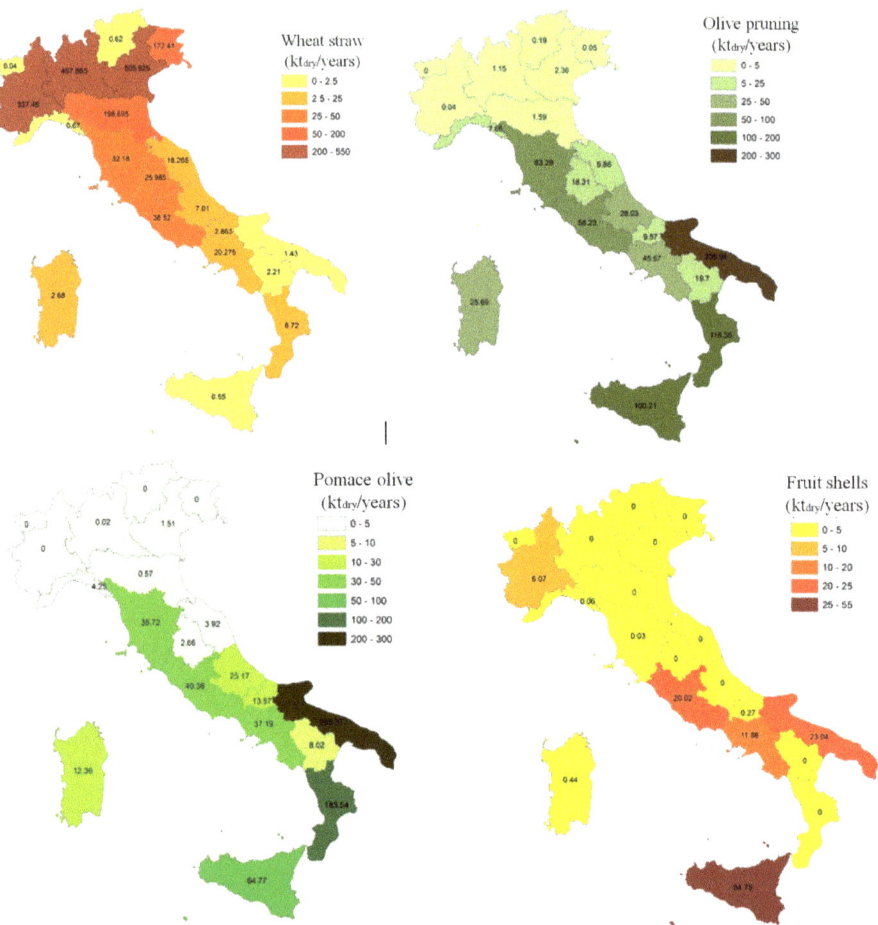

Figure 1. Diffusion of the biomass waste on Italian soil analyzed in this study [43].

2.2. Process Modeling

For the model developed in this study, the Aspen Plus process model simulator was used. The following assumptions were considered for the simulation:

- The process is steady state and isothermal [48–50];
- Drying and pyrolysis take place instantaneously, and the volatile products mainly consist of H_2, CO, CO_2, CH_4, and H_2O [25,46];

- Char is 100% carbon [51]; and
- All gases behave ideally.

Equations (1)–(8) are the chemical reactions considered in this work for the gasification process [17,26,52] and are listed in Table 4.

Table 4. Chemical reactions involved in the process.

Reaction Number	Reaction Equation	Reaction Name	Heat of Reaction ΔH (kJ/mol)
1	$C + O_2 \rightarrow CO_2$	Carbon combustion	−393.0
2	$C + 0.5O_2 \rightarrow CO$	Carbon partial oxidation	−112.0
3	$C + H_2O \leftrightarrow CO + H_2$	Water gas reaction	+131.0
4	$CO + H_2O \leftrightarrow CO_2 + H_2$	Water gas-shift reaction	−41.0
5	$H_2 + 0.5O_2 \rightarrow H_2O$	Hydrogen partial combustion	−242.0
6	$CH_4 + H_2O \rightarrow CO + 3H_2$	Steam reforming of methane	+206.0
7	$H_2 + S \rightarrow H_2S$	H_2S formation	−20.2
8	$0.5N_2 + 1.5H_2 \rightarrow NH_3$	NH_3 formation	−46.0
9	$H_2 + 2Cl \rightarrow 2\ HCl$	HCl formation	−92.31

The Boudouard reaction was not considered in this simulation as it does not achieve kinetic equilibrium and causes destabilization in reactor behavior [53].

For the simulation, we analyzed two different configurations:

(1) Air was used as the only gasifying agent, and the relative Aspen Plus flowsheet is shown in Figure 2;

(2) Steam was used as the gasifying agent, and the associate Aspen Plus flowsheet is shown in Figure 3.

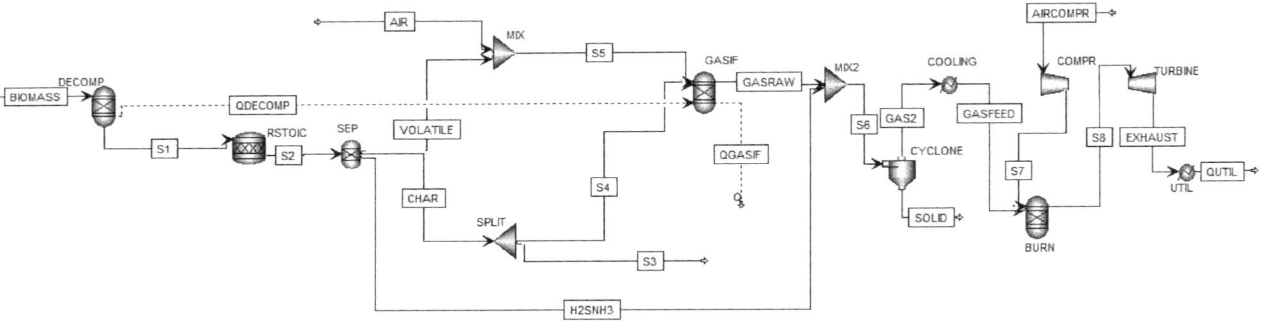

Figure 2. Flowchart of the plant evaluated in this study (the hatched streams are the heat streams, representing thermal recoveries; the continuous streams are the material streams), considering air as the gasifying agent.

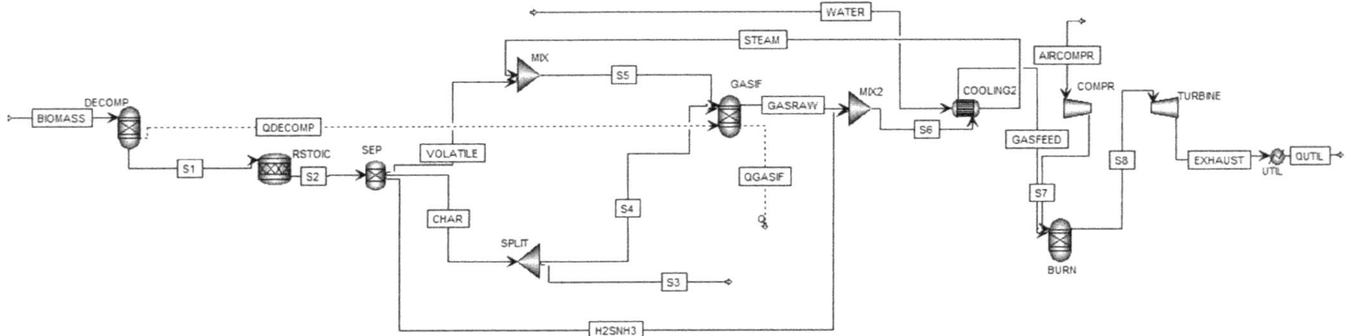

Figure 3. Flowchart of the plant evaluated in this study (the hatched streams are the heat streams, representing thermal recoveries; the continuous streams are the material streams), considering steam as the gasifying agent.

The stream BIOMASS was specified as a non-conventional stream, as defined by its proximate and ultimate analysis. The BIOMASS stream goes to the RYELD reactor used to simulate the decomposition of the unconventional feed into its conventional components (carbon, hydrogen, oxygen, sulfur, nitrogen, and ash by specifying the yield distribution according to the biomass ultimate analysis in Table 3).

Off products from DECOMP moves in the RSTOIC block to simulate the production of HCl, NH_3, and H_2S , through the reactions (7) and (8), by considering the fractional conversion for H_2S and HCl as equal to 1 and NH_3 as equal to 0.5 [54], because, in order to have more realistic results, these components cannot be modeled with a chemical equilibrium. The resulting stream S2 moves into a separator SEP, which divides the stream into three sub-streams: volatile part VOLATILE, char part CHAR, and a stream composed of NH_3, HCl, and H_2S , called H_2SNH_3. The VOLATILE stream, after mixing with oxidizing fluid, goes in the gasifier GASIF. The CHAR stream is split into two sub-streams: S3, which represents the un-reacted char, and S4, which represents the char that reacted in the gasifier.

To simulate the gasification process in Aspen Plus, we used a RGibbs reactor, called GASIF in the flowsheets of Figures 2 and 3, modeled with the restricted chemical equilibrium, which allowed us to describe the syngas composition more accurately than the equilibrium models. Equations (1)–(6) of Table 4 are the chemical reactions considered in this work for the gasification process. The restricted chemical equilibrium can be obtained by choosing the calculation option "Restrict chemical equilibrium-specify temperature approach or reactions" in Aspen Plus and specifying the zero temperature approach for each reaction in the gasifier model. In this way, the RGibbs evaluates the chemical equilibrium constant for each reaction at the reactor temperature, thereby giving the equilibrium gas composition [55,56]. In the present model, tar formation and catalyst deactivation were not taken into account. The block CYCLONE represents the simulation of gas cleaning, where solid parts are separated from gas. We only considered this step of cleaning because an engine needs gas at low temperature to have low density, so this type of cleaning is sufficient for our purposes. Therefore, the stream GAS2 goes into a cooler called COOLING in Figures 2 and 3 that represents the cooling of syngas before its entrance into the internal combustion engine, which was simulated with a gas turbine and composed of the block COMPR and the block TURBINE. At the end, the EXHAUST stream goes into a cooler UTIL to achieve the chosen utilization temperature.

The configuration presented in Figure 3 is the same as that of Figure 2 except for the utilization of steam instead of air: the stream WATER goes into the cooler, called COOLING2, which is a counter-current heat exchanger where syngas loses temperature and water is lifted up to the saturation temperature that becomes the STEAM stream. Therefore, the following Tables 5 and 6 showing the ASPEN units and the system operating conditions, are valid for the flowchart presented in Figure 2 and in Figure 3.

Table 5. Description of ASPEN Plus flowsheet unit operation presented in Figure 2.

ASPEN Plus Name	Block ID	Description
RYIELD	DECOMP	Yield reactor—converts the non-conventional stream "BIOMASS" into its conventional components
RSTOIC	RSTOIC	Rstoic reactor—simulates the production of HCl, NH_3 and H_2S
SEP	SEP	Separator—separates the biomass in three streams: volatile, char and a stream of NH_3 and H_2S
MIXER	MIX	Mixer—mixes oxidizing fluid with VOLATILE stream, that represents combustible fluid
MIXER	MIX2	Mixer—mixes the gas from gasifier with NH_3, HCl and H_2S
FSPLIT	SPLIT	Splitter—splits char unreacted (S3) from char to burn (S4)

Table 5. Cont.

ASPEN Plus Name	Block ID	Description
RGIBBS	GASIF	Gibbs free energy reactor—simulates drying, pyrolysis, partial oxidation and gasification and restricts chemical equilibrium of the specified reactions to set the syngas composition by specifying a temperature approach for individual reactions
CYCLONE	CYCLONE	Cyclone—simulates gas-solid separation
HEATER	COOLING	Heater—lowers the temperature between GASIF and ICE
COMPR	COMPR	Compressor—used to simulate internal combustion engine
COMPR	TURBINE	Turbine—used to simulate internal combustion engine
RGIBBS	BURN	Combustion chamber—used to simulate gas turbine combustion
HEATER	UTIL	Heater—lowers the temperature of exhausted fumes to utilization temperature

Table 6. System operating conditions.

Plant Unit	Process Parameters	Value
GASIF	Temperature	800 °C
	Pressure	1 bar
COOLING	Temperature syngas (out)	30 °C
	Pressure syngas (out)	1 bar
UTIL	Temperature (out)	80 °C

3. Internal Combustion Engine Simulation

The ICE was simulated as a gas turbine in this paper. The process parameters are shown in Table 7 and discussed in Section 4.7 for the reference case of hazelnut shells and olive pruning feeding. The gas turbine engine was fed with syngas at an ambient temperature (30 °C) that was compressed by up to 20 bar pressure before entering the turbine [39,40].

Table 7. Gas turbine's cycle operating conditions.

Process Parameters	Value
Temperature, syngas (in)	30 °C
Temperature, air (in, compressor)	20 °C
Equivalence ratio [35]	3
Isentropic expansion coefficient	90%
Isentropic compression coefficient	90%
Pressure, fumes (out, turbine)	1 bar

4. Results and Discussion

In this simulation, we considered 1 MWth as the input size and the HHV of each of the four biomass wastes analyzed and the feed was fixed in this way: for the hazelnut shells, the input flow settled at the constant flow rate of 180 kg/h; for the olive pruning, the input flow settled at the constant flow rate of 170 kg/h; for the olive pomace, the input flow settled at the constant flow rate of 153 kg/h; and for wheat straw, it was settled at the constant flow rate of 179 kg/h.

In the first configuration with air, the gasification agent considered was at the constant flow rate of 159 kg/h at 25 °C and 1 bar.

Focusing on the syngas composition out of the gasifier, a sensitivity study was carried out by varying:

- The gasifier operating temperature to verify the influence of gasification temperature on the syngas composition, from 785 to 870 °C, in case of air as oxidant agent;

- The ER, to analyze the system reaction changing the input flow of air by varying the equivalent ratio from 0.2 to 0.6, but keeping the gasification temperature constant at 800 °C in order to evaluate the decrease in the energy needed for the gasification reaction (thermal energy that has to be added);
- The gasifier operating temperature and ER simultaneously to evaluate the LHV of the syngas and the cold gas efficiency η_{CG}, which represents the fraction of energy in the biomass feed that can be acquired as energy from the use of the produced syngas. The cold gas efficiency was calculating using the following equation:

$$\eta_{CG} = \frac{M_{syn} \cdot LHV_{syn}}{M_{biomass} \cdot LHV_{biomass}}, \quad (1)$$

  where M_{syn} and $M_{biomass}$ are the mass of the produced syngas and the original biomass, respectively; LHV_{syn} and $LHV_{biomass}$ are the LHV of the produced syngas and the original biomass, respectively.
- The steam to biomass (S/B) ratio, in the configuration of Figure 3, to study the possible improvements of the plant efficiency when more steam was delivered to the gasifier.

4.1. Syngas Composition

At the gasification temperature of 800 °C and with the input flow rate declared above, with air as the gasifying agent, the simulation was conducted in Aspen Plus, as shown in Figure 2. The compositions of the product syngas for each biomass waste analyzed are shown in Table 8.

Table 8. Composition of the syngas in %dry mole fraction.

Component ($\%_{dry\ mole\ fraction}$)	Hazelnut Shells	Olive Pruning	Olive Pomace	Wheat Straw
H_2	20.7	20.4	23.1	21.9
CO	14.8	14.6	16.1	15.1
CO_2	13	12.7	10.1	11.7
H_2O	19.6	18.6	14.3	18.5
CH_4	2	2	4.3	2.7
HCl	0.17	0.38	0.05	0.03

4.2. Effect of Gasification Temperature

The syngas composition, in the stream GASRAW as defined in Figure 2, was obtained by varying the gasification temperature between 785 and 870 °C. The sensitivity analysis conducted for the hazelnut shells is shown in Figure 4a, for the olive pruning in Figure 4b, for olive pomace in Figure 4c, and for wheat in Figure 4d.

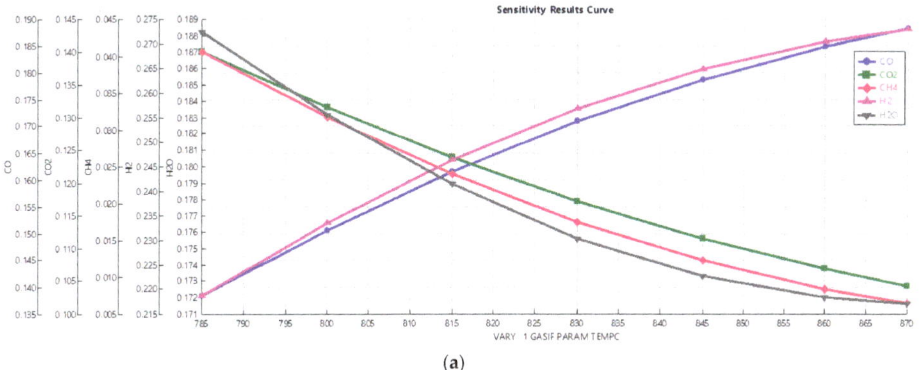

(a)

Figure 4. *Cont.*

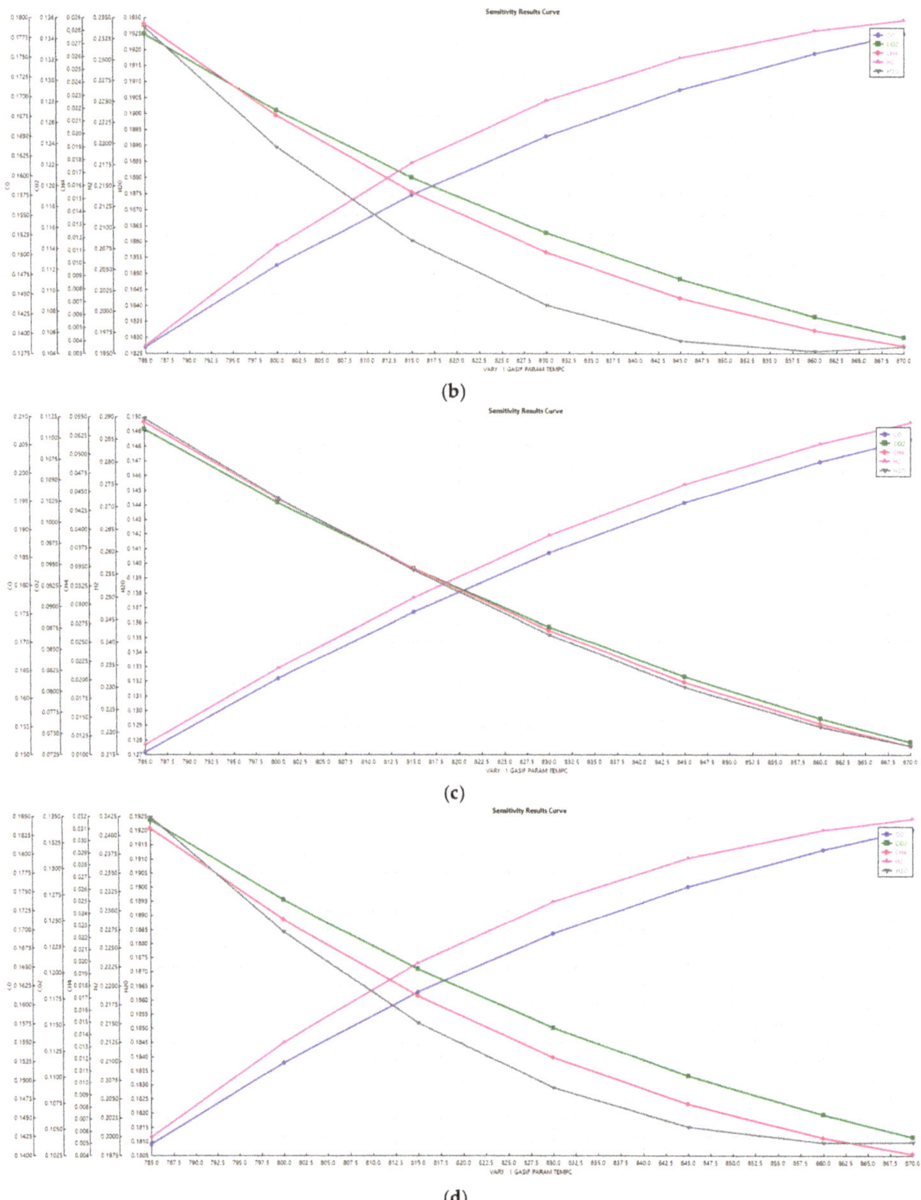

(b)

(c)

(d)

Figure 4. (**a**) Effect of gasification temperature on the syngas composition from hazelnut shells. (**b**) Effect of gasification temperature on the syngas composition from olive pruning. (**c**) Effect of gasification temperature on the syngas composition from olive pomace. (**d**) Effect of gasification temperature on the syngas composition from wheat straw.

From Figure 4a–d, it can be observed that the concentrations of CO and H_2O increased with an increase in temperature, instead the concentrations of CO_2 and CH_4 decrease with increasing in temperature. Similar trends were reported in [55]. The endothermic reactions (3) and (6) reported in Table 4 favor their forward reaction with increasing gasification temperature and will result in an increase of the concentration of CO and H_2 and a decrease of CO_2 and CH_4. However, the decrease of CH_4 is mostly determined by the effect of steam methane reforming, which is prevalent at high temperature.

4.3. Effect of ER

The effect of ER on syngas composition was investigated. Figure 5a–d show the trend of syngas composition by varying ER from 0.2 to 0.6 and maintaining the gasification temperature at 800 °C.

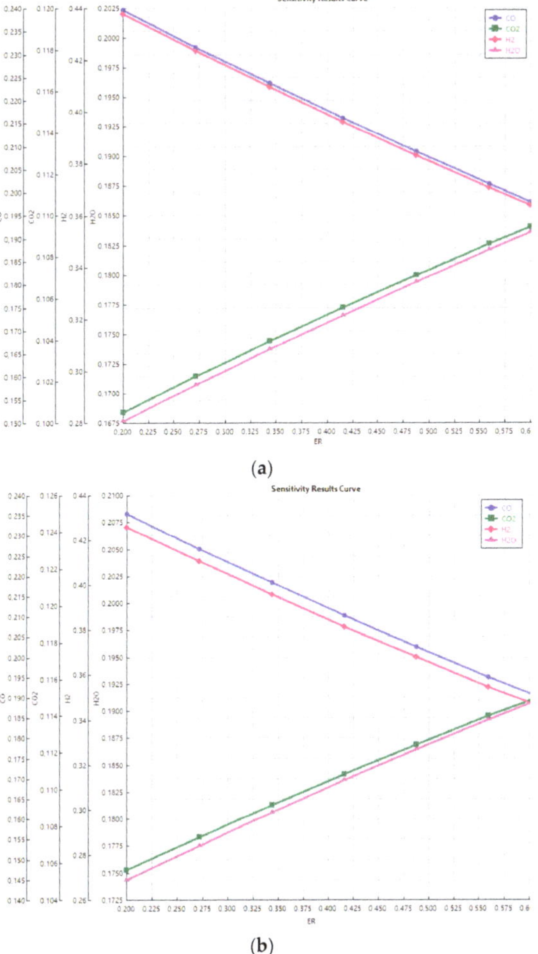

(a)

(b)

Figure 5. *Cont.*

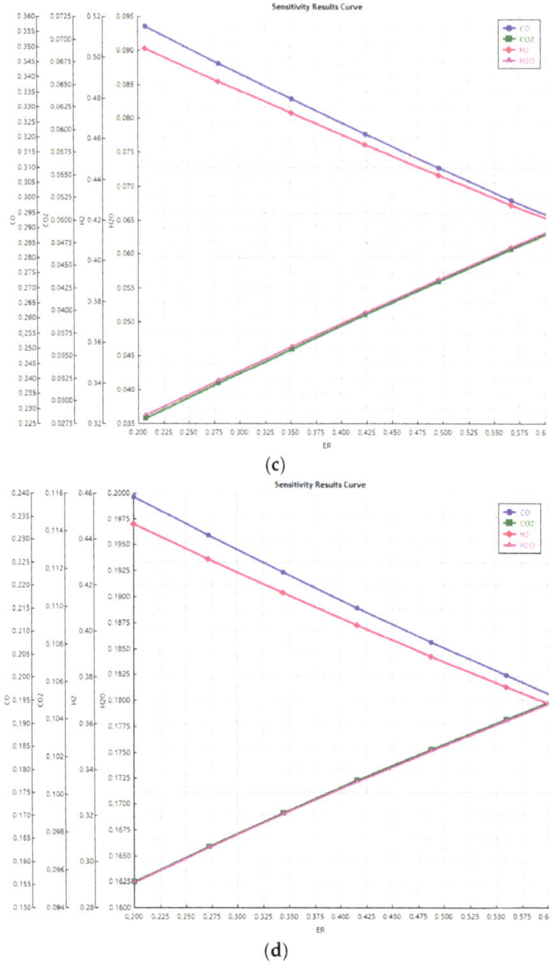

Figure 5. (**a**) Effect of air equivalent ratio on the syngas composition from hazelnut shells. (**b**) Effect of air equivalent ratio on the syngas composition from olive pruning. (**c**) Effect of air equivalent ratio on the syngas composition from olive pomace. (**d**) Effect of air equivalent ratio on the syngas composition from wheat straw.

The trend obtained showed good agreement with the results in the literature. With the increase in ER, the yields of CO_2 and H_2O increased, and the yields of H_2 and CO decreased. In order to evaluate the thermodynamic balance into the gasifier, Figure 6 shows the gasifier heat required and the LHV of the syngas produced (stream GASRAW), using olive pruning as an example because the others showed a similar trend. The heat required and LHV decreased as the ER increased, as foreseen from the previous figures and from the increase in the oxidant. The LHV varied between 5 and 4 MJ/Nm^3, while the heat demand Q varied between 257 and 185 MJ/h. In the example of olive pruning, given the similar results for the other biomass wastes, the gas yield was 1.7 Nm^3/kg and the biomass inlet was 170 kg/h, so the variation of 1 MJ/Nm^3 of the LHV corresponded to a variation of 100 MJ/h while the Q variation was 72 MJ/h. As the LHV decreased faster than Q with the increase of the ER and a loss of LHV accounted for more than a decrease of heat demand, the optimum value was lowest at ER = 0.2, when considering the overall energy balance. Indeed, at ER = 0.2, the corresponding values of Q and

LHV were the highest (260 MJ/h and 5 MJ/Nm^3, respectively), and with an increase in the ER, there was a decrease in efficiency given that the lower LHV was not compensated for by the decrease in the heat demand.

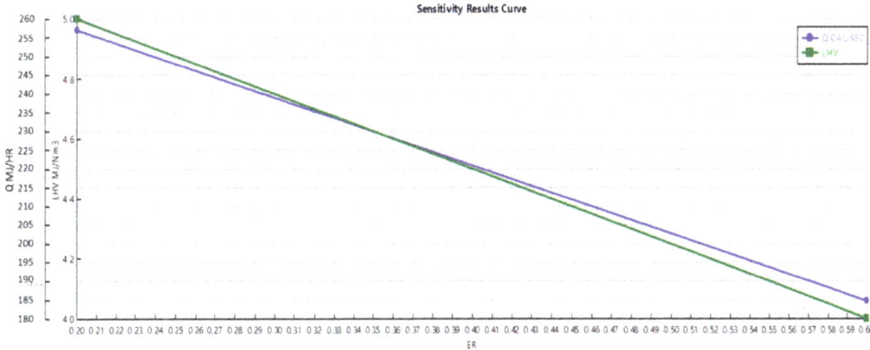

Figure 6. Gasifier heat demand and LHV vs. ER.

4.4. Cold Gas Efficiency and LHV vs. Gasification Temperature and ER

In Figure 7a–d, it can be seen that the value of the cold gas efficiency, named CGEFF, and the LHV (MJ/Nm^3) on the y-axis, was obtained by varying the gasifier temperature corresponding to the parametric curves representing the ER.

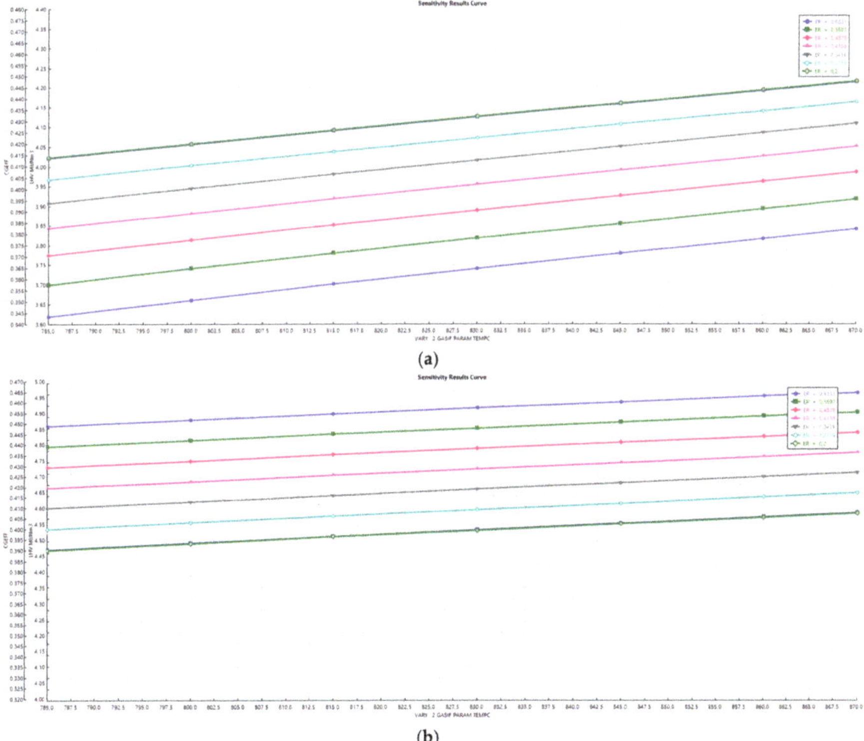

(a)

(b)

Figure 7. *Cont.*

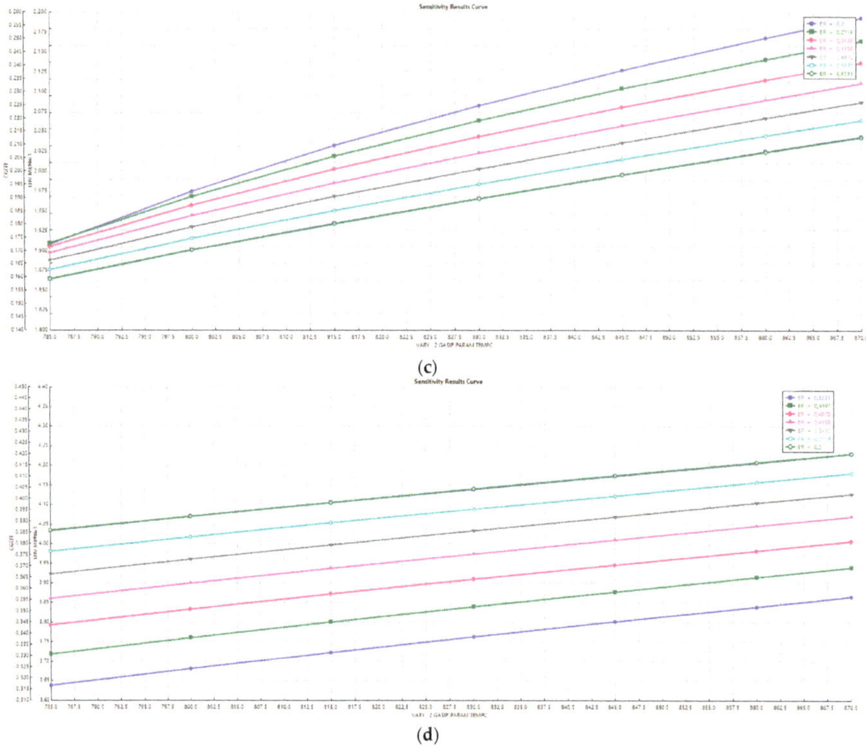

Figure 7. (**a**) Cold gas efficiency and LHV for hazelnut shells. (**b**) Cold gas efficiency and LHV for olive pruning. (**c**) Cold gas efficiency and LHV for olive pomace. (**d**) Cold gas efficiency and LHV for wheat straw.

As shown in Figure 7a–d, the cold gas efficiency and the LHV decreased as the ER increased; according to Figure 6, they showed increasing behavior as the temperature rose, so higher values of ER are not useful because the lower value of LHV means that lower heat can be generated through gas combustion, which leads to lower net power from the turbines. The best combination of LHV and cold gas efficiency for each biomass waste was:

- Hazelnut shells, η_{CG} = 42% and LHV = 4 MJ/Nm^3 at 870 °C and ER = 0.2;
- Olive pruning, η_{CG} = 46.4% and LHV = 5 MJ/Nm^3 at 870 °C and ER = 0.2;
- Olive pomace, η_{CG} = 26% and LHV = 2.2 MJ/Nm^3 at 870 °C and ER = 0.2;
- Wheat straw, η_{CG} = 41% and LHV = 4 MJ/Nm^3 at 870 °C and ER = 0.2.

However, the necessary ER calculated for the total combustion considered an excess of air of 10%, which was equal to 0.27. Therefore, the best values of LHV and cold gas efficiency obtained by moving the parametric line representing ER = 0.27 in Figure 7a–d are:

- Hazelnut shells, η_{CG} = 43.5% and LHV = 4.15 MJ/Nm^3 at 870 °C and ER = 0.27;
- Olive pruning, η_{CG} = 45.5% and LHV = 4.9 MJ/Nm^3 at 870 °C and ER = 0.27;
- Olive pomace, η_{CG} = 24.5% and LHV = 2.16 MJ/Nm^3 at 870 °C and ER = 0.27;
- Wheat straw, η_{CG} = 41% and LHV = 4.16 MJ/Nm^3 at 870 °C and ER = 0.27.

4.5. *Effect of Steam to Biomass (S/B) Ratio*

Considering the configuration shown in Figure 3 where the oxidant was only steam, a sensitivity analysis was carried out by varying the S/B parameter between 0.2 to 1.35. The S/B ratio is important to identify the quantitative effects of the addition of steam on the performance of the gasifier. Figure 8a–d show the effect of the S/B ratio on the syngas composition at a gasification temperature of 800 °C for the biomass wastes analyzed.

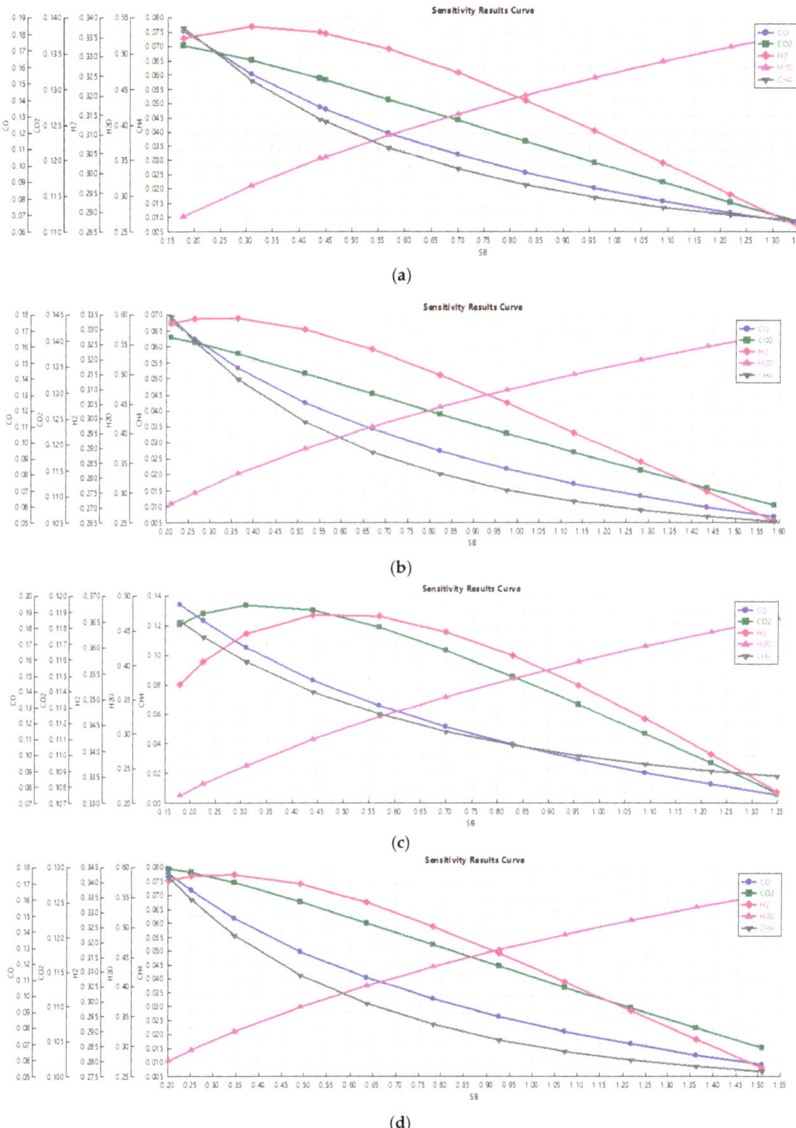

Figure 8. (**a**) Effect of the S/B ratio on the syngas composition from hazelnut shells. (**b**) Effect of the S/B ratio on the syngas composition from olive pruning. (**c**) Effect of the S/B ratio on the syngas composition from pomace olive. (**d**) Effect of the S/B ratio on the syngas composition from wheat straw.

It was observed that the concentration of H_2 increased with the increasing S/B ratio until it reached a maximum; then the concentration decreased. The hydrogen peak was almost at the beginning, which was due to the absence of air and the use of a variable external source of heat. In particular, Figure 8 shows that there was a lower regime of steam to biomass in order to reduce the heat demand, which, as shown in Figure 8, increased with the increase of S/B.

In order to evaluate the thermodynamic balance into the gasifier, Figure 9 shows the gasifier heat required and the LHV of the syngas produced (stream GASRAW). This has been shown only for the example of hazelnut shells as the other sources showed a similar trend. The heat required Q and LHV increased as the S/B increased, as foreseen from the previous figures and from the increase in the oxidant. The LHV varied between 6.5 and 9 MJ/Nm^3 while the heat demand Q varied between 550 and 1550 MJ/h. In the case of the hazelnut shells, which was similar to that of the other waste sources, the gas yield was 1.56 Nm^3/kg and the biomass inlet was 180 kg/h, so the variation of 1 MJ/Nm^3 of the LHV corresponded to a variation of 350 MJ/h. Moreover, as shown in Figure 9, the curve representing the LHV had a higher slope and was always stronger with respect to the heat demand Q. For this reason, the optimum had the lowest value of S/B after the intersection point of the two curves. Considering the overall energy balance, a good value of S/B could be 0.2. However, as S/B increased, the LHV also increased. Therefore, each time, a careful evaluation is needed in order to determine the aim of the research. If, for example, the goal was to improve the H_2 production or the increment of the LHV value, great heat required for the gasifier could be accepted.

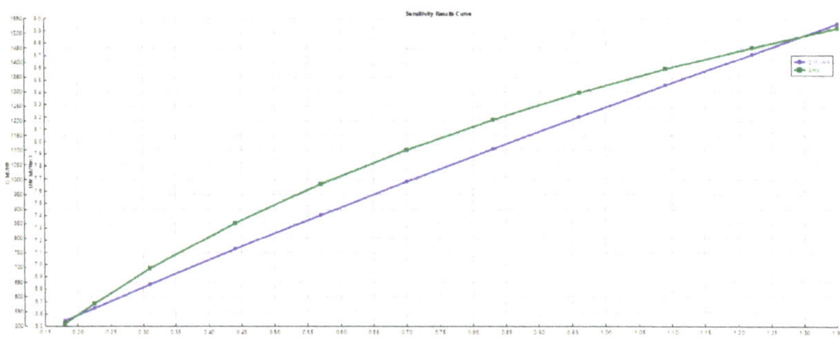

Figure 9. Gasifier heat demand vs. S/B considering hazelnut shells.

4.6. *Cold Gas Efficiency and LHV vs. Gasifier Temperature and S/B*

Referring to the configuration shown in Figure 3 where the gasifying agent is steam, the values of the cold gas efficiency and the LHV obtained by varying the gasifier temperature corresponding to the parametric curves representing the S/B are shown in Figure 10a–d.

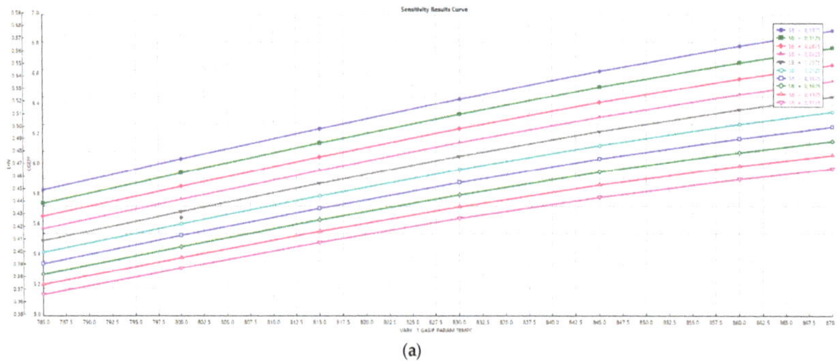

(a)

Figure 10. *Cont.*

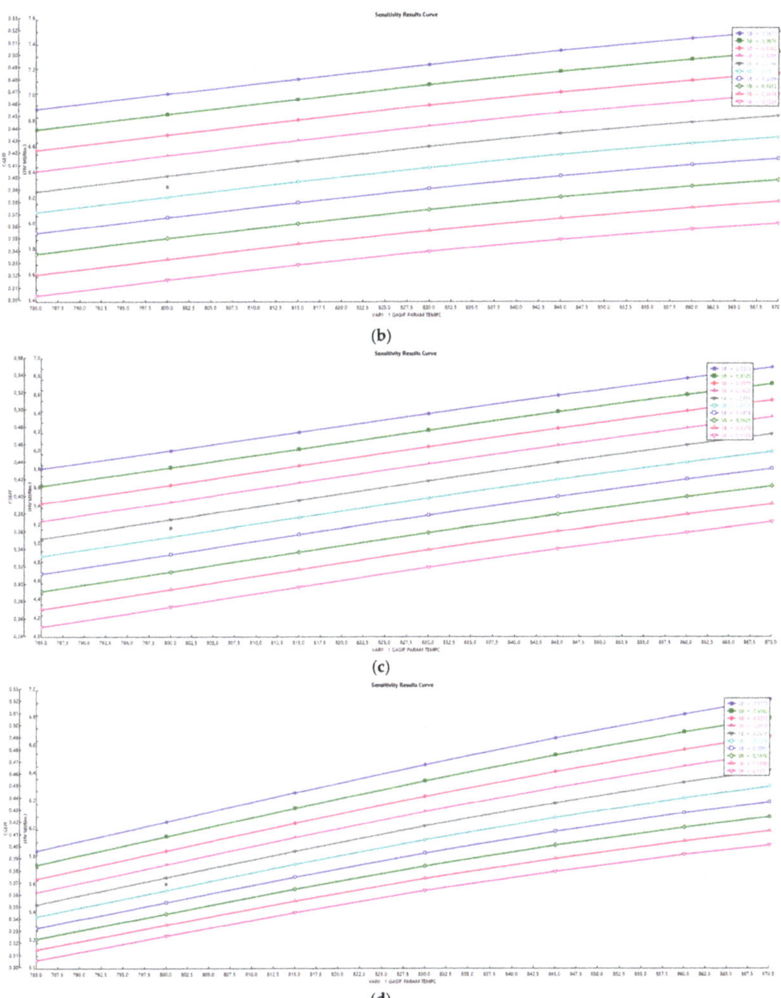

Figure 10. (**a**) Cold gas efficiency and LHV for hazelnut shells. (**b**) Cold gas efficiency and LHV for olive pruning. (**c**) Cold gas efficiency and LHV for olive pomace. (**d**) Cold gas efficiency and LHV for wheat straw.

Considering that the simulation was conducted assuming that S/B = 0.33 and that the increase of S/B means an increase of the heat required, we chose to stay with a low value of steam to biomass. Figure 10a–d show a decrease in the cold gas efficiency and the LHV with the increase in temperature and decrease of the S/B ratio. A comparison between Figure 10a–d shows that the cold gas efficiency was higher for hazelnut shells than for the othr biomass wastes and its maximum value was 58% at 870 °C with a S/B = 0.33. The highest value of LHV and cold gas efficiency for each biomass waste type was:

- Hazelnut shells, η_{CG} = 58% and LHV = 6.9 MJ/Nm^3 at 785 °C and S/B = 0.33;
- Olive pruning, η_{CG} = 55% and LHV = 6.9 MJ/Nm^3 at 785 °C and S/B = 0.33
- Olive pomace, η_{CG} = 54% and LHV = 6.7 MJ/Nm^3 at 785 °C and S/B = 0.33;
- Wheat straw, η_{CG} = 51% and LHV = 6.8 MJ/Nm^3 at 785 °C and S/B = 0.33.

4.7. Internal Combustion Engine Performance

As a result of the consideration explained in Sections 4.4 and 4.6 by taking into account the highest value of LHV and cold gas efficiency, we chose to analyze the ICE behavior using the example of olive pruning for the configuration of air gasification and the example of hazelnut shells for the configuration of steam gasification. For the two cases under observation, the following Table 9 quotes the electrical efficiency and the cogeneration efficiency, by bringing the exhaust fumes at the utilization temperature of 80 °C and a pressure drop in the turbine of 10 kPa.

The cogeneration efficiency is defined as follows:

$$\eta_{CHP} = \frac{N_{TURB} + Q_{EXCH} + Q_{EX}}{LHV_{BIOM} \cdot M_{BIOM} + Q_{INPUT}}, \tag{2}$$

where N_{TURB} is the effective electrical power of the turbine, Q_{EXCH} is the heat of the exchangers, Q_{EX} is the heat produced to bring the exhausted fumes to 80 °C, LHV_{BIOM} is the lower heat value of the biomass, M_{BIOM} is the mass of the biomass and Q_{INPUT} is the heat associate to the Gibbs reactor.

The electrical efficiency is defined as:

$$\eta_{el} = \frac{N_{TURB}}{LHV_{BIOM} \cdot M_{BIOM}}. \tag{3}$$

Table 9. Cold Gas, Electrical and Cogenerative efficiencies of the analyzed biomass waste.

Biomass	Gasification Agent	Gasification Temperature (°C)	Gas Yield (Nm^3/kg)	LHV (MJ/Nm^3)	η_{CG} (%)	η_{el} (%)	η_{CHP} (%)
Olive pruning	Air, ER = 0.27	785	1.7	4.2	35	26	41
Hazelnut shells	Steam, S/B = 0.4	785	1.56	7.25	60	30	64

5. Conclusions

An Aspen Plus model was developed for the gasification of biomass waste and for power generation from syngas. The most available biomass wastes on Italian soil were investigated to select those most suitable for the gasification process. The main parameters governing the gasification process of biomass waste in a bubbling fluidized bed gasifier using air and steam as the oxidizing agents were discussed. The effect of gasification temperature, ER, and S/B ratio was analyzed and the results showed that it was more useful to work at high temperature, low ER, and with a S/B of around 0.33. The value of the cold gas efficiency and the LHV achieved for each biomass waste in different configurations and operative conditions were studied. The best syngas compositions to feed the ICE were:

- In the case of air gasification with olive pruning and an ER = 0.27, it had 26% electrical efficiency, 46.5% cold gas efficiency, and 41% cogeneration global efficiency;
- In the case of steam gasification with hazelnut shells and a S/B = 0.33, it had 30% electrical efficiency, 58% cold gas efficiency, and 64% cogeneration global efficiency.

These results confirm that the gasifier/ICE is an attractive technique when considering the environmental benefits and the electrical efficiency obtained.

Author Contributions: E.B.: concept, methodology and revision of the simulation results; A.C.: biomass analysis in particular regarding the agricultural biomass waste typologies, properties and availability; V.M.: software simulation, draft writing; M.V.: management of the resources, data curation, validation, writing, review and editing.

Funding: This research was chiefly funded by "HBF2.0" research project within *Ricerca di Sistema* of Italian Ministry of the Economic Development, grant number CCSEB_00224 and partially funded by MIUR (Italian MInistry for education, University and Research), Law 232/2016, "Department of excellence".

Conflicts of Interest: The authors declare no conflict of interest.

References

1. Grafakos, S.; Enseñado, E.; Flamos, A.; Rotmans, J. Mapping and Measuring European Local Governments' Priorities for a Sustainable and Low-Carbon Energy Future. *Energies* **2015**, *8*, 11641–11666. [CrossRef]
2. Anifantis, A.S.; Colantoni, A.; Pascuzzi, S. Thermal energy assessment of a small scale photovoltaic, hydrogen and geothermal stand-alone system for greenhouse heating. *Renew. Energy* **2017**, *103*, 115–127. [CrossRef]
3. Manzone, M.; Airoldi, G.; Balsari, P. Energetic and Economic Evaluation of a Poplar Cultivation for the Biomass Production in Italy. *Biomass Bioenergy* **2009**, *33*, 1258–1264. [CrossRef]
4. Testa, R.; Di Trapani, A.M.; Foderà, M.; Sgroi, F.; Tudisca, S. Economic Evaluation of Introduction of Poplar as Biomass Crop in Italy. *Renew. Sustain. Energy Rev.* **2014**, *38*, 775–780. [CrossRef]
5. Zeman, F. *Metropolitan Sustainability: Understanding and Improving the Urban Environment*; Woodhead Publishing: Cambridge, UK, 11 September 2012; ISBN 9780857096463.
6. Hill, J.; Nelson, E.; Tilman, D.; Polasky, S.; Tiffany, D. Environmental, Economic, and Energetic Costs and Benefits of Biodiesel and Ethanol Biofuels. *Proc. Natl. Acad. Sci. USA* **2006**, *103*, 11206–11210. [CrossRef] [PubMed]
7. Ahmadi, P.; Dincer, I.; Rosen, M.A. Development and Assessment of an Integrated Biomass-Based Multi-Generation Energy System. *Energy* **2013**, *56*, 155–156. [CrossRef]
8. González-García, S.; Bacenetti, J. Exploring the Production of Bio-Energy from Wood Biomass. Italian Case Study. *Sci. Total Environ.* **2019**, *647*, 158–168. [CrossRef] [PubMed]
9. McKendry, P. Energy Production from Biomass (Part 3): Gasification Technologies. *Bioresour. Technol.* **2002**, *83*, 55–63. [CrossRef]
10. Orecchini, F.; Bocci, E. Biomass to Hydrogen for the Realization of Closed Cycles of Energy Resources. *Energy* **2007**, *32*, 1006–1011. [CrossRef]
11. Caputo, A.C.; Palumbo, M.; Pelagagge, P.M.; Scacchia, F. Economics of Biomass Energy Utilization in Combustion and Gasification Plants: Effects of Logistic Variables. *Biomass Bioenergy* **2005**, *28*, 35–51. [CrossRef]
12. Cuellar, A.D.; Herzog, H. A Path Forward for Low Carbon Power from Biomass. *Energies* **2015**, *8*, 1701–1715. [CrossRef]
13. Yao, Z.; You, S.; Ge, T.; Wang, C.H. Biomass Gasification for Syngas and Biochar Co-Production: Energy Application and Economic Evaluation. *Appl. Energy* **2018**, *209*, 43–55. [CrossRef]
14. Bocci, E.; Sisinni, M.; Moneti, M.; Vecchione, L.; Di Carlo, A.; Villarini, M. State of Art of Small Scale Biomass Gasification Power Systems: A Review of the Different Typologies. *Energy Procedia* **2014**, *45*, 247–256. [CrossRef]
15. Chen, H.; Zhang, X.; Wu, B.; Bao, D.; Zhang, S.; Li, J.; Lin, W. Analysis of Dual Fluidized Bed Gasification Integrated System with Liquid Fuel and Electricity Products. *Int. J. Hydrog. Energy* **2016**, *41*, 11062–11071. [CrossRef]
16. Kabalina, N.; Costa, M.; Yang, W.; Martin, A. Production of Synthetic Natural Gas from Refuse-Derived Fuel Gasification for Use in a Polygeneration District Heating and Cooling System. *Energies* **2016**, *9*, 1080. [CrossRef]
17. Nikoo, M.B.; Mahinpey, N. Simulation of Biomass Gasification in Fluidized Bed Reactor Using ASPEN PLUS. *Biomass Bioenergy* **2008**, *32*, 1245–1254. [CrossRef]
18. Dogru, M.; Howarth, C.R.; Akay, G.; Keskinler, B.; Malik, A.A. Gasification of Hazelnut Shells in a Downdraft Gasifier. *Energy* **2002**, *27*, 415–427. [CrossRef]
19. Sharma, S.; Sheth, P.N. Air-Steam Biomass Gasification: Experiments, Modeling and Simulation. *Energy Convers. Manag.* **2016**, *110*, 307–318. [CrossRef]
20. Rajabi Hamedani, S.; Villarini, M.; Colantoni, A.; Moretti, M.; Bocci, E. Life Cycle Performance of Hydrogen Production via Agro-Industrial Residue Gasification—A Small Scale Power Plant Study. *Energies* **2018**, *11*, 675. [CrossRef]
21. Azimov, U.; Tomita, E.; Kawahara, N.; Harada, Y. Effect of Syngas Composition on Combustion and Exhaust Emission Characteristics in a Pilot-Ignited Dual-Fuel Engine Operated in PREMIER Combustion Mode. *Int. J. Hydrog. Energy* **2011**, *36*, 11985–11996. [CrossRef]

22. Shilling, N.Z.; Lee, D.T. *IGCC—Clean Power Generation Alternative for Solid Fuels*; PowerGen Asia: Ho Chi Minh City, Vietnam, 2003.
23. Pilatau, A.Y.; Viarshyna, H.A.; Gorbunov, A.V.; Nozhenko, O.S.; Maciel, H.S.; Baranov, V.Y.; Mucha, O.V.; Maurao, R.; Lacava, P.T.; Liapeshko, I.; et al. Analysis of Syngas Formation and Ecological Efficiency for the System of Treating Biomass Waste and Other Solid Fuels with CO_2 recuperation Based on Integrated Gasification Combined Cycle with Diesel Engine. *J. Braz. Soc. Mech. Sci. Eng.* **2014**, *36*, 673–679. [CrossRef]
24. Couto, N.; Rouboa, A.; Silva, V.; Monteiro, E.; Bouziane, K. Influence of the Biomass Gasification Processes on the Final Composition of Syngas. *Energy Procedia* **2013**, *36*, 596–606. [CrossRef]
25. Thapa, S.; Bhoi, P.R.; Kumar, A.; Huhnke, R.L. Effects of Syngas Cooling and Biomass Filter Medium on Tar Removal. *Energies* **2017**, *10*, 349. [CrossRef]
26. Doherty, W.; Reynolds, A.; Kennedy, D. Aspen Plus Simulation of Biomass Gasification in a Steam Blown Dual Fluidised Bed. In *Materials and Processes for Energy: Communicating Current Research and Technological Developments*; Méndez-Vilas, A., Ed.; Formatex Research Centre: Badajoz, Spain, 2013.
27. Pradhan, A.; Baredar, P.; Kumar, A. Syngas as An Alternative Fuel Used in Internal Combustion Engines: A Review. *J. Pure Appl. Sci. Technol.* **2015**, *5*, 51–66.
28. Marculescu, C.; Cenuşă, V.; Alexe, F. Analysis of Biomass and Waste Gasification Lean Syngases Combustion for Power Generation Using Spark Ignition Engines. *Waste Manag.* **2016**, *47*, 133–140. [CrossRef] [PubMed]
29. Monteiro, E.; Sotton, J.; Bellenoue, M.; Moreira, N.A.; Malheiro, S. Experimental Study of Syngas Combustion at Engine-like Conditions in a Rapid Compression Machine. *Exp. Therm. Fluid Sci.* **2011**, *35*, 1473–1479. [CrossRef]
30. Liu, C.; Song, H.; Zhang, P.; Wang, Z.; Wooldridge, M.S.; He, X.; Suo, G. A Rapid Compression Machine Study of Autoignition, Spark-Ignition and Flame Propagation Characteristics of $H_2/CH_4/CO/$Air Mixtures. *Combust. Flame* **2018**, *188*, 150–161. [CrossRef]
31. Villarini, M.; Bocci, E.; Di Carlo, A.; Savuto, E.; Pallozzi, V. The Case Study of an Innovative Small Scale Biomass Waste Gasification Heat and Power Plant Contextualized in a Farm. *Energy Procedia* **2015**, *82*, 335–342. [CrossRef]
32. Ando, Y.; Yoshikawa, K.; Beck, M.; Endo, H. Research and Development of a Low-BTU Gas-Driven Engine for Waste Gasification and Power Generation. *Energy* **2005**, *30*, 2206e18. [CrossRef]
33. Pushp, M.; Mande, S. *Development of 100% Producer Gas Engine and Field Testing with Pid Governor Mechanism for Variable Load Operation*; SAE Pap.; SAE: Warrendale, PA, USA, 2008.
34. Rakopoulos, C.D.; Michos, C.N.; Giakoumis, E.G. Availability Analysis of a Syngas Fueled Spark Ignition Engine Using a Multi-Zone Combustion Model. *Energy* **2008**, *33*, 1378–1398. [CrossRef]
35. Martínez, J.D.; Mahkamov, K.; Andrade, R.V.; Silva Lora, E.E. Syngas Production in Downdraft Biomass Gasifiers and Its Application Using Internal Combustion Engines. Renew. *Energy* **2012**, *38*, 1–9.
36. Boloy, R.A.M.; Silveira, J.L.; Tuna, C.E.; Coronado, C.R.; Antunes, J.S. Ecological Impacts from Syngas Burning in Internal Combustion Engine: Technical and Economic Aspects. Renew. Sustain. *Energy Rev.* **2011**, *15*, 5194–5201. [CrossRef]
37. Mohon Roy, M.; Tomita, E.; Kawahara, N.; Harada, Y.; Sakane, A. Performance and Emission Comparison of a Supercharged Dual-Fuel Engine Fueled by Producer Gases with Varying Hydrogen Content. *Int. J. Hydrog. Energy* **2009**, *34*, 7811–7822. [CrossRef]
38. Warnecke, R. Gasification of Biomass: Comparison of Fixed Bed and Fluidized Bed Gasifier. *Biomass Bioenergy* **2000**, *18*, 489–497. [CrossRef]
39. Baratieri, M.; Baggio, P.; Bosio, B.; Grigiante, M.; Longo, G.A. The Use of Biomass Syngas in IC Engines and CCGT Plants: A Comparative Analysis. *Appl. Therm. Eng.* **2009**, *29*, 3309–3318. [CrossRef]
40. Damartzis, Th.; Michailos, S.; Zabaniotou, A. Energetic Assessment of a Combined Heat and Power Integrated Biomass Gasification–Internal Combustion Engine System by Using Aspen Plus®. *Fuel Process. Technol.* **2012**, *95*, 37–44. [CrossRef]
41. Pallozzi, V.; Di Carlo, A.; Bocci, E.; Villarini, M.; Foscolo, P.U.; Carlini, M. Performance Evaluation at Different Process Parameters of an Innovative Prototype of Biomass Gasification System Aimed to Hydrogen Production. *Energy Convers. Manag.* **2016**, *130*, 34–43. [CrossRef]
42. González-Vázquez, M.P.; García, R.; Pevida, C.; Rubiera, F. Optimization of a Bubbling Fluidized Bed Plant for Low-Temperature Gasification of Biomass. *Energies* **2017**, *10*, 306. [CrossRef]

43. Castelli, S. *Biomasse ed energia: Produzione, gestione e processi di trasformazione*; Maggioli Editore: Milan, Italy, 2011; ISBN 8838765278/9788838765278.
44. Biagini, E.; Barontini, F.; Tognotti, L. Gasification of Agricultural Residues in a Demonstrative Plant: Vine Pruning and Rice Husks. *Bioresour. Technol.* **2015**, *194*, 36–42. [CrossRef] [PubMed]
45. Colantoni, A.; Longo, L.; Gallucci, F.; Monarca, D. Pyro-Gasification of Hazelnut Pruning Using a Downdraft Gasifier for Concurrent Production of Syngas and Biochar. *Contemp. Eng. Sci.* **2016**, *9*, 1339–1348. [CrossRef]
46. Lapuerta, M.; Hernández, J.J.; Pazo, A.; López, J. Gasification and Co-Gasification of Biomass Wastes: Effect of the Biomass Origin and the Gasifier Operating Conditions. *Fuel Process. Technol.* **2008**, *89*, 828–837. [CrossRef]
47. Nordin, A. Chemical Elemental Characteristics Fuels. *Biomass Bioenergy* **1994**, *6*, 339–347. [CrossRef]
48. Ye, G.; Xie, D.; Qiao, W.; Grace, J.R.; Lim, C.J. Modeling of Fluidized Bed Membrane Reactors for Hydrogen Production from Steam Methane Reforming with Aspen Plus. *Int. J. Hydrog. Energy* **2009**, *34*, 4755–4762. [CrossRef]
49. Lv, P.; Xiong, Z.; Chang, J.; Wu, C.; Chen, Y.; Zhu, J. An Experimental Study on Biomass Air–Steam Gasification in a Fluidized Bed. *Bioresour. Technol.* **2004**, *95*, 95–101. [CrossRef] [PubMed]
50. Sadaka, S.S.; Ghaly, A.E.; Sabbah, M.A. Two Phase Biomass Air-Steam Gasification Model for Fluidized Bed Reactors: Part I—Model Development. *Biomass Bioenergy* **2002**, *22*, 439–462. [CrossRef]
51. Demirbaş, A. Carbonization Ranking of Selected Biomass for Charcoal, Liquid and Gaseous Products. *Energy Convers. Manag.* **2001**, *42*, 1229–1238. [CrossRef]
52. Begum, S.; Rasul, M.G.; Akbar, D.; Ramzan, N. Performance Analysis of an Integrated Fixed Bed Gasifier Model for Different Biomass Feedstocks. *Energies* **2013**, *6*, 6508–6524. [CrossRef]
53. Franco, C.; Pinto, F.; Gulyurtlu, I.; Cabrita, I. The Study of Reactions Influencing the Biomass Steam Gasification Process. *Fuel* **2003**, *82*, 835–842. [CrossRef]
54. Torres, W.; Pansare, S.S.; Goodwin, J.G. Hot Gas Removal of Tars, Ammonia, and Hydrogen Sulfide from Biomass Gasification Gas. *Catal. Rev.* **2007**, *49*, 407–456. [CrossRef]
55. Pala, L.P.R.; Wang, Q.; Kolb, G.; Hessel, V. Steam Gasification of Biomass with Subsequent Syngas Adjustment Using Shift Reaction for Syngas Production: An Aspen Plus Model. Renew. *Energy* **2017**, *101*, 484–492. [CrossRef]
56. Liao, C.H.; Summers, M.; Seiser, R.; Cattolica, R.; Herz, R. *Simulation of a Pilot-Scale Dual-Fluidized-Bed Gasifier for Biomass*; Environmental Progress and Sustainable Energy, John Wiley & Sons, Ltd.: Hoboken, NJ, USA, 2014; pp. 732–736.

Article

Performance Assessment of Front-Mounted Beet Topper Machine for Biomass Harvesting

Volodymyr Bulgakov [1], Simone Pascuzzi [2,*], Semjons Ivanovs [3], Francesco Santoro [2], Alexandros Sotirios Anifantis [2] and Ievhen Ihnatiev [4]

1 Department of Mechanics, Faculty of Construction and Design, National University of Life and Environmental Sciences of Ukraine, 03041 Kyiv, Ukraine; vbulgakov@meta.ua
2 Department of Agricultural and Environmental Science, University of Bari Aldo Moro, Via Amendola 165/A, 70126 Bari, Italy; francesco.santoro@uniba.it (F.S.); alexandrossotirios.anifantis@uniba.it (A.S.A.)
3 Faculty of Engineering, Latvia University of Life Sciences and Technologies, LV-2130 Jelgava, Latvia; semjons@apollo.lv
4 Department of Machine-Using in Agriculture, Dmytro Motornyi Tavria State Agrotechnological University, 18, Khmelnytskiy Ave., 72312 Melitopol, Ukraine; yevhen.ihnatiev@tsatu.edu.ua
* Correspondence: simone.pascuzzi@uniba.it; Tel.: +39-080-544-2214

Received: 10 June 2020; Accepted: 4 July 2020; Published: 8 July 2020

Abstract: Sugar beet is an extensive crop of great agronomic value with significant productive and economic returns and Ukraine's sugar beet accounts for about 5.1% of the overall world production. Sugar beets and the by-products resulting from its manufacturing transformation are a significant renewable energy resource. A new high-quality performance prototype of a sugar beet top harvester, front mounted on a tractor, was built by the authors in Ukraine. The aim of this study is to evaluate the main performance parameters related to the operation of this new machine. Field tests were carried out linking the prototype to a wheel tractor, whilst suitable sensors measured the significant kinematic and dynamic parameters, allowing experimental data collection to assess the machine's performance parameters. The entire technological process of harvesting and transporting the beet tops to the beet top storage unit required power ranging from 6.42 to 17.65 kW. At the topmost tested forward speed, the required tractor traction force was less than 1.9 kN with the power required by the shaft that drives the screw conveyor ranging from 3.1 to 4.6 kW. This value was the lowest for a speed of the tractor–beet top harvesting machine aggregate ranging from 0.9 to 1.2 m·s^{-1}.

Keywords: sugar beet; beet top cutting; tractor–harvester aggregate; biomass

1. Introduction

World sugar beet production amounts to approximately 2.75·10^8 ton with a devoted area of 2.7·10^8 m^2 [1]. In terms of both production and harvested area, referring to overall world data, the EU contribution is 43.5% and 36.3%, respectively, with the most contributive countries being France and Germany with 14.4% and 10.1%, respectively, and 9.5% and 8.6%, respectively, while Italy contributes only 0.7% for both ratios. Referring instead to Ukraine, the previously referred ratios are 5.1% and 5.7%, respectively [1]. However, it should be mentioned that beet and sugar production regulation within the EU is based on the Common Market Organization (CMO) and that, in 2006, the CMO was completely amended, leading to a strong reduction in EU sugar production [2]. Some countries such as France and Germany, considered more suited to beet production and more efficient from an industrial point of view, have been little affected by the changes, whereas others such as Italy have suffered significant consequences. Italy, at the time active in the sector with 19 sugar industries and approximately 1.4·10^9 m^2 of beet-cultivated area, after the reform gradually reduced their industrial plants from 19 to 4 and their overall beet-cultivated area decreased considerably [2]. Sugar beet is an extensive crop of great agronomic value with significant productive and economic returns; moreover, it has always been considered an "improvement crop" from which all crops in succession benefit [3–6]. Sugar beet and the by-products resulting from its industrial transformation are a noteworthy renewable energy resource [7–10], e.g., pulps can be used in biogas and electricity production [11–15], and beet leaves and tops are currently used as a fundamental component in the food rations of animal farms, as they are rich in nutrients and they can be employed as a substrate in anaerobic digestion for renewable energy production, due to the high content of both sugar and almost completely digestible fibers [16,17]. Furthermore, large amounts of the sugar industry's different kinds of generated waste, such as sugar beet pulp and leaves, can be employed as precious substrates in the production of biotechnology cellular proteins, enzymes, organic acids, etc. [18–20]. For all the aforesaid reasons, a high performance and quality of sugar beet top harvesting that can be achieved only using specialized machines clearly appears to be of paramount importance [21–24]. The beet harvesting machines commonly used in Europe do not meet these performance standards, in particular referring to the beet tops which are not collected after cutting but simply crushed and spread on the soil in such a way that makes them unusable also for animal fodder [25–27]. Furthermore, even in the case of high-performance beet harvesting machines, some specific conditions, such as those in Ukraine, may arise from which many unresolved problems still could derive. For example, one of these problems is a deterioration in the quality of the collected beets, especially in difficult harvesting conditions, such as high soil hardness or excessive humidity, the irregularity and non-linearity of the crop rows, excessive weeds, and so on [28–33]. Therefore, the scientific and research community continue to search for design and technical solutions in order to meet these operative needs and, at the same time, reduce the energy required by the digging and harvesting process of the beets, so increasing the productivity and reliability of the machines [34–39]. The study, design and prototyping of a new sugar beet top harvester in Ukraine that, when front mounted on the tractor, considerably increases beet top harvesting performances has to be considered in this framework [40–42]. According to experimental data obtained from field tests, the aim of this study was to evaluate the main performance parameters related to the operation of this new machine.

2. Materials and Methods

2.1. The Beet Topper Machine

The research focuses on a new three-row beet top harvesting machine equipped with improved working devices which allow the machine to cut the beet tops and transport them into a loading chute [43–46]. In particular, the developed front-mounted beet top harvester is founded on the concept of a mower-shredder and is mounted on a wheeled tractor (Figure 1a), from which it receives motion and power by means of a cardan universal joint (7 in Figure 1b) linked to the power take-off of the tractor itself. The beet top harvester continuously cuts rosette leaves during its forward motion, regulating the cutting height of a cylindrical rotor (4 in Figure 1b) equipped with arc-shaped knives by means of pneumatic feeler wheels (3 in Figure 1b), ensuring an effective result in the cutting of beet tops [47]. Once cut, the beet tops are transferred onto a screw conveyor that ensures their transportation into a loading chute (6 in Figure 1b) and subsequent delivery into a trailer running alongside through a chute. The main technical characteristics of this new beet top harvester are: (i) three sugar beet root crops rows working at a width of 1.35 m, (ii) a forward speed of up to 2.1 $m{\cdot}s^{-1}$, (iii) a mass equal to 850 kg, (iv) a working capacity within a range of 1.0–1.2 $ha{\cdot}h^{-1}$ [41].

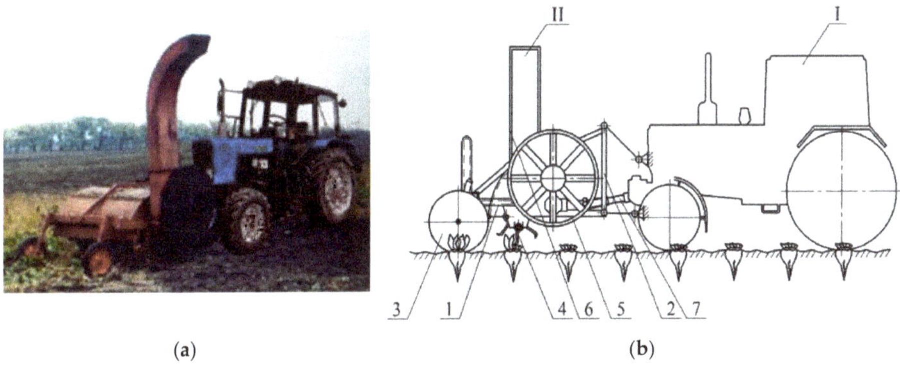

(a) (b)

Figure 1. Tractor–harvester aggregate: (**a**) photo taken during experimental field test phase, (**b**) schematic representation: I—Wheeled, row-crop integrated tractor; II—Front-mounted beet topper machine: 1—frame, 2—point hitch, 3—pneumatic feeler wheel, 4—rotary beet top cutting device, 5—screw conveyor, 6—loading chute, 7—cardan universal joint.

2.2. The Field Tests

The experimental tests took place in the Vasilkovsky district, Kiev Region, and were aimed at assessing the main performance parameters of the new three-row beet top harvester in the typical operating conditions of the harvesting phase. The beet topper machine was joined at the front of a wheeled MTZ-80 tractor using a semi-mounted coupling through the three-point hitch, deriving the needed motion from the tractor power take-off (PTO) (Figure 1). Throughout the whole duration of the experimental tests, the tractor rear axle was the only drive axle with the front one disabled. The following performance parameters during the carried-out tests were taken into account [48,49]: (i) the tractor traction power N, kW; (ii) the tractor traction force R, N; (iii) the required torque at the tractor PTO T_{PTO}, N·m; and (iv) the required power at the tractor PTO N_{PTO}, kW, which is related to T_{PTO} and to the PTO shaft angular speed ω_{PTO} through the equation $N_{PTO} = T_{PTO} \cdot \omega_{PTO}$. The tractor traction power (N) was measured using sensors able to measure torque and the angular speeds of the tractor rear left and right drive axle shafts, whereas the tractor traction force (R) was given by [50,51]:

$$R = \frac{N}{V} \quad (1)$$

where V, m·s^{-1} is the aggregate tractor–beet top harvester forward speed.

The tractor's left and right rear drive axle shaft's torques and angular speeds were measured by means of Zemic BA350KA (Zemic Europe B.V.—Etten-Leur, The Netherlands) contactless rotary torque transducers, capable of gauging both the torque strain in the shafts, via an on-shaft microprocessor circuit, and shaft rotational speed, and whose main features are: 1000 Ω nominal resistance and a −30 to +80 °C working temperature range.

The actual tractor–beet top harvester aggregate forward speed was measured through a track measuring wheel equipped with an Autonics PR12-4DN stationary proximity sensor (Autonics, Busan, South Korea) whose main characteristics are: cylindrical round (PR Series) type, 12–24 V DC voltage, M12 sensing side diameter, 4 mm sensing distance, 500 Hz response frequency (Figure 2).

Figure 2. Tractor–harvester aggregate actual speed measuring wheel.

The torque T_{PTO} and power N_{PTO} required by the PTO were evaluated through the same aforesaid Zemic BA350KA (Zemic Europe B.V.—Etten-Leur, The Netherlands) sensor arranged on the tractor PTO (Figure 3).

Figure 3. Required torque and power sensor arranged at the tractor power take-off (PTO).

All used sensors were connected to a laptop via an L-CARD model E14-140-M (Moscow, Russian Federation) converter whose main characteristics are: a 48 MHz 32 bit processor, 8 differential (16 if common ground is used) input channels.

The tests were carried out maintaining both the engine crankshaft and tractor PTO speeds at 1200 rpm and 540 rpm, respectively, and considering three different forward speeds of the tractor–harvester aggregate, obtained by means of three different gear-range lever combinations: (i) 1st gear and the high speed ratio range lever engaged with an estimated forward speed of V = 1.46 m·s^{-1} (5.27 km·h^{-1}); (ii) 2nd gear and the low speed ratio range lever engaged with an estimated forward speed of V = 1.88 m·s^{-1} (6.78 km·h^{-1}); and (iii) 3rd gear and the high speed ratio range lever engaged with an estimated forward speed of V = 2.49 m·s^{-1} (8.97 km·h^{-1}). In these conditions, tests were carried out within the manufacturers' recommended forward tractor–harvester aggregate speed range of 0.9–2.50 m·s^{-1} (3.2–9.0 km·h^{-1}) and for each gear-range lever combination, three repetitions were carried out while measuring actual tractor–harvester aggregate speed.

Furthermore, for each chosen gear-range lever combination, three different screw conveyor angular speeds were considered (5 in Figure 1b). In particular, whilst maintaining a constant at 57 rad·s^{-1} (540 rpm), the angular speed of both the tractor PTO and the shaft that drives the cylindrical rotor equipped with beet top cutting knives, three different angular speeds of screw conveyor driving were considered: (i) $\omega_{po1} = 57$ rad·s^{-1}; (ii) $\omega_{po2} = 39$ rad·s^{-1}; and (iii) $\omega_{po3} = 0$ rad·s^{-1} (Figure 4).

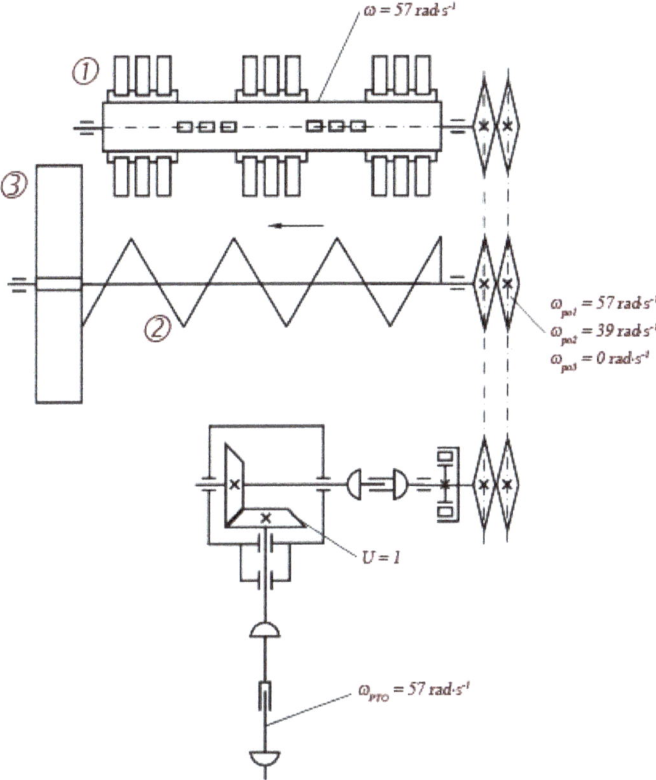

Figure 4. Kinematic diagram of the tested three row beet top harvester: 1—rotary beet top cutting device, 2—screw conveyor, 3—loading chute feeder.

2.3. Data Analysis

The experimental data were processed using Microsoft Excel software in order to carry out a regression analysis of each studied performance parameter, by means of the least-squares method [52]. As is known, this criterion is a technique for fitting the "best" curve to the sample $\hat{x}$, $\hat{y}$ observations. It involves minimizing the sum of the squared (vertical) deviations of points from the curve:

$$Min \sum (\hat{y}_i - y_i)^2 \tag{2}$$

where:

$\hat{y}_i$ refers to the actual observations
y_i refers to the corresponding fitted values, so that $(\hat{y}_i - y_i) = e_i$, the residual [52].

The data were processed using different regression functions (linear, polynomial, power and exponential), calculating the corresponding coefficients of the determination R^2 and residuals. Among these regression functions, only the 2nd order polynomial was considered, because for all the examined performance parameters, it allowed the achievement of the highest value of R^2 and the lowest residuals. The 2nd order polynomial was given by:

$$p(x) = a_0 + a_1 x + a_2 x^2 \tag{3}$$

where the coefficients a_0, a_1 and a_2 were calculated by the solution of the following matrix system:

$$\hat{V}^t\hat{V}A = \hat{V}^t\hat{Y} \tag{4}$$

where:

$\hat{V}$ is a Vandermonde matrix, which contains the observation values $\hat{V}_{i,j} = \hat{x}_{i-1}^{j-1}$;
$\hat{V}^t$ is the transpose matrix of $\hat{V}$;
A is the column vector of the terms a_i;
$\hat{Y}$ is the column vector of the observations $\hat{y}_i$.

Practically, considering the 9 couples of experimental measures for each studied performance parameter, the linear matrix system (4) gives:

$$\hat{V}^t\hat{V} = \begin{pmatrix} 1 & 1 & 1 & \dots & 1 \\ \hat{x}_0 & \hat{x}_1 & \hat{x}_2 & \dots & \hat{x}_8 \\ \hat{x}_0^2 & \hat{x}_1^2 & \hat{x}_2^2 & \dots & \hat{x}_8^2 \end{pmatrix} \begin{pmatrix} 1 & \hat{x}_0 & \hat{x}_0^2 \\ 1 & \hat{x}_1 & \hat{x}_1^2 \\ 1 & \hat{x}_2 & \hat{x}_2^2 \\ \dots & \dots & \dots \\ 1 & \hat{x}_8 & \hat{x}_8^2 \end{pmatrix} = \begin{pmatrix} z_{1,1} & z_{1,2} & z_{1,3} \\ z_{2,1} & z_{2,2} & z_{2,3} \\ z_{3,1} & z_{3,2} & z_{3,3} \end{pmatrix} \tag{5}$$

and the note term of the system (4) gives:

$$\hat{V}^t\hat{Y} = \begin{pmatrix} 1 & 1 & 1 & \dots & 1 \\ \hat{x}_0 & \hat{x}_1 & \hat{x}_2 & \dots & \hat{x}_8 \\ \hat{x}_0^2 & \hat{x}_1^2 & \hat{x}_2^2 & \dots & \hat{x}_8^2 \end{pmatrix} \begin{pmatrix} \hat{y}_0 \\ \hat{y}_1 \\ \hat{y}_2 \\ \vdots \\ \hat{y}_8 \end{pmatrix} = \begin{pmatrix} d_1 \\ d_2 \\ d_3 \end{pmatrix}. \tag{6}$$

Finally, substituting (5) and (6) in (4), it gives the following square matrix system:

$$\begin{pmatrix} z_{1,1} & z_{1,2} & z_{1,3} \\ z_{2,1} & z_{2,2} & z_{2,3} \\ z_{3,1} & z_{3,2} & z_{3,3} \end{pmatrix} \begin{pmatrix} a_0 \\ a_1 \\ a_2 \end{pmatrix} = \begin{pmatrix} d_1 \\ d_2 \\ d_3 \end{pmatrix} \tag{7}$$

whose solution allows the assessment of the coefficients a_0, a_1 and a_2 and then attainment of the 2nd order least-squares polynomial (2).

At the end of the calculations, for each performance parameter, the residuals average e was evaluated through the following equation:

$$e = \frac{1}{9}\sum_{i=1}^{9}(\hat{y}_i - y_i) \tag{8}$$

3. Results and Discussion

In Table 1, all the obtained experimental data corresponding to each of the considered gear-range lever combinations, screw conveyor angular speeds and repetitions are reported as follows: (i) tractor–harvester aggregate forward speed V, m·s^{-1}; (ii) tractor PTO required torque T_{PTO}, N·m; (iii) traction force R, N; (iv) tractor PTO required power N_{PTO}, kW; and (v) total required power measured at the tractor rear drive axle N, kW.

Table 1. Beet top harvester obtained experimental data.

Tractor Gear-Range Lever	V, m·s^{-1}	T_{PTO}, N·m	R, N	N_{PTO}, kW	N, kW
		ω_{po1} = 57 rad·s^{-1}			
	0.990	91.35	1092.2	5.165	1.081
1st gear-high speed ratio	0.960	97.15	1244.6	5.493	1.194
	0.954	92.8	914.4	5.247	1.272
	1.464	97.15	1447.8	5.493	2.119
2nd gear-low speed ratio	1.380	101.5	1625.6	5.739	2.242
	1.524	117.5	1727.2	6.644	2.631
	2.118	168.2	1803.4	9.511	3.818
3rd gear-high speed ratio	2.196	169.7	1930.4	9.596	4.238
	2.022	162.4	1727.2	9.183	3.491
		ω_{po2} = 39 rad·s^{-1}			
	1.014	60.9	1244.6	3.444	1.261
1st gear-high speed ratio	1.146	71.05	1193.8	4.017	1.367
	1.176	66.7	1346.2	3.772	1.582
	1.698	81.2	1574.8	4.591	2.673
2nd gear-low speed ratio	1.536	71.05	1295.4	4.017	1.989
	1.722	78.3	1701.8	4.427	2.929
	2.124	111.7	2108.2	6.316	4.476
3rd gear-high speed ratio	2.220	156.6	1549.4	8.855	3.438
	2.058	142.1	1955.8	8.035	4.024
		ω_{po3} = 0 rad·s^{-1}			
	1.140	50.75	1041.4	2.870	1.186
1st gear-high speed ratio	1.236	60.9	1320.8	3.444	1.632
	1.314	66.7	1117.6	3.772	1.468
	1.692	71.05	1473.2	4.017	2.492
2nd gear-low speed ratio	1.746	58.00	1295.4	3.280	2.261
	1.614	59.45	1625.6	3.362	2.623
	2.256	89.9	1752.6	5.083	3.953
3rd gear-high speed ratio	2.166	97.15	1981.2	5.493	4.290
	2.148	85.55	1752.6	4.837	3.763

ω_{po1}, ω_{po2}, ω_{po3}: screw conveyor drive shaft angular speeds.

Table 2 reports second order polynomial regressions with the corresponding R^2 values and residuals for the assessed performance parameters; the response variables are the torque T_{PTO} and the power N_{PTO}, the tractor traction force R, and the total tractor power N, respectively. The explanatory variable is the aggregate forward speed. These regression functions are plotted in Figures 5–7.

Table 2. Second order regression results of different V dependent parameters.

		Screw Conveyor Drive Shaft Angular Speed	
	ω_{po1} = 57 rad·s^{-1}	**ω_{po2} = 39 rad·s^{-1}**	**ω_{po3} = 0 rad·s^{-1}**
R *	$R = -264.003 \cdot V^2 + 1433.064 \cdot V$ $R^2 = 0.8661 - e = -3.048$	$R = -238.756 \cdot V^2 + 1366.048 \cdot V$ $R^2 = 0.6589 - e = 1.246$	$R = -121.706 \cdot V^2 + 1090.533 \cdot V$ $R^2 = 0.7845 - e = 0.538$
T_{PTO}	$T_{PTO} = 54.410 \cdot V^2 - 104.234 \cdot V + 143.073$ $R^2 = 0.9709 - e = -0.001$	$T_{PTO} = 90.495 \cdot V^2 - 224.716 \cdot V + 202.692$ $R^2 = 0.8969 - e = 0.001$	$T_{PTO} = 32.815 \cdot V^2 - 78.774 \cdot V + 104.929$ $R^2 = 0.8238 - e = 0.000$
N_{PTO}	$N_{PTO} = 3.0769 \cdot V^2 - 5.8944 \cdot V + 8.0899$ $R^2 = 0.97092 - e = 0.0000$	$N_{PTO} = 5.1190 \cdot V^2 - 12.7129 \cdot V + 11.4660$ $R^2 = 0.89686 - e = -0.0001$	$N_{PTO} = 1.8552 \cdot V^2 - 4.4539 \cdot V + 5.9337$ $R^2 = 0.82376 - e = -0.0001$
N *	$N = 0.4746 \cdot V^2 + 0.8380 \cdot V$ $R^2 = 0.98197 - e = -0.0037$	$N = 0.5623 \cdot V^2 + 0.6372 \cdot V$ $R^2 = 0.89571 - e = -0.0026$	$N = 0.6672 \cdot V^2 + 0.3490 \cdot V$ $R^2 = 0.94897 - e = -0.0001$

* Provided that the regression lines pass through the coordinates' origin.

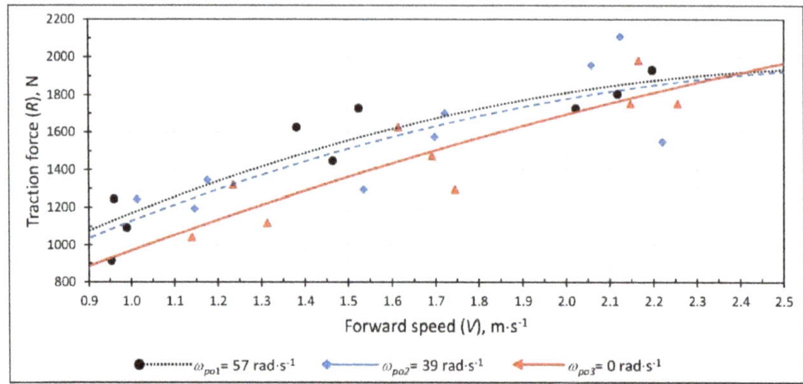

Figure 5. Tractor traction force R vs. aggregate tractor–beet top harvester forward speed V for different values of screw conveyor drive shaft angular speed.

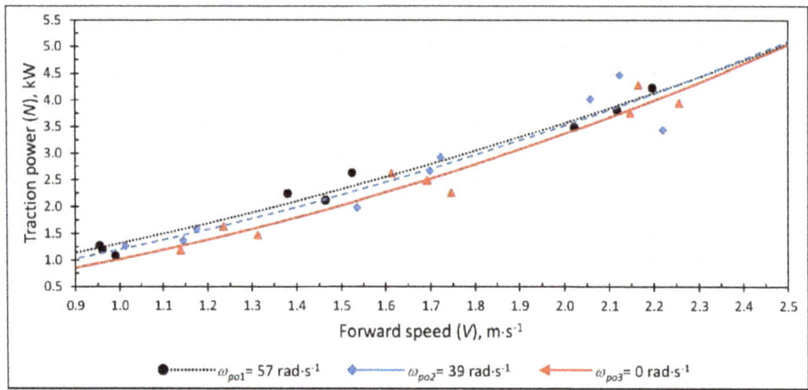

Figure 6. Tractor traction power N vs. aggregate tractor–beet top harvester forward speed V for different values of screw conveyor drive shaft angular speed.

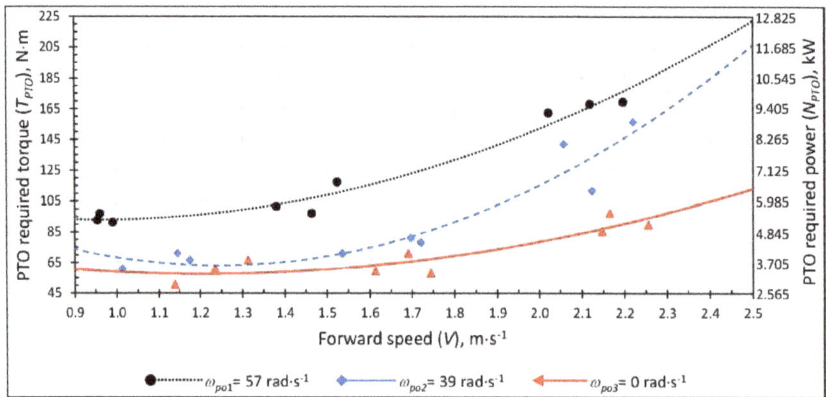

Figure 7. PTO required torque T_{PTO} and power N_{PTO} vs. aggregate tractor–beet top harvester forward speed V for different screw conveyor drive shaft angular speeds.

The traction force required by the tractor was not so much influenced by the screw conveyor drive shaft angular speed and, even if its increase was almost 94% in the considered aggregate forward speed range (Figure 5), in absolute terms the increase was of only almost 1 kN. This behavior can be explained by considering that the tractor traction force is mainly the sum of the wheels rolling resistance and the push force required by the front beet topper machine. As is known, the rolling resistance of tires on a surface is mainly connected to the hysteresis in tire materials caused by the deflection of the tire casing during rolling as well as by its operating conditions, such as surface conditions, inflation pressure, rolling speed, temperature, and so on [49,50]. Nevertheless, in the considered speed range, that is within 0.9 and 2.5 m·s^{-1}, the tire's rolling resistances can be considered almost constant and the low absolute increase in the traction force is probably due to the increase in the flow rate of tops that have to be processed by the cutting apparatus as the aggregate forward speed increases [27,29]. Therefore, according to other studies, the tractor traction force is affected by the speed of the aggregate, rather than by the screw conveyor drive shaft angular speed variations [23,31]. In the above-mentioned speed range, the traction force varies from 1.1 to 1.9 kN, considering $\omega_{po1} = 57$ rad·s^{-1}.

Figure 6 shows how the traction power measured at the tractor's rear drive axles increased within the range of 0.81–5.0 kW as the aggregate forward speed increased and also, for this parameter, no significant dependency related to screw conveyor drive shaft angular speed variations appeared.

Referring to the relations between the required torque, the power at the tractor PTO shaft and the aggregate forward speed evaluated for different screw conveyor drive shaft angular speeds, as shown in Figure 7, there was limited growth of the interested parameters for an increase of the forward speed up to about 1.5 m·s^{-1}. For higher speed values, an important increase of the required torque and power can be observed. Furthermore, the screw conveyor drive shaft angular speed had a significant effect on the PTO required torque and power values. At an angular speed of $\omega_{po1} = 57$ rad·s^{-1}, the required torque (power) was 93.3 N·m (5.3 kW), and at a forward speed of 0.9 m·s^{-1}, this increased by 139% to 222.6 N·m (12.6 kW) at a speed of 2.5 m·s^{-1}. Whereas when the screw conveyor is turned off $\omega_{po3} = 0$ rad·s^{-1}, the required torque (power) increased by 86.7% from 60.6 N·m (3.4 kW) at a forward speed of 0.9 m·s^{-1} to 113.1 N·m (6.4 kW) at a speed of 2.5 m·s^{-1}. This behavior is clearly connected to the greater product flow rate that the cutting apparatus must process as forward speed increases.

Taking into account both the $N(V)$ and $N_{PTO}(V)$ relationships between the tractor power, PTO required power and aggregate forward speed represented in Figures 6 and 7, it is possible to assess that the entire technological process of the harvesting and transporting to the storage unit of the beet tops requires a power which ranges from 6.42 to 17.65 kW (with $\omega_{po1} = 57$ rad·s^{-1}). However, in normal operating conditions with a forward speed ranging from 1.7 to 2 m·s^{-1}, the total required power ranges from 10 to 12 kW.

During the tests, the lower screw conveyor drive shaft angular speed ($\omega_{po2} = 39$ rad·s^{-1}) ensured the proper performance of the harvesting machine. Nevertheless, in the case of increased humidity or an excessive amount of weeds on the field, congestions and obstructions occurred in the screw conveyor operation. According to the experimental results, the suitable screw conveyor drive shaft angular speed has to be in the range of 50 to 60 rad·s^{-1} so the chosen angular speed $\omega_{po1} = 57$ rad·s^{-1} can be a guarantee of reliability in all operating conditions.

The executed field tests carried out using the new beet top harvesting machine also highlighted that its average energy costs, related to a single work row (N_{PTO} = 3.1 kW and N = 1.4 kW), are significantly lower than the corresponding performance parameter values of beet top harvesting machines currently in use on Ukrainian farms.

4. Conclusions

Sugar beet is a temperate climate crop which is grown profitably in almost all areas of the world with latitudes over than 30° where the winters are not very hard. Obviously, cultivation systems and material inputs must be adjusted according to the climate and soil characteristics, taking into account that the quality of the beet deeply affects the operative efficiency of the process carried out inside a sugar beet factory. During the harvesting operations, the cut off and collection of the beet tops must be performed properly with suitable machines. The new front-mounted beet topper machine analyzed in this study is able to process three-rows of beets simultaneously, under conditions of high quality performance of the technological process. The results of the executed test highlighted its good performance, pointing out that the tractor power and its traction force, as well as the torque and the power required at its power take-off, are on average 1.2 to 1.5 times lower than the corresponding performance parameters of the beet top harvesters currently employed in Ukraine. Nevertheless, further technical improvement of the screw conveyor system is under study in order to make better the efficiency of the system that allows the transport and loading of cut beet tops. Further experimental campaigns will be then necessary to re-verify the performance parameters analyzed in this study.

Author Contributions: Conceptualization, V.B., S.P., S.I., and I.I.; methodology, V.B., S.P., S.I., F.S., A.S.A. and I.I.; formal analysis, investigation and data curation, S.I., F.S., A.S.A. and I.I.; writing—original draft preparation, V.B., S.P., S.I., and I.I.; writing—review and editing, V.B. and S.P.; supervision, V.B. and S.P. All authors have read and agreed to the published version of the manuscript.

Funding: This research received no external funding.

Conflicts of Interest: The authors declare no conflict of interest.

References

1. FAO. Food and Agriculture Organization of the United Nations 2018. Available online: http://www.fao.org/faostat/en/#data/QC (accessed on 3 July 2020).
2. Stevanato, P.; Chiodi, C.; Broccanello, C.; Concheri, G.; Biancardi, E.; Pavli, O.; Skaracis, G. Sustainability of the sugar beet crop. *Sugar Tech.* **2019**, *21*, 703–716. [CrossRef]
3. Bulgakov, V.M. *Beet Harvesting Machines*; Agrarian Science: Kiev, Ukraine, 2011; p. 351.
4. Khvostov, V.A.; Reingart, E.S. *Machines for Harvesting Root Crops and Onions (Theory, Design, Calculation)*; VISHOM: Moscow, Russia, 1995; p. 391.
5. Pogorelyj, L.; Tatjanko, N. *Machines for Harvesting Sugar Beet*; Fenix: Kyiv, Ukraine, 2014; p. 369.
6. Silva, R.P.; Rolim, M.M.; Gomes, I.F.; Pedrosa, E.M.R.; Tavares, U.E.; Santos, A.N. Numerical modeling of soil compaction in a sugarcane crop using the finite element. *Soil Tillage Res.* **2018**, *181*, 1–10. [CrossRef]
7. Bulgakov, V.M.; Adamchuk, V.V.; Nozdrovicky, L.; Boris, M.M.; Ihnatiev, Y.I. Properties of the sugar beet tops during the harvest. In Proceedings of the 6th International Conference on Trends in Agricultural Engineering 2016, Prague, Czech Republic, 7–9 September 2016; pp. 102–108.
8. Anifantis, A.S.; Pascuzzi, S.; Scarascia-Mugnozza, G. Geothermal source heat pump performance for a greenhouse heating system: An experimental study. *J. Agric. Eng.* **2016**, *47*, 164–170. [CrossRef]
9. Pascuzzi, S.; Santoro, F. Evaluation of farmers' OSH hazard in operation nearby mobile telephone radio base stations. In Proceedings of the 16th International Scientific Conference "Engineering for Rural Development", Jelgava, Latvia, 24–26 May 2017; Volume 16, pp. 748–755. [CrossRef]
10. Hamedani, S.R.; Villarini, M.; Colantoni, A.; Carlini, M.; Cecchini, M.; Santoro, F.; Pantaleo, A. Environmental and economic analysis of an anaerobic co-digestion power plant integrated with a compost plant. *Energies* **2020**, *13*, 2724. [CrossRef]
11. Pascuzzi, S.; Bulgakov, V.; Santoro, F.; Anifantis, A.S.; Ivanovs, S.; Holovach, I. A Study on the drift of spray droplets dipped in airflows with different directions. *Sustainability* **2020**, *12*, 4644. [CrossRef]
12. Pantaleo, A.; Villarini, M.; Colantoni, A.; Carlini, M.; Santoro, F.; Hamedani, S.R. Techno-economic modeling of biomass pellet routes: Feasibility in Italy. *Energies* **2020**, *13*, 1636. [CrossRef]
13. Santoro, F.; Anifants, A.S.; Ruggiero, G.; Zavadskiy, V.; Pascuzzi, S. Lightning protection systems suitable for stables: A case study. *Agriculture* **2019**, *9*, 72. [CrossRef]
14. Pascuzzi, S.; Santoro, F. Analysis of possible noise reduction arrangements inside olive oil mills: A Case Study. *Agriculture* **2017**, *7*, 88. [CrossRef]
15. Anifantis, A.S.; Camposeo, S.; Vivaldi, G.A.; Santoro, F.; Pascuzzi, S. Comparison of UAV photogrammetry and 3D modeling techniques with other currently used methods for estimation of the tree row volume of a super-high-density olive orchard. *Agriculture* **2019**, *9*, 233. [CrossRef]

16. Guerrieri, A.S.; Anifantis, A.S.; Santoro, F.; Pascuzzi, S. Study of a large square baler with innovative technological systems that optimize the baling effectiveness. *Agriculture* **2019**, *9*, 86. [CrossRef]
17. Cerruto, E.; Manetto, G.; Santoro, F.; Pascuzzi, S. Operator dermal exposure to pesticides in tomato and strawberry greenhouses from hand-held sprayers. *Sustainability* **2018**, *10*, 2273. [CrossRef]
18. Berlowska, J.; Binczarski, M.; Dziugan, P.; Wilkowska, A.; Kregiel, D.; Witonska, I. Sugar beet pulp as a source of valuable biotechnological products. In *Handbook of Food Bioengineering*; Elsevier: Amsterdam, The Netherlands, 2018; pp. 359–392.
19. Manetto, G.; Cerruto, E.; Pascuzzi, S.; Santoro, F. Improvements in citrus packing lines to reduce the mechanical damage to fruit. *Chem. Eng. Trans.* **2017**, *58*, 391–396. [CrossRef]
20. Pascuzzi, S.; Santoro, F. Analysis of the almond harvesting and hulling mechanization process: A case study. *Agriculture* **2017**, *7*, 100. [CrossRef]
21. Pascuzzi, S.; Santoro, F. Exposure of farm workers to electromagnetic radiation from cellular network radio base stations situated on rural agricultural land. *Int. J. Occup. Saf. Ergo.* **2020**, *21*, 351–358. [CrossRef] [PubMed]
22. Bulgakov, V. Haulm Harvesting Machine. Ukrainian Patent 83051, 6 October 2008.
23. Bulgakov, V.; Pascuzzi, S.; Ivanovso, S.; Nadykto, V.; Nowak, J. Kinematic discrepancy between driving wheels evaluated for a modular traction device. *Biosyst. Eng.* **2020**, *196*, 88–96. [CrossRef]
24. Bulgakov, V.; Pascuzzi, S.; Adamchuk, V.; Ivanovs, S.; Pylypaka, S. A theoretical study of the limit path of the movement of a layer of soil along the plough mouldboard. *Soil Tillage Res.* **2019**, *195*, 104406. [CrossRef]
25. Butenin, N.V.; Lunts, J.L.; Merkin, D.R. *The Course of Theoretical Mechanics*; Dynamics Science: Moscow, Russia, 1985; p. 495.
26. Pascuzzi, S. A multibody approach applied to the study of driver injures due to a narrow-track wheeled tractor rollover. *J. Agric. Eng.* **2015**, *46*, 105–144. [CrossRef]
27. Vasilenko, P.M. *Introduction to Agricultural Mechanics*; Agriculture: Kiev, Ukraine, 1996; p. 252.
28. Pascuzzi, S.; Anifantis, A.S.; Santoro, F. The concept of a compact profile agricultural tractor suitable for use on specialised tree crops. *Agriculture* **2020**, *10*, 123. [CrossRef]
29. Reznik, N.E. *The Theory of Cutting with a Blade and the Basics of Calculating the Cutting Apparatus*; Mechanical Engineering: Moscow, Russia, 1975; p. 311.
30. Kalpakjian, S.; Schmid, S.R. *Manufacturing Engineering and Technology*, 6th ed.; Prentice Hall: New York, NY, USA, 2010; p. 1176.
31. Bulgakov, V.; Golovach, I.; Ivanovs, S.; Ihnatiev, Y. Theoretical simulation of parameters of cleaning sugar beet heads from remnants of leaves by flexible blade. In Proceedings of the 16th International Scientific Conference "Engineering for Rural Development", Jelgava, Latvia, 24–26 May 2017; Volume 16, pp. 288–295.
32. Vasilenko, I.F. *Theory of Cutting Machines Reaping Machines*; Works of the VISHOM: Moscow, Russia, 1937; Volume 5, pp. 7–14.
33. Bulgakov, V.; Adamchuk, V.; Ivanovs, S.; Ihnatiev, Y. Theoretical investigation of aggregation of top removal machine frontally mounted on wheeled tractor. In Proceedings of the 16th International Scientific Conference Engineering for Rural Development, Jelgava, Latvia, 24–26 May 2017; Volume 16, pp. 273–280.
34. Bulgakov, V.; Pascuzzi, S.; Beloev, H.; Ivanovs, S. Theoretical investigations of the headland turning agility of a trailed asymmetric implement-and-tractor aggregate. *Agriculture* **2019**, *9*, 224. [CrossRef]
35. Bulgakov, V.; Pascuzzi, S.; Ivanovs, S.; Kaletnik, G.; Yanovich, V. Angular oscillation model to predict the performance of a vibratory ball mill for the fine grinding of grain. *Biosyst. Eng.* **2018**, *171*, 155–164. [CrossRef]
36. Bulgakov, V.; Pascuzzi, S.; Nadykto, V.; Ivanovs, S. A mathematical model of the plane-parallel movement of an asymmetric machine-and-tractor aggregate. *Agriculture* **2018**, *8*, 151. [CrossRef]
37. Pascuzzi, S.; Cerruto, E. An innovative pneumatic electrostatic sprayer useful for tendone vineyards. *J. Agric. Eng.* **2015**, *46*, 123–127. [CrossRef]
38. Pascuzzi, S.; Cerruto, E.; Manetto, G. Foliar spray deposition in a "tendone" vineyard as affected by airflow rate, volume rate and vegetative development. *Crop. Prot.* **2017**, *91*, 34–48. [CrossRef]
39. Pascuzzi, S.; Santoro, F.; Manetto, G.; Cerruto, E. Study of the correlation between foliar and patternator deposits in a "Tendone" vineyard. *Agric. Eng. Int. CIGR J.* **2018**, *20*, 97–107.
40. Bulgakov, V.; Adamchuk, V.; Nozdrovicky, L.; Ihnatiev, Y. Theory of vibrations of sugar beet leaf harvester front-mounted on universal tractor. *Acta Technol. Agric.* **2017**, *4*, 96–103. [CrossRef]

41. Ihnatiev, Y. Theoretical research and development of new design of beet tops harvesting machinery. In Proceedings of the V International Scientific-Technical Conference Agricultural Machinery, Varna, Bulgaria, 21–24 June 2017; Volume 1, pp. 46–48.
42. Tatyanko, N.V. The calculation of the working bodies for cutting machines of sugar beet tops. *Tract. Agric. Mach.* **1962**, *11*, 18–21.
43. Bulgakov, V.; Pascuzzi, S.; Anifantis, A.S.; Santoro, F. Oscillations analysis of front-mounted beet topper machine for biomass harvesting. *Energies* **2019**, *12*, 2774. [CrossRef]
44. Bulgakov, V.; Pascuzzi, S.; Santoro, F.; Anifantis, A.S. Mathematical model of the plane-parallel movement of the self-propelled root-harvesting machine. *Sustainability* **2018**, *10*, 3614. [CrossRef]
45. Pascuzzi, S. Outcomes on the spray profiles produced by the feasible adjustments of commonly used sprayers in "tendone" vineyards of Apulia (Southern Italy). *Sustainability* **2016**, *8*, 1307. [CrossRef]
46. Pascuzzi, S. The effects of the forward speed and air volume of an air-assisted sprayer on spray deposition in "tendone" trained vineyards. *J. Agric. Eng.* **2013**, *3*, 125–132. [CrossRef]
47. Bosoy, E.S. *Cutting Apparatus of the Harvesting Machines*; Theory and Calculation; Mechanical Engineering: Moscow, Russia, 1967; p. 167.
48. Dospehov, B. *Methodology of Field Experiments*; Nauka: Moscow, Russia, 1985; p. 351. (In Russian)
49. Kolychev, E.I. On the choice of the design impact case in the study of the smoothness of the movement of tractors and agricultural machinery. *Tract. Agric. Mach.* **1976**, *3*, 9–11.
50. Morozov, B.I.; Gringauz, N.M. Calculation of the movement of the wheeled vehicle on a rough road. *Mech. Electrif. Social. Agric.* **1969**, *1*, 11–14.
51. Bulgakov, V.; Pascuzzi, S.; Adamchuk, V.; Kuvachov, V.; Nozdrovicky, L. Theoretical study of transverse offsets of wide span tractor working implements and their influence on damage to row crops. *Agriculture* **2019**, *9*, 144. [CrossRef]
52. Carnahan, B.; Luther, H.A.; Wilkes, J.O. *Applied Numerical Methods*; John Wiley & Sons: New York, NY, USA, 1969; p. 604.

Review

Pellet Production from Woody and Non-Woody Feedstocks: A Review on Biomass Quality Evaluation

Rodolfo Picchio [1,*], Francesco Latterini [2], Rachele Venanzi [1], Walter Stefanoni [2], Alessandro Suardi [2], Damiano Tocci [1] and Luigi Pari [2]

1 Department of Agricultural and Forest Sciences (DAFNE), Tuscia University, 01100 Viterbo, Italy; venanzi@unitus.it (R.V.); toccidamiano91@gmail.com (D.T.)

2 CREA Research Centre for Engineering and Agro-Food Processing, Via della Pascolare, 16, 00015 Monterotondo (Rome), Italy; francesco.latterini@crea.gov.it (F.L.); walter.stefanoni@crea.gov.it (W.S.); alessandro.suardi@crea.gov.it (A.S.); luigi.pari@crea.gov.it (L.P.)

* Correspondence: r.picchio@unitus.it; Tel.: +39-0761-357-400

Received: 17 April 2020; Accepted: 1 June 2020; Published: 8 June 2020

Abstract: Forest and agricultural biomass are important sources of renewable and sustainable fuel for energy production. Their increasing consumption is mainly related to the increase in global energy demand and fossil fuel prices but also to the limited availability of petroleum and the lower environmental impact of these biomass compared with other non-renewable fuels. In particular, the pellet sector has seen important developments in terms of both production and the number of installed transformation plants. In addition, pellet production from non-woody biomass is increasing in importance. One of the fundamental aspects for the correct and sustainable use of a biofuel is evaluation of its quality. This is even more important when dealing with pellet production, considering the broad spectrum of possible raw materials for pelletizing. Considering the significant number of papers dealing with pellet quality evaluation and improvement in the last decade, this review aims to give the reader an overall view of the most current knowledge about this large and interesting topic. We focused on pellets of agricultural and forestry origin and analyzed papers regarding the specific topic of pellet quality evaluation and improvement from the last five years (2016–2020). In particular, the review findings are presented in the following order: the influence of different agro-forest management systems on pellet quality; analysis of pellets from pure feedstocks (no blending or binders); the influence of blending and binders on pellet quality; and the influence of pre and post treatments. Finally, a brief discussion about actual research lacks in this topic and the possibilities for future research are presented. It is important to underline that the present review is focused on the influence of the biomass characteristics on pellet quality. The effects of the process parameters (die temperature, applied pressure, holding time) on pellet features are not considered in this review, because that is another very large topic deserving a dedicated paper.

Keywords: woody pellet; agropellet; quality; standards; blending

1. Introduction

The constant decreasing availability of fossil fuels as energy sources and the huge environmental problems related to their use, have led to great interest in renewable energy sources [1]. One of these interesting sources is undoubtedly biomass. The main problems related to the use of biomass for energy production are linked to its irregular shape, low bulk density, and high moisture content which create difficulties in handling, transport, and storage [2–6]. A recent trend to remedy the critical issues in the use of biomass for energy purposes is densification and standardization to exploit a homogeneous and easy-to-use solid biofuel, which is also characterized by a higher energy density [7]. Currently, there are various densification processes available, but the most commonly used is that of pelletization.

Biomass use in form of pellets generates a biofuel that is more cost-effective than the direct use of non-modified biomass residues for energy production. This process, called extrusion, consists of applying high pressure and high temperatures to semi-dry biomass pre-processed in dust, sawdust or shavings, passing it through a hole of a few millimeters in size, and producing small cylinders that are cut to the desired length and then cooled [8]. This process increases the bulk density of biomass, thus reducing handling, transport, and storage costs [9–13]. The great importance of the pelletization process for energy production is highlighted by the data from a 2019 statistical report on pellets [14], which underlines the increasing amount of pellet production worldwide (Figure 1).

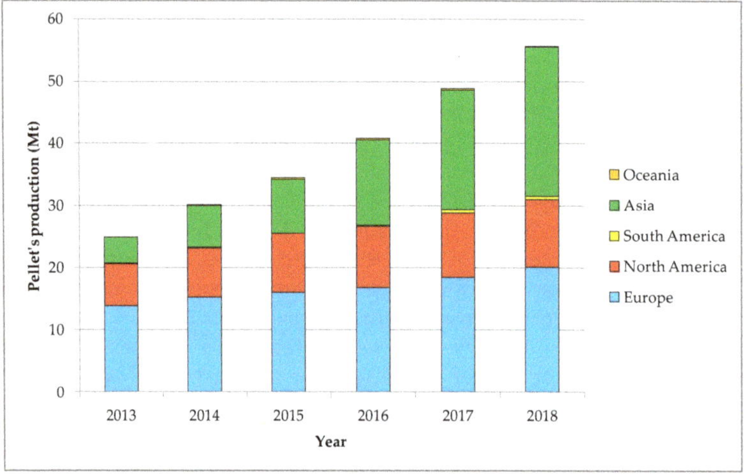

Figure 1. Worldwide pellet production from 2013 to 2018 [14].

One of the fundamental aspects for the correct and sustainable use of a biofuel is the evaluation of its quality. Pellet quality is affected by multiple parameters linked to both the properties of the raw material used, for example, particle size distribution, moisture content, and chemical composition, and to the operating conditions, like the die temperature, applied pressure, and holding time [15–17]. In addition, pellets can be produced from various feedstocks, mainly related to agricultural and forest activities [18–21]. Considering the broad spectrum of possible raw materials for pelletizing, quality evaluation of pellets is even more important.

In particular, of the many types of biomass, wood represents the main feedstock for pellet production [21]. Sawdust is an ideal substrate, as it is untreated and even minor contaminants are removed through bark removal and washing of saw logs prior to sawing [8]. However, considering the increasing demand for wood pellets and the limited supply of sawmill residue, there has been growing interest and exploration in the production of pellets from other processes and sources [22]. Recently, pellets produced from forest logging residue or dedicated agroforestry plantations have been used [23].

Non-woody biomass, like agricultural residue, is one of the most important alternative feedstocks for pelletization [8]. This kind of biomass could be of significant use thanks to the large amount available, the low price, and the importance of agricultural waste re-use as by-products in maintaining a green and circular economy [24]. However, in most cases, the quality of agropellets is less than that of forest pellets, and the main disadvantages are related to the low bulk density, high ash content, and low calorific value [25]. One possible solution is mixing woody and non-woody biomass. In fact, using various blends generally leads to an increase in pellet quality [10].

Moreover, in recent years, several studies have focused on improving the pellet quality, in particular regarding durability and bulk density, through the use of organic components, for example, sewage sludge and inorganic binders [26–28].

Another solution to improve the quality is pre-treatment of the feedstock or post-treatment of the pellet. The main aim of pre-treatment is the structural change in biomass feedstock to make it more suitable for pelletization. The two most well-known pre-treatments are steam explosion treatment and torrefaction [29]. Steam pre-treatment, also known as autohydrolysis or steam explosion, consists of exposing feedstock, firstly, to steam (typically at 140–260 °C) and then, suddenly, to atmospheric pressure [29]. The first step leads to the hydrolysis of hemicellulose and to the activation of lignin, while the second phase (pressure drop) causes biomass fragmentation [30,31]. This type of treatment has been found to be particularly effective for agricultural residue and hardwood [32]. The main aim of steam explosion is to increase mechanical strength due to the activation of inherent lignin [29]. On the other hand, torrefaction is a thermochemical treatment which can be applied to both the feedstock (pre-treatment) and to the already formed pellet (post-treatment) [29]. This type of treatment is performed in an anoxic environment or under a very low oxygen content (<6 vol% O_2) in a temperature range of between 200 and 300 °C [33,34]. When applied to the feedstock, this process results, generally, in more grindable material with a higher energy density, increased homogeneity, a higher thermal stability, and greater hydrophobicity [29]. The main problem linked with feedstock torrefaction is the high friction in the die channel which causes an increase in energy consumption during pellet formation [35]. Moreover, pelletization of torrefied biomass is more difficult and the correct formation of pellets is not always achieved [36]. In order to avoid these criticalities, this process is also used as a post-treatment with the aim of increasing the heating value and hydrophobicity of the pellet [29].

Considering the significant number of papers dealing with pellet quality evaluation and improvement in the last decade, this review aims to give to the reader an overall view of the most current knowledge on this large and interesting topic. We focused on pellets of agricultural and forestry origin and analyzed papers from the last five years (2016–2020) that investigated the specific topic of pellet quality evaluation and improvement.

First, the methodology used to construct the review is given. Then, an overview of the pellet quality standards classification is reported. After this, the review findings are presented and commented in the following order: influence on pellet quality of different agro-forest management systems; analysis of pellets from pure feedstocks (no blending or binders); the influence of blending and binders on pellet quality; and the influence of pre and post treatments. Finally, some ideas for future research directions were shown and some conclusions were formulated.

It is important to underline that the present review is focused on the influence of the biomass characteristics on pellet quality. The effects of the process parameters (die temperature, applied pressure, holding time) on pellet features are not considered in this review, because that is another very large topic which deserves a dedicated review.

2. Materials and Methods

This bibliographical search was developed using Boolean operators and implementing a symbolic logic system that creates relationships between concepts and words. The use of Boolean searching to carry out a systematic review allows one to analyze all studies in a specific research field [37]. Research was performed using the databases Scopus, ISI Web of Knowledge, and Google Scholar. The first keyword used was "pellet", limiting the research only to the "Energy" subject area. In this way 8364 findings were produced. Subsequently, the research was refined by limiting the findings only to the period 2016–2020. The number of findings decreased to 2323. Then, the keyword "quality" was used to further refine the papers. Consequently, the number of findings decreased to 626.

After this, authors performed paper selection by reading the title and abstract of each article, and 71 articles were identified as suitable for the present review. A total of 43.66% of the analyzed papers dealt with the use of pure feedstock for pellet production; 21.13% were about the influence of blending on quality; 18.31% concerned the effect of binders; 15.49% dealt with pre- or post- treatment consequences; and only one paper (1.41%) was about the influence of agrosystem management on pellet quality (Figure 2). Moreover, in Figure 2 an overview about the main topics of all papers dealing with pellet quality on Scopus repository is given to the reader. As it is possible to notice, there is a good balance among the percentage of papers dealing with forest biomass, agricultural residues, and both these are used as feedstock for pellet production.

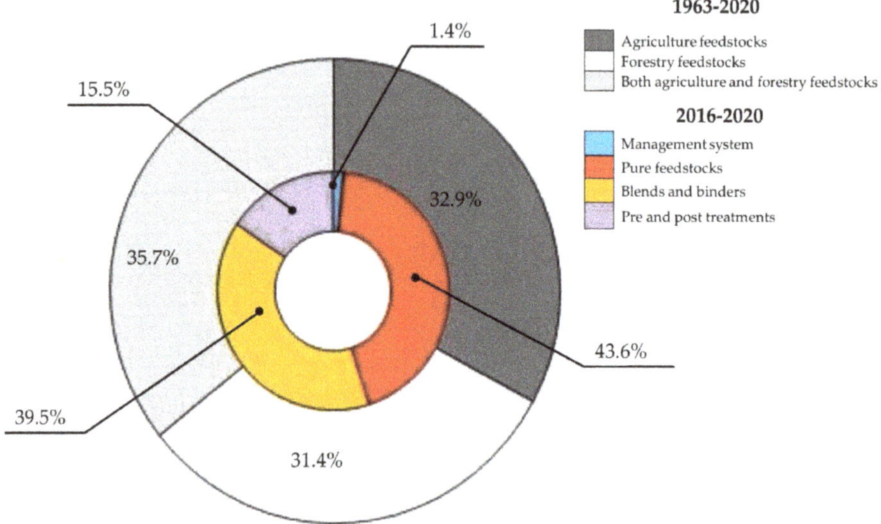

Figure 2. Percentages of the various identified main topics in relation to the total number of analyzed papers.

3. Pellet Quality Standards

International pellet quality standards were developed by the International Organization for Standardization (ISO). In particular, standards for the use of pellets as biofuel include EN ISO 17225-1 for general quality requirements, EN ISO 17225-2 for graded wood pellets for industrial and domestic use, and EN ISO 17225-6 for graded non-woody pellets. In the past, several European countries have developed regulations and standards for pellet quality certification, in particular, the Austrian standard ÖNORM M 7135, the Swedish standard SS 187120, the German standards DIN 51731 and DIN EN 15270, the Italian standard CTI-R04/05, and the French recommendation ITEBE [38].

ISO fuel specification standards (EN ISO 17225 series) were published in May 2014, and this series replaced EN 14961. National standardization bodies in Europe published the EN ISO 17225 series as a national standard at the end of November 2014 [39].

The graded wood pellet standard (EN ISO 17225-2) is related to the use of pellets for industrial and non-industrial use. Non-industrial use means fuel intended to be used in smaller appliances, such as in households and small commercial and public sector buildings [39]. According to this standard, the best quality class is A1, which represents virgin woods and chemically untreated wood residue that are low in ash and nitrogen. Pellets with a slightly higher ash content and nitrogen content are described as A2. Finally, there is property class B. This class includes chemically treated industrial wood by-products and residue [39]. Within ISO 17225-2, a classification of pellets for industrial use is also reported. This classification provides three different quality classes (I1, I2, and I3), which present slightly more restrictive requirements in comparison to classes A1, A2, and B for pellets for domestic use. The advent of this standard at European level represented an important step for the sector, ensuring greater transparency on a product along the full evolution and allowing greater uniformity with world markets.

Non-woody pellet standards (ISO 17225-6) relate to pellets made from blends and mixtures, including herbaceous, fruit, or aquatic biomass. This standard provides two classification tables, one for herbaceous and fruit biomass and blends and one for straw, miscanthus, and reed canary grass pellets. In general, non-woody pellets have high ash, chlorine, nitrogen, and sulfur contents [39] as well as a lower heating value (LHV). Consequently, their required standard is less restrictive, while considering the achievement of lower quality levels than wood pellets, in this sense, a greater attention would be desirable about clearly presenting qualitative differences and suggestions for use. Considering the high dynamism of this sector, it is important that standards for solid biofuels should be continuously under development. In particular, they should be quite strict especially in relation to the use of heterogeneous materials. For example, the US EPA (Environmental Protection Agency) imposed detailed restrictions on the type of fuels, including density, dimensions, inorganic fines (≤1%), chlorides, ash content (<2%), interdiction of demolition or construction waste, and trace metal concentrations less than 100 mg/kg.

4. Influence of Agro-Forest Management System on Pellet Quality

The only paper that was found to deal with the influence of different management systems on pellet quality was that presented by Civitarese et al. [1]. In their study, a comparison of pellet quality was reported, taking into consideration three different ways of managing dedicated plantations for biomass production. In particular, the authors analyzed pellets from poplar plantations managed through short rotation forestry (SRF), medium rotation forestry with a six-year rotation period (MRF6), and medium rotation forestry with a nine-year rotation period (MRF9). The obtained results showed that all investigated categories presented a shortfall (in relation to ISO standards) regarding bulk density. The SRF pellets also showed problems with excessive ash content, and the durability of MRF9 was not sufficient to reach ISO quality standards. For the other investigated parameters, for example the lower heating value (LHV) or heavy metal content, all materials reached the A1 class standard according to EN ISO 17225-2. According to the authors' findings it seems that the best rotation cycle for poplar SRC with reference to pellet production is a 6 year period. Those with periods of three years indeed showed a major ash content, because of the higher percentage of bark, while 9 years material showed higher heavy metal content, moreover it is important to underline that such a long rotation period implies a more complex and specialized mechanization level for harvesting operation considering the high diameter of the stems to be cut. From the findings it is clear that the management system could influence pellet quality and consequently the energetic performance, for example: (i) pellet density increases char combustion time; (ii) pellets composed of bark had up to 50% longer char combustion time. These performances could be directly linked to changes in extractives, ash, and lignin, characteristics that depend upon tree vigor class. Generally, low vigor trees have higher extractives, ash, and lignin contents than the vigorous trees.

5. Pure Feedstock for Pellet Production

Before showing the main findings about the characteristics of pellet produced from new feedstocks, it is necessary to give the reader a view of the characteristics of pellet made on existing and established feedstocks, i.e., hardwood tree species [40]. Such information is given in Table 1, in which characteristics of pellet from pine, spruce, and hemlock are reported. Moreover, in Table 1 the quality requirements for wood pellet for A1, A2 and B class are also shown.

Table 1. Characteristics of pellet produced from pine, hemlock and spruce and ISO 17225-2 standards requirements for A1, A2, and B classes.

	LHV MJ kg^{-1}	Bulk Density ($kg\ m^{-3}$)	Moisture (%)	Durability (%)	Ash Content (%)	N (%)	S (%)	Cl (%)	Feedstock
[41]	—	603.00	8.50	98.50					*Pinus spp*
[42]	18.50	709.00	9.60	97.80	0.70	0.06	0.02	0.01	*Pinus spp*
[43]	18.13	540.90	—	—	—	—	—	—	*Pinus spp*
[44]	—	680.00	8.86	—	0.78	1.43	0.02	0.02	*Pinus spp*
[45]	16.90	—	7.30	—	1.30	0.50	<0.01	—	*Pinus spp*
[46]	—	—	—	—	0.10	—	—	—	*Picea spp*
[47]	18.69	709.46	7.84	93.60	0.70	0.09	<0.01	<0.01	*Picea spp*
[47]	18.28	756.23	5.79	96.80	0.35	0.07	0.01	<0.001	*Tsuga spp*
EN ISO 17225-2 A1	≥16.50	≥600	≤10.00	≥97.50	≤0.70	≤0.30	≤0.04	≤0.02	
EN ISO 17225-2 A2	≥16.50	≥600	≤10.00	≥97.50	≤1.20	≤0.50	≤0.05	≤0.02	
EN ISO 17225-2 B	≥16.50	≥600	≤10.00	≥96.50	≤2.00	≤1.00	≤0.05	≤0.03	

The production of forest residue pellets from the thinning of Mediterranean pine stands (*Pinus halepensis* Mill. and *Pinus pinaster* Aiton), which is generally an uneconomic intervention [48], generally showed a good quality according to ISO standards. In particular, the highest quality was shown for pellets from larger diameter logs and debarked ones, while pellets from branches showed a lower quality [49]. This aspect is linked to the higher percentage of bark present in branches which has negative influence on ash content.

Another interesting raw material for pellet production is beech (*Fagus sylvatica* L.) wood. Pellets produced from this species showed very good quality, reaching the A1 standard for most parameters and A2 for ash content [50]. This could be very interesting for the beech wood value chain, considering the large number of beech forests throughout Europe and taking into consideration that the possibilities of this forest-wood chain are currently not being completely exploited [51]. Obviously, economic evaluations are needed to put into practice what is written above.

The major problem for other types of forest residual biomass seemed to be an excessive ash content. Chestnut (*Castanea sativa* Mill.) pellets only reached the B class ISO standard [52], and the same was observed for birch (*Betula spp*) sawdust [53]. An even greater ash content was shown by willow (*Salix spp*), poplar (*Populus spp*), and scots pine (*Pinus sylvestris* L.) bark pellet [50].

In relation to the pellet quality from tropical wood species, Parra Artemio et al. [54] investigated pellet samples from *Albizia amara* Roxb., *Ebenopsis ebano* (Berland.) Barneby, and *Havardia pallens* Benth. The results were very interesting with good results for ash content, durability, and calorific value. However, other tests are needed to improve the knowledge on the pellet quality of tropical woody species considering the importance that the establishment of a pellet value chain could have for the economy of developing countries in such areas of the world.

Another interesting type of feedstock for pellet production could be shrub wood. This is mainly of interest for arid or semi-arid zones, where, after wildfire prevention interventions, there could be a consistent amount of biomass from shrubby species [55].

By focusing on pellet quality from shrubby biomass, it is possible to notice the generally good quality in terms of bulk density, durability, and LHV. Instead, the main problems are linked to the excessive or consistent ash content as well as excess sulfur and chlorine concentrations [52,55]. The best pellet quality was shown by *Rhododendron ponticum* L. and *Genista cinerascens* L. [52,55]. Additionally, in this case, the major cause of the high ash content is probably linked to the high percentage of bark present in such species as a consequence of low stem diameter. A possible solution to using such feedstock for pellet production, thus evaluating a raw material whose disposal after forest restoration intervention is actually a cost, could be a certain percentage of blending with softwood sawdust in order to reduce the ash content.

The analysis of pellets from energy crops indicated a generally good quality in relation to the ISO standard for agropellets. Aragon-Garita [56] investigated the quality of various energy crops in Costa Rica, in particular *Gynerium sagittatum* Aubl., *Phyllostachys aurea* Riviere, *Arundo donax* L., *Pennisetum purpureum* Schumach, and *Sorghum bicolor* L. All pellets reached the requirements of the quality standard for LHV and almost every type, except *Arundo donax* L., also showed good quality in terms of ash content. However, *Pennisetum purpureum* Schumach and *Sorghum bicolor* L. contained excessive moisture content. The aspect of excessive moisture is a key issue for herbaceous energy crops management and the possible solutions are, on the one hand, a well-planned supply chain which ensures a drying period in optimal conditions, thus avoiding dry matter losses, and on the other hand the usage of harvesting systems which allow simpler on-field drying, for example mower-conditioners or dedicated headers. Another study highlighted the reasonably good performance of the Reed Canary Grass (*Phalaris arundinacea* L.) pellet, which reached an overall quality class of B according to ISO 17225-6 [50]. Finally, the *Jatropha curcas* L. pellet showed good results concerning bulk density and LHV but had a slightly excessive amount of moisture [57,58]. However, deeper investigation is needed to assess the overall feasibility of using this feedstock for pellet production.

Considering the pellet quality from agricultural residue feedstock, the first thing that can be observed is that it represents the topic of the majority of the investigated papers. This shows the great importance given by scientific research to the issue of agricultural residue valorization through densification and use as biofuels [59–62].

One of the most frequently investigated feedstocks is rice (*Oryza* spp.) cultivation residue [63], particularly straw, husk, and chaff. Yang et al. [64] analyzed pellets from straw and husks and detected excess chlorine in both types as well as excessive ash in the rice husk pellet. The same problems of an excessive ash content, together with low durability and low LHV, were detected for rice husk pellets by Rios-Badran et al. [65]. Regarding rice chaff, problems related to durability and an excess chlorine content were shown [66].

Corn (*Zea mays* L.) residue is another feasible feedstock for pellet production. Analysis conducted in the last five years showed a generally good quality for the pellets produced from corn cobs (obviously considering agropellet standards as a reference) but with slightly conflicting evidence. In fact, Djatkov [67] showed good quality in terms of durability and mechanical resistance parameters, while Miranda [68] highlighted the presence of excess chlorine and a low durability. Corn stover seems to be a feedstock with lower potential [67].

As a general comment to the reported above issue of pellets production from agricultural residues, it is important to underline that it is probably almost impossible to improve the quality of pellet produced from such feedstocks "as is" because of the intrinsic characteristics of these materials. On the other hand, it is fundamental to value these residues considering the large amount of them and the need for renewable energy production. Pelletization represents a valid solution to reach this aim, since one of the main problems related to residues is the low bulk density. Scientific efforts are needed in order to develop plants able to face the main flaws of such kinds of pellet also in the long run. Another possible solution is, even in this case, blending with materials with higher lignin content such as coniferous wood. Finally, a key issue could be developing and/or give major space to technologies able to produce pellet directly on field, in order to further lower supply chain costs.

Some commonly investigated feedstocks in the reference period (2016–2020) that did not perform well are oil palm (*Elaeis* spp.) residues, such empty fruit bunches, fruit mesocarps, leaves, and fronds. All of these feedstocks showed problems linked with their ash content, durability, bulk density, and LHV [25,69,70].

Coffee residue (spent coffee grounds and coffee husks) could be another possible suitable material for pelletization, but more studies of this feedstock are needed. Jeguirim et al. [71] analyzed spent coffee ground and coffee husk pellets, finding that they were high quality in terms of moisture, ash content, and LHV. However, Park et al. [66] reported a consistent lack in quality regarding the bulk density, durability, chlorine content, and copper content of spent coffee ground pellets. These controversial results claim for further scientific investigation on this field, in order to better define the possibility of using this important residue for energy production.

Pelletization of another important agriculture residue, biomass from pruning operations, was investigated by several authors in the reference period (2016–2020). The main problems with this type of biomass, in particular for olive grove and vineyard pruning, are its excessive ash content and, most of all, its excessive copper content, which is probably linked to the use of phytosanitary treatments with copper-based products [72,73]. Apple pruning seems, instead, to be a better material for pelletization, with an ash content and LHV that met the ISO standards; however, problems linked to a low bulk density have been detected [74]. Regarding the issue of pellet production from pruning residues, a key aspect to be further investigated is the economical sustainability of the supply chain, considering that pruning collection represents a certain cost by itself [75,76], also in this case an interesting solution could be the development of mobile technologies for pellets production at the farm, thus limiting supply chain costs [73].

Finally, many papers published in the last 5 years have focused on the use of other alternative residual biomass for pelletization. In order to give the reader a comprehensive view of these, Table 2 shows the main positive and negative aspects of each of these alternative feedstocks.

Table 2. Main postive and negative aspects found in the literature for relatively new feedstocks for pellet production.

Feedstock	Reference	Positive Aspects	Negative Aspects
Wheat straw	[50]	Lower Heating Value (LHV), ash, all chemical parameters except for chlorine	Chlorine
Scenedesmus microalgae	[68]	Durability and LHV	Ash, Nitrogen, Chlorine
Garden waste	[77]	LHV	Ash, Moisture
Soybean	[78]	—	Ash, LHV (very important shortfall)
Sugarcane bagasse	[78]	—	Ash, LHV (very important shortfall)
Cherry stones	[79]	LHV, all chemical parameters	Bulk density
Fallen leaves	[80]	Durability	Bulk density, LHV
Chamomile wastes	[53]	LHV, and Ash (mostly LHV)	Bulk density

As it is possible to see in Table 2, the main problems for the majority of these materials are their ash content (microalgae, garden wastes, soybean, sugarcane bagasse) and bulk density (cherry stones, fallen leaves and chamomile wastes); however, some of them also show a shortfall related to their LHV (soybean, sugarcane bagasse, fallen leaves) and chlorine content (wheat straw, microalgae).

Regarding feedstocks that present shortfalls related to physicochemical variables, for example, LHV or ash content, it is difficult to improve these negative aspects without using binders, blending, or pre-/post-treatments. In contrast, for feedstocks that are lacking in quality in relation to ISO standards in terms of mechanical characteristics, it could be possible to improve these aspects by enhancing the pelletization process, for example, by applying major pressure or a different temperature. Therefore, it could be interesting to produce cherry stone pellets or chamomile waste under different pelletization conditions and test whether the bulk density of the obtained material meets the quality standards.

It is interesting to note that considering raw materials mainly as "watertight compartments", and analyzing in detail their energetic characteristics, was the first step in the production of high-quality pellets. Initially, blending these raw materials for the production of high-quality pellets was a topic overlooked by researchers and more developed at a technical and commercial level. Current research developments in the sector allowed integrated analyzes aimed at the production of high-quality pellets. The data presented in the next chapter show the important results reached transferring the results of pure research to technical applications.

6. Effects of Blending and Binders on Pellet Quality

One of the possible solutions to improve pellet quality is co-pelletization, i.e., blending two or more feedstocks. The general trend is to use woody materials, for example sawdust, to improve the overall quality of pellets from alternative feedstocks [29]. An increase in the lignin content of wood is associated with an improvement in pellet quality, mostly regarding the heating value, bulk density, durability, and ash content [29,81].

In recent years, co-pelletizing has been one of the most commonly investigated methods. Mixing reed canary grass, timothy hay (*Phleum pratense* L.), and switchgrass (*Panicum virgatum* L.) with pine and spruce sawdust resulted in good quality pellets. In particular, an increase in LHV and reduced ash, chlorine, and nitrogen contents were detected in comparison to pure herbaceous feedstocks which also required only 2 kJ of energy for the pelletization almost as low as woody biomass [82]. Mixture of pine and spruce sawdust with bamboo improved the ash content (lower than 8%) and bulk density which increased from 0.54 to 0.60 g·cm^{-3} despite the slight reduction in LHV [83]. Regarding blending ratios, the addition of 30%–40% pine or spruce sawdust consistently improved wheat straw and maize pellets [10,72].

Higher proportions of pine sawdust are necessary for other feedstocks. Garcia et al. [84] mixed pine sawdust with many agricultural alternative raw materials and evaluated the quality of the obtained pellet according to ISO standards for industrial pellets. Blends of pine sawdust with almond shell and olive stone contents of up to 30 wt%, as well as a pine cone leaf content of up to 15 wt% produced I1 pellets. Blends of pine sawdust with coffee dregs, coffee husks, and grape pomace proportions of up to 10 wt%; blends with hazelnut shell, miscanthus, pine kernel shell, and switchgrass contents of up to 15 wt%; and blends with a pine cone leaf content of between 15 and 30 wt% generated I3 class pellets. Classification was not possible for cocoa shells mixed with pine sawdust due to the low bulk density of the pellets [84].

In addition to coniferous species, birch, osier, and alder wood showed good performances following blending; specifically, they were mixed with reed canary grass to improve the LHV of the produced pellet [85]. Less satisfactory results were reported for aspen wood [85].

Olive pruning residual biomass is another woody material that was shown to be feasible for the improvement of pellet quality. Mixing pruning residues with 25% stage 2 olive pomace and 50% stage 3 olive pomace produced a good quality pellet. In particular, a consistent increase in durability and a lower ash content (up to 2.4%) were shown in comparison with pellets from pure olive pomace although the extremely high values of ash-forming elements such as Fe, Mg, and K [86].

Interesting findings were reported by Hosseinizad et al. [87] and Cui et al. [88] regarding the blending of woody biomass with *Chlorella* spp. microalgae. Mixing microalgae with sawdust resulted in a consistently lower energy requirement for the pelletization process and a substantial increase in the mechanical characteristics of pellets.

Regarding non-woody blends, recent scientific findings have reported a generally low efficiency in comparison to woody–non-woody blends. Wang et al. [89] reported an increase only in mechanical characteristics by mixing wheat and rice straw, while the LHV did not improve substantially. Similar results were presented by Lisowski et al. [90] in relation to hay and straw mixing—a mechanical improvement without positive effects on LHV.

Binders are organic or inorganic substances that can be added, generally in a lower quantity, to feedstock for pelletization, in order to improve the mechanical characteristics of the pellet. In addition, binders can reduce the energy required for pelletization.

Sewage sludge is a possible binder, and scientific evidence has been reported regarding its capacity to improve the mechanical characteristics of pellets but, on the other hand, it increases the ash content from 4.28% to 13.01% when the fir–sludge ratio shifts from 25 to 50 wt% [91]. At the same straw–sludge ratios, the residues increase from 15.86 to 20.04 wt%.

Abedi et al. [92,93] used lignin and proline as binders for oat (*Avena sativa* L.) pellets and showed a positive effect on the LHV but a slightly negative effect on the ash content. The use of oat hull as cobinder at the concentrations of 10% and 50%, increases the HHV of sawdust pellet by 0.8 MJ/kg [93].

Another interesting bio-oil for pelletization is apple tree pyrolysis, which has been shown to improve the hydrophobicity of pellets. In fact, moisture content varied from 0% to 4.74% in 30 hours test [94].

Potato starch is another common binder that can reduce the energy needed for pellet formation; on the other hand, it increases the moisture content and substantially decreases the LHV [95,96]. Concentrations of 10%, 20%, and 30% provided ash content of 1.45%, 1.50%, and 1.59% and calorific value of 18.2, 18.1, and 18.0 MJ kg^{-1}, respectively.

The effects of paraffin, corn starch, and dolomite on the quality of wheat straw pellets were investigated by Gageanu et al. [97,98]. They found beneficial effects of these binders on the pellet length, surface, and shape. These additives were also beneficial for decreasing the moisture content. The bulk density significantly increased for samples obtained using additives. LHV registered a small decrease when paraffin and corn starch were used as additives to wheat straw pellets. The ash content was also positively influenced by using additives.

Other binders which have been shown to increase the mechanical characteristics of pellets are sugar beet molasses [99], Persea kurzii kosterm powder [100], carboxymethyl cellulose [101], calcium carbonate [102], and cow dung [103]. Cashew nut shell decreases the mechanical characteristics but increases the LHV of pellet of approximately 1 and 0.5 cal·g^{-1} in comparison with the use of Persea and dammar as binders, respectively. On the other hand, the use of cashew nut shells causes the production of unpleasant smoke and tar deposits [100].

Gerhig et al. [104] tested the use of kaolin to improve the quality of pellets from willow using short rotation coppices. Kaolin did not have a positive influence on the LHV or the ash content. Conflicting results were reported for chemical characteristics with decreases in sulfur and chlorine but substantial increases in Al and Pb. Moreover, CO emissions decreased, but SO_2 emissions were significant.

Finally, Cheng et al. [27] analyzed the use of coal tar (CTR) as binder. They reported a better mechanical resistance and an increased LHV of 19.32, 21.35 and 21.00 MJ kg^{-1} in wheat straw pellets, sawdust pellets and moso bamboo pellets, respectively, when 35 wt% CTR was applied. At the same concentration, also LHV of lignite pellets increased from 13.52 (without CTR) to 18.61 MJ·kg^{-1}.

Obviously, a proximate analysis is needed for a deeper evaluation of this binder.

An overall summary on the literature findings about the effects of blending or binders on pellet quality is given in Table 3.

From Table 3, it is possible to see that the majority (71.9%) of analyses conducted in the last years focused on only four variables (LHV, ash content, bulk density, and durability). Table 3 also reveals that either blending or binder addition led to an increase in mechanical characteristics in all cases. However, the results regarding the improvement of ash content and, particularly, LHV were not always positive.

Finally, it is necessary to give some information about the impact of blending and binding processes on pellets production costs. As found by Garcia [105] adding 20% glycerol to torrefied pine pellet increased production cost from 142.50 to 237 EUR t^{-1}. However, a decrease of 10%–20% of storage and transport costs was detected, and this aspect is very interesting mostly for industrial use of pellet [105]. On the other hand natural blending of wooden biomass and agricultural residues seems to be a solution with lower costs, in particular a mixture of 50% maize residues and 50% *Pinus radiata* sawdust showed 42.8% lower production costs than pure pine pellet [72]. The research has reached a high degree of improvement but now it will be necessary, in the qualification of the pellet, to consider also parameters related to the sustainability of production, the only real way to evaluate management systems, feedstock, and treatments in an integrated way.

Table 3. Overall summary of the effect of blending or binders on pellet quality. Green cue balls indicate a positive effect of the blending/binder on the variable, red cue balls indicate a negative effect on the variable, and white cue balls represent that the variable was not investigated in the study or that no effect was found by the authors. LHV: lower heating value; BD: bulk density; Dur: durability; Ash: ash content; RE: energy required for pelletization; Hyd: hydrophobicity; Moi: moisture. "Analysis %" represents the ratio between the number of papers that analyzed variable "x" and the total number of analyses conducted on all variables in every investigated paper. "Success %" represents the ratio between the number of papers that showed a good effect of the blend/binder on variable "x" and the total number of papers that analyzed variable "x".

Reference	Treatment		LHV	BD	Dur	Ash	RE	Hyd	Moi	Cl	N	S	Al	Pb	CO	SO_2
	Blending	Binding														
[27]		X	🟢	🟢	🟢	○	○	○	○	○	○	○	○	○	○	○
[82]	x		🟢	○	○	🟢	🟢	○	○	🟢	🟢	○	○	○	○	○
[83]	x		🔴	🟢	○	🟢	○	○	○	○	○	○	○	○	○	○
[85]	x		🟢	○	○	🟢	○	○	○	🟢	○	○	○	○	○	○
[86]	x		○	○	🟢	🟢	○	○	○	○	○	○	○	○	○	○
[87]	x		○	🟢	🟢	○	🟢	○	○	○	○	○	○	○	○	○
[88]		X	○	🟢	🟢	○	🟢	○	○	○	○	○	○	○	○	○
[89]	x		🔴	🟢	🟢	○	○	○	○	○	○	○	○	○	○	○
[90]		X	🔴	🟢	🟢	○	○	○	○	○	○	○	○	○	○	○
[91]	x		○	🟢	🟢	🔴	○	○	○	○	○	○	○	○	○	○
[92]	x		🟢	○	○	🔴	○	○	○	○	○	○	○	○	○	○
[93]	x		🟢	○	○	🔴	○	○	○	○	○	○	○	○	○	○
[94]		X	○	○	○	○	○	🟢	○	○	○	○	○	○	○	○
[95]	x		🔴	○	○	○	🟢	○	🔴	○	○	○	○	○	○	○
[96]		X	🔴	○	○	○	🟢	○	🔴	○	○	○	○	○	○	○
[97]		X	🔴	🟢	○	🟢	○	○	🟢	○	○	○	○	○	○	○
[98]		X	🔴	🟢	○	🟢	○	○	🟢	○	○	○	○	○	○	○
[99]		X	○	🟢	○	○	○	○	○	○	○	○	○	○	○	○
[100]		X	○	🟢	🟢	○	○	○	○	○	○	○	○	○	○	○
[101]		X	○	🟢	🟢	○	○	○	○	○	○	○	○	○	○	○
[102]		X	○	🟢	🟢	○	○	○	○	○	○	○	○	○	○	○
[103]		X	○	🟢	🟢	○	○	○	○	○	○	○	○	○	○	○
[104]	x		🔴	○	○	🔴	○	○	○	🟢	○	🟢	🔴	🔴	🟢	🔴
Analysis %			19.4	20.9	16.4	14.9	7.5	1.5	6.0	4.5	1.5	1.5	1.5	1.5	1.5	1.5
Success %			30.8	100.0	100.0	60.0	100.0	100.0	50.0	100.0	100.0	100	0.0	0.0	100.0	0.0

7. Pre- and Post-Treatments

The last section of the present review deals with the influences of pre- and post-treatments on the pellet quality. In this review, only one paper about steam explosion treatment influence was detected.

The mentioned above study analyzed, in detail, how steam treatment changes the biomass structure in order to determine the parameter with the biggest influence on pellet improvement [106]. The authors found that the structural changes caused by steam treatment, such as lignin relocation, hemicellulose hydrolysis, and size reduction, all aid in particle binding during pelletization. However, they have different contributions. In particular, the most important effect of steam treatment is lignin modification, while particle size distribution changes are not so important in improving the pelletization process and pellet quality [106]. About economic aspects of steam explosion treatment Pirraglia et al. [107] found an impact on overall production costs of 13%.

Consistently, significant attention has been given to the torrefaction process, particularly as a pre-treatment. Feedstock torrefaction was very effective in improving pellet parameters, mostly those linked with energetic efficiency [108] but also mechanical characteristics [109,110]. Pellets produced from torrefied biomass were of a high quality according to ISO standards for industrial pellets when this treatment was applied to pruning residual biomass of olives and almonds. In particular, these pellets had an LHV that was 25%–30% higher than that of raw biomass and reached I1 or I2 classes for size, moisture, bulk density, and heating value, but only obtained a classification of I3 for ash content [111].

This feedstock treatment was useful for improving the quality of pellets from blended material or from feedstock with the use of binders. Torrefied wood pine, with 20 wt.% glycerol and with 10 wt.% grape pomace and 10 wt.% glycerol met the quality standards for industrial pellet [105].

Less satisfactory results were shown in the reference period (2016–2020) for the use of torrefaction as a post-treatment. Manoucherijizad et al. [112] showed this post-treatment at a temperature range from 230 to 290 °C improved the heating value and hydrophobicity of the torrefied wood pellets as compared with raw wood pellets. On the other hand, the hardness and durability of these pellets need to be improved to prevent dust formation during long-distance transport and safe storage operations.

Finally, it is important to notice that several recent studies focused on the issue of torrefaction process improvement. One possible solution is the use of ultrasonic vibration when pelletizing torrefied biomass. Song et al. [113] found that torrefied wheat straw biomass could be densified into good quality pellets with the assistance of ultrasonic vibration, whereas with the same pelleting pressure but without ultrasonic vibration, good pellets could barely be made.

Another useful way to improve the performance of this pre-treatment is to use "pressurized steam torrefaction". Kudo et al. [114] carried out the torrefaction of hardwood at 180–250 °C in the presence of saturated steam. This treatment considerably improved the pelletability of biomass, producing pellets with a tensile strength 5.2 times higher than that of the original biomass. This improvement in pelletability was comparable to that of wet torrefaction (process in compressed water). Meanwhile, pressurized steam torrefaction showed higher energy densification than wet process [114].

An interesting approach which was shown to be a feasible way to improve the quality of pellets produced from feedstocks affected by an excessive ash content is the combination of water leaching and torrefaction.

In 2019, Gong et al. [115] applied a first step of water leaching on empty fruit bunches from oil palm, rice straws, and sugarcane bagasse. Subsequently, torrefaction was performed at 200 °C for 5 min. Leaching allowed the removal of the majority of ash and also practically reduced the chlorine and potassium concentrations. After this treatment, heating values increased by 4.42% in the empty fruit bunches, 4.68% in the rice straw, and 5.30% in the sugarcane bagasse [115].

The last method to improve the efficiency of this process, among those investigated by scientific research in the last five years, is the application of binders to biomass before torrefaction and pelletization. According to Rejdak et al. [116], a very good performance was shown when modified wheat starch was used as a binder, which resulted in improvements in the mechanical characteristics of torrefied pellets. The same author found that a blend of natural wheat starch, molasses, and sodium lignosulfonate resulted in a pellet with good efficiency with slightly lower mechanical characteristics in comparison to one produced with modified wheat starch [116].

Focusing on the economic assessment of the torrefaction process of densified biomass, literature findings reported that the torrefaction process increases the production cost by about 10%. Pellet production costs indeed 47-64 EUR t^{-1} using sawdust as feedstock [117] and the price rises to 88–160 or 50–70 EUR t^{-1} starting from forest residues [118–121]; while prices for torrefied pellet production are in the range of 136–169 EUR t^{-1} [122–124]. On the other hand, it is important to underline that cost per energy unit of torrefied pellet can be lower than "normal" pellet's one. Indeed Yun et al. [125] reported a price ranging from 14.10 to 17.05 EUR GJ^{-1} for not treated pellet's production, while 11.87–13.71 EUR GJ^{-1} for torrefied pellet. In particular, the torrefied process with the best economic performance resulted to be torrefaction before grinding.

Considering what showed in this chapter, the two main (pre or post) treatments that currently seem to produce real quality improvements are the steam explosion system and torrefaction. Assessing the data reported in the cited researches, starting with the same qualitative characteristics of the pellet produced, in terms of sustainability, torrefaction is the most effective treatment. However, doubts remain about the real scale production chains that could be activated with these treatment methods and their efficacy and sustainability in comparison with traditional pelletization.

8. Some Ideas for Future Research Directions

When it comes to be consistent with a range of qualitative aspects related to pellet, the quality of starting material is crucial to the final characteristics of the product. Therefore, minimum standards of quality are strongly required in order to provide the market with a reliable product. Unfortunately, this threshold might not be met in future because of the ongoing changes in the field of renewable energy. In fact, according to the most recent regulations set by the European policy making (cit RED2), the contribution of agricultural residues to the production of bioenergy will be enhanced, so the use of non-woody biomass could be more and more relevant. Therefore, the future scenario of pelletization may change alongside the different kind of biomass available, thus a better understanding about the effects of using blends and binders to enhance the quality of pellet is fundamental. However, the whole binding mechanisms involved in the process result wide and difficult to be comprehended and properly managed. So far, very few efforts have been dedicated for the investigation of relationship between particle binding and pellet quality, which represents a gap in the literature demanding for study.

Remaining focused on the initial theme, there is a necessity to develop new mathematical models to assess the performance of pellets under different operating conditions. Currently, researchers found significant prediction model equations, but they are not dimensionally homogenous and cannot be generalized and used for further new materials or the production of new shapes of pellets. In line with this, proposing new significant empirical equations that describe any property of the pellet as a function of other features is an essential issue for pellets production improvement, especially if related to the original feedstock. This could be achieved by predicting functional relationships between the different physical and mechanical characteristics of pellets and the original raw materials, on the basis of the regression analysis of the experimental data.

Another emergent topic in tune with the relation between the pellet quality and its raw origin concerns the necessity to understand the influence of the diverse biomass feedstocks, mixed biomass with non-biodegradable wastes such as plastic, pre-treatment methods and the interaction of process parameters on the fuel pellet quality. Considering the potential markets of "biomass" as a substituent to wood and fossil fuels, future researches should be addressed to the assessment of fuel pellets quality and environmental impacts due to its production and use.

In accordance with this last point, and consistent with the other treated issues and concerns about biomass for energy use, the research should try to broaden the analyzes, increasingly focusing at life cycle scale, with the aim to include all the sustainability pillars as much as possible. This is one of the main objectives and challenges for researchers, called upon to clarify and understand how to approach to the stakeholders involved in the fuel biomass supply chain worldwide.

As highlighted in the other chapters, it would be considered useful and appropriate in the pellet qualification to add some indicators or indices linked to the sustainability of production, parameters that more and more often could make the difference to guide the choice of the global market or of the individual consumer.

Another aspect related to pellet qualification is a proper mechanical characterization. Studies on the compressive mechanical strength of pellets, and how this relates to bioenergy storage, transport, and related processes, are necessary. Low mechanical characteristics can affect pellet integrity during transport and handling, this could be key in ensuring a standardized product for processing. This minimizes transport costs and reduces the risk of fires through dust explosions.

9. Conclusions

This review focused on the most recent (2016–2020) scientific contributions regarding wood or agropellet quality. Only a few cases referred to papers published before 2016. The issue of good quality is fundamental for every biofuel and is probably even more important for pellets because of the large number of different possible feedstocks for pelletization.

It is possible to summarize the main findings of this review as follows:

1. Very few studies have investigated the relationship between different agroforest management systems and the quality of obtained pellets. This could be interesting focus for future research to give a better understanding of this topic.

2. As reported in many other studies, including the most recent ones, wood pellets have a higher quality than agropellets, particularly in terms of their bulk density, ash content, heating value, and chemical composition.
3. However, agropellets are an interesting way to valorize agricultural waste—mostly for industrial use.
4. Blending and using binders are possible methods to improve pellet quality, but their use must be evaluated on a case-by-case basis.
5. There have been very interesting findings regarding the blending of woody biomass and microalgae, showing a consistent improvement in pellet quality.
6. Torrefaction seems to be the most investigated treatment to improve pellet quality, and recent studies tried to further improve such processes (ultrasonic torrefaction, pressurized steam torrefaction).

Author Contributions: Conceptualization, R.P. and F.L.; methodology, R.P., F.L., R.V.; formal analysis, F.L., W.S., A.S., D.T.; investigation, R.P., F.L.; resources, R.P., L.P.; data curation, F.L., W.S., A.S., R.V.; writing—original draft preparation, R.P., F.L., L.P.; writing—review and editing, R.P., F.L., R.V., L.P.; supervision, R.P., L.P.; funding acquisition, F.L., W.S., A.S., L.P. All authors have read and agreed to the published version of the manuscript.

Funding: The work was performed in the framework of the European project AGROinLOG "Demonstration of innovative integrated biomass logistics centers for the Agro-industry sector in Europe". This project has received funding from the European Union's Horizon 2020 research and innovation program under Grant Agreement No 727961.

Acknowledgments: This work was supported by the Italian Ministry for education and the University and Research (MIUR) for financial support (Law 232/2016, Italian University Departments of excellence)—UNITUS-DAFNE WP3.

Conflicts of Interest: The authors declare no conflict of interest. The funders had no role in the design of the study; in the collection, analyses, or interpretation of data; in the writing of the manuscript, or in the decision to publish the results.

References

1. Civitarese, V.; Acampora, A.; Sperandio, G.; Assirelli, A.; Picchio, R. Production of wood pellets from poplar trees managed as coppices with Different harvesting cycles. *Energies* **2019**, *12*, 2973–2988. [CrossRef]
2. Verma, V.K.; Bram, S.; De Ruyck, J. Small scale biomass heating systems: Standards, quality labelling and market driving factors—An EU outlook. *Biomass Bioenergy* **2009**, *33*, 1393–1402. [CrossRef]
3. Kaliyan, N.; Vance Morey, R. Factors affecting strength and durability of densified biomass products. *Biomass Bioenergy* **2009**, *33*, 337–359. [CrossRef]
4. Nunes, L.J.R.; Matias, J.C.O.; Catalão, J.P.S. Mixed biomass pellets for thermal energy production: A review of combustion models. *Appl. Energy* **2014**, *127*, 135–140. [CrossRef]
5. Puig-Arnavat, M.; Shang, L.; Sárossy, Z.; Ahrenfeldt, J.; Henriksen, U.B. From a single pellet press to a bench scale pellet mill—Pelletizing six different biomass feedstocks. *Fuel Process. Technol.* **2016**, *142*, 27–33. [CrossRef]
6. Rentizelas, A.A.; Tolis, A.J.; Tatsiopoulos, I.P. Logistics issues of biomass: The storage problem and the multi-biomass supply chain. *Renew. Sustain. Energy Rev.* **2009**, *13*, 887–894. [CrossRef]
7. Sánchez, J.; Curt, M.D.; Sanz, M.; Fernández, J. A proposal for pellet production from residual woody biomass in the island of Majorca (Spain). *AIMS Energy* **2015**, *3*, 480. [CrossRef]
8. Whittaker, C.; Shield, I. Factors affecting wood, energy grass and straw pellet durability—A review. *Renew. Sustain. Energy Rev.* **2017**, *71*, 1–11. [CrossRef]
9. Gilbert, P.; Ryu, C.; Sharifi, V.; Swithenbank, J. Effect of process parameters on pelletisation of herbaceous crops. *Fuel* **2009**, *88*, 1491–1497. [CrossRef]
10. Stasiak, M.; Molenda, M.; Bańda, M.; Wiącek, J.; Parafiniuk, P.; Gondek, E. Mechanical and combustion properties of sawdust—Straw pellets blended in different proportions. *Fuel Process. Technol.* **2017**, *156*, 366–375. [CrossRef]
11. Zhou, Y.; Zhang, Z.; Zhang, Y.; Wang, Y.; Yu, Y.; Ji, F.; Ahmad, R.; Dong, R. A comprehensive review on densified solid biofuel industry in China. *Renew. Sustain. Energy Rev.* **2016**, *54*, 1412–1428. [CrossRef]

12. Kaliyan, N.; Morey, R.V. Natural binders and solid bridge type binding mechanisms in briquettes and pellets made from corn stover and switchgrass. *Bioresour. Technol.* **2010**, *101*, 1082–1090. [CrossRef] [PubMed]
13. Van der Stelt, M.J.C.; Gerhauser, H.; Kiel, J.H.A.; Ptasinski, K.J. Biomass upgrading by torrefaction for the production of biofuels: A review. *Biomass Bioenergy* **2011**, *35*, 3748–3762. [CrossRef]
14. Calderòn, C.; Colla, M.; Jossart, J.-M.; Hemeleers, N.; Cancian, G.; Aveni, N.; Caferri, C. Bioenergy Europe, Statistical report on pellet. In Proceedings of the European Biomass Conference and Exhibition, Stockholm Sweden, 12–15 June 2017.
15. Samuelsson, R.; Thyrel, M.; Sjöström, M.; Lestander, T.A. Effect of biomaterial characteristics on pelletizing properties and biofuel pellet quality. *Fuel Process. Technol.* **2009**, *90*, 1129–1134. [CrossRef]
16. Larsson, S.H.; Thyrel, M.; Geladi, P.; Lestander, T.A. High quality biofuel pellet production from pre-compacted low density raw materials. *Bioresour. Technol.* **2008**, *99*, 7176–7182. [CrossRef]
17. Lestander, T.A.; Finell, M.; Samuelsson, R.; Arshadi, M.; Thyrel, M. Industrial scale biofuel pellet production from blends of unbarked softwood and hardwood stems—the effects of raw material composition and moisture content on pellet quality. *Fuel Process. Technol.* **2012**, *95*, 73–77. [CrossRef]
18. Nizamuddin, S.; Mubarak, N.M.; Tiripathi, M.; Jayakumar, N.S.; Sahu, J.N.; Ganesan, P. Chemical, dielectric and structural characterization of optimized hydrochar produced from hydrothermal carbonization of palm shell. *Fuel* **2016**, *163*, 88–97. [CrossRef]
19. Toscano, G.; Riva, G.; Pedretti, E.F.; Corinaldesi, F.; Mengarelli, C.; Duca, D. Investigation on wood pellet quality and relationship between ash content and the most important chemical elements. *Biomass Bioenergy* **2013**, *56*, 317–322. [CrossRef]
20. Ahn, B.J.; Chang, H.; Lee, S.M.; Choi, D.H.; Cho, S.T.; Han, G.; Yang, I. Effect of binders on the durability of wood pellets fabricated from Larix kaemferi C. and Liriodendron tulipifera L. sawdust. *Renew. Energy* **2014**, *62*, 18–23. [CrossRef]
21. Križan, P.; Matú, M.; Šooš, L'.; Beniak, J. Behavior of beech sawdust during densification into a solid biofuel. *Energies* **2015**, *8*, 6382–6398.
22. Stelte, W.; Sanadi, A.R.; Shang, L.; Holm, J.K.; Ahrenfeldt, J.; Henriksen, U.B. Recent developments in biomass pelletization–A review. *BioResources* **2012**, *7*, 4451–4490.
23. Picchio, R.; Spina, R.; Sirna, A.; Monaco, A.L.; Civitarese, V.; Giudice, A.D.; Suardi, A.; Pari, L. Characterization of woodchips for energy from forestry and agroforestry production. *Energies* **2012**, *5*, 3803–3816. [CrossRef]
24. Mostafa, M.E.; Hu, S.; Wang, Y.; Su, S.; Hu, X.; Elsayed, S.A.; Xiang, J. The significance of pelletization operating conditions: An analysis of physical and mechanical characteristics as well as energy consumption of biomass pellets. *Renew. Sustain. Energy Rev.* **2019**, *105*, 332–348. [CrossRef]
25. Tenorio, C.; Moya, R.; Valaert, J. Characterisation of pellets made from oil palm residues in Costa Rica. *J. Oil Palm Res.* **2016**, *28*, 198–210. [CrossRef]
26. Li, H.; Jiang, L.-B.; Li, C.-Z.; Liang, J.; Yuan, X.-Z.; Xiao, Z.-H.; Xiao, Z.-H.; Wang, H. Co-pelletization of sewage sludge and biomass: The energy input and properties of pellets. *Fuel Process. Technol.* **2015**, *132*, 55–61. [CrossRef]
27. Cheng, J.; Zhou, F.; Si, T.; Zhou, J.; Cen, K. Mechanical strength and combustion properties of biomass pellets prepared with coal tar residue as a binder. *Fuel Process. Technol.* **2018**, *179*, 229–237. [CrossRef]
28. Kijo-Kleczkowska, A.; Środa, K.; Kosowska-Golachowska, M.; Musiał, T.; Wolski, K. Combustion of pelleted sewage sludge with reference to coal and biomass. *Fuel* **2016**, *170*, 141–160. [CrossRef]
29. Azargohar, R.; Nanda, S.; Dalai, A.K. Densification of agricultural wastes and forest residues: A review on influential parameters and treatments. *Recent Adv. Biofuels Bioenergy Util.* **2018**, 27–51. [CrossRef]
30. Kumar, P.; Barrett, D.M.; Delwiche, M.J.; Stroeve, P. Methods for pretreatment of lignocellulosic biomass for efficient hydrolysis and biofuel production. *Ind. Eng. Chem. Res.* **2009**, *48*, 3713–3729. [CrossRef]
31. Ramos, L.P. The chemistry involved in the steam treatment of lignocellulosic materials. *Quim. Nova* **2003**, *26*, 863–871. [CrossRef]
32. Sun, Y.; Cheng, J. Hydrolysis of lignocellulosic materials for ethanol production: A review. *Bioresour. Technol.* **2002**, *83*, 1–11. [CrossRef]
33. Stelte, W.; Holm, J.K.; Sanadi, A.R.; Barsberg, S.; Ahrenfeldt, J.; Henriksen, U.B. Fuel pellets from biomass: The importance of the pelletizing pressure and its dependency on the processing conditions. *Fuel* **2011**, *90*, 3285–3290. [CrossRef]

34. Mei, Y.; Liu, R.; Yang, Q.; Yang, H.; Shao, J.; Draper, C.; Zhang, S.; Chen, H. Torrefaction of cedarwood in a pilot scale rotary kiln and the influence of industrial flue gas. *Bioresour. Technol.* **2015**, *177*, 355–360. [CrossRef] [PubMed]
35. Li, H.; Liu, X.; Legros, R.; Bi, X.T.; Lim, C.J.; Sokhansanj, S. Pelletization of torrefied sawdust and properties of torrefied pellets. *Appl. Energy* **2012**, *93*, 680–685. [CrossRef]
36. Kumar, L.; Koukoulas, A.A.; Mani, S.; Satyavolu, J. Integrating torrefaction in the wood pellet industry: A critical review. *Energy Fuels* **2017**, *31*, 37–54. [CrossRef]
37. Picchio, R.; Proto, A.R.; Civitarese, V.; Di Marzio, N.; Latterini, F. Recent Contributions of Some Fields of the Electronics in Development of Forest Operations Technologies. *Electronics* **2019**, *8*, 1465. [CrossRef]
38. García-Maraver, A.; Popov, V.; Zamorano, M. A review of European standards for pellet quality. *Renew. Energy* **2011**, *36*, 3537–3540. [CrossRef]
39. Alakangas, E. Solid biofuels for energy: A lower greenhouse gas alternative. *Green Energy Technol.* **2014**, *28*, 1–3.
40. Rhén, C.; Gref, R.; Sjöström, M.; Wästerlund, I. Effects of raw material moisture content, densification pressure and temperature on some properties of Norway spruce pellets. *Fuel Process. Technol.* **2005**, *87*, 11–16. [CrossRef]
41. Masche, M.; Puig-Arnavat, M.; Jensen, P.A.; Holm, J.K.; Clausen, S.; Ahrenfeldt, J.; Henriksen, U.B. From wood chips to pellets to milled pellets: The mechanical processing pathway of Austrian pine and European beech. *Powder Technol.* **2019**, *350*, 134–145. [CrossRef]
42. Monedero, E.; Portero, H.; Lapuerta, M. Combustion of poplar and pine pellet blends in a 50 kw domestic boiler: Emissions and combustion efficiency. *Energies* **2018**, *11*, 1580. [CrossRef]
43. Santos, L.B.; Striebeck, M.V.; Crespi, M.S.; Capela, J.M.V.; Ribeiro, C.A.; De Julio, M. Energy evaluation of biochar obtained from the pyrolysis of pine pellets. *J. Anal. Calorim.* **2016**, *126*, 1879–1887. [CrossRef]
44. Liu, Z.; Jiang, Z.; Fei, B.; Cai, Z.; Liu, X. Comparative properties of bamboo and pine pellets. *Wood Fiber Sci.* **2014**, *46*, 510–518.
45. Rabaçal, M.; Fernandes, U.; Costa, M. Combustion and emission characteristics of a domestic boiler fired with pellets of pine, industrial wood wastes and peach stones. *Renew. Energy* **2013**, *51*, 220–226. [CrossRef]
46. Stelte, W.; Clemons, C.; Holm, J.K.; Sanadi, A.R.; Ahrenfeldt, J.; Shang, L.; Henriksen, U.B. Pelletizing properties of torrefied spruce. *Biomass Bioenergy* **2011**, *35*, 4690–4698. [CrossRef]
47. Brackley, A.M.; Parrent, D.J. Production of wood pellets from Alaska-grown white spruce and hemlock. *Usda. Serv.* **2011**. [CrossRef]
48. Savelli, S.; Cavalli, R.; Baldini, S.; Picchio, R. Small scale mechanization of thinning in artificial coniferous plantation. *Croat. J. Eng.* **2010**, *31*, 11–21.
49. Lerma-Arce, V.; Oliver-Villanueva, J.V.; Segura-Orenga, G. Influence of raw material composition of Mediterranean pinewood on pellet quality. *Biomass Bioenergy* **2017**, *99*, 90–96. [CrossRef]
50. Agar, D.A.; Rudolfsson, M.; Kalén, G.; Campargue, M.; Da Silva Perez, D.; Larsson, S.H. A systematic study of ring-die pellet production from forest and agricultural biomass. *Fuel Process. Technol.* **2018**, *180*, 47–55. [CrossRef]
51. Picchio, R.; Spina, R.; Calienno, L.; Venanzi, R.; Lo Monaco, A. Forest operations for implementing silvicultural treatments for multiple purposes. *Ital. J. Agron.* **2016**, *11*, 156–161.
52. Gündüz, G.; Saraçoğlu, N.; Aydemir, D. Characterization and elemental analysis of wood pellets obtained from low-valued types of wood. *Energy Sources* **2016**, *38*, 2211–2216. [CrossRef]
53. Zawiślak, K.; Sobczak, P.; Kraszkiewicz, A.; Niedziółka, I.; Parafiniuk, S.; Kuna-Broniowska, I.; Tanaś, W.; Żukiewicz-Sobczak, W.; Obidziński, S. The use of lignocellulosic waste in the production of pellets for energy purposes. *Renew. Energy* **2020**, *145*, 997–1003. [CrossRef]
54. Artemio, C.P.; Maginot, N.H.; Serafín, C.U.; Rahim, F.P.; Guadalupe, R.Q.J.; Fermín, C.M. Physical, mechanical and energy characterization of wood pellets obtained from three common tropical species. *PeerJ* **2018**, *2018*, 1–16. [CrossRef] [PubMed]
55. Bados, R.; Esteban, L.S.; Pérez, P.; Mediavilla, I.; Fernández, M.J.; Barro, R.; Corredor, R.; Carrasco, J.E. Study of the production of pelletized biofuels from Mediterranean scrub biomass. In Proceedings of the European Biomass Conference and Exhibition, Stockholm Sweden, 12–15 June 2017.

56. Aragón-Garita, S.; Moya, R.; Bond, B.; Valaert, J.; Tomazello Filho, M. Production and quality analysis of pellets manufactured from five potential energy crops in the Northern Region of Costa Rica. *Biomass Bioenergy* **2016**, *87*, 84–95. [CrossRef]
57. Ramírez, V.; Martí-Herrero, J.; Romero, M.; Rivadeneira, D. Energy use of Jatropha oil extraction wastes: Pellets from biochar and Jatropha shell blends. *J. Clean. Prod.* **2019**, *215*, 1095–1102. [CrossRef]
58. Romuli, S.; Karaj, S.; Correa, C.R.; Kruse, A.; Müller, J. Physico-mechanical properties and thermal decomposition characteristics of pellets from Jatropha curcas L. residues as affected by water addition. *Biofuels* **2019**, 1–8. [CrossRef]
59. Alarcon, M.; Santos, C.; Cevallos, M.; Eyzaguirre, R.; Ponce, S. Study of the Mechanical and Energetic Properties of Pellets Produce from Agricultural Biomass of Quinoa, Beans, Oat, Cattail and Wheat. *Waste Biomass Valori* **2017**, *8*, 2881–2888. [CrossRef]
60. Burg, P.; Masan, V.; Ludin, D. Possibilities of using grape marc for making fuel pellets. *Eng. Rural Dev.* **2017**, *16*, 1333–1338.
61. Stelte, W.; Barsberg, S.T.; Clemons, C.; Morais, J.P.S.; de Freitas Rosa, M.; Sanadi, A.R. Coir Fibers as Valuable Raw Material for Biofuel Pellet Production. *Waste Biomass Valori* **2019**, *10*, 3535–3543. [CrossRef]
62. Royo, J.; Canalís, P.; Quintana, D.; Díaz-Ramírez, M.; Sin, A.; Rezeau, A. Experimental study on the ash behaviour in combustion of pelletized residual agricultural biomass. *Fuel* **2019**, *239*, 991–1000. [CrossRef]
63. Marrugo, G.; Valdés, C.F.; Gómez, C.; Chejne, F. Pelletizing of Colombian agro-industrial biomasses with crude glycerol. *Renew. Energy* **2019**, *134*, 558–568. [CrossRef]
64. Yang, I.; Kim, S.-H.; Sagong, M.; Han, G.S. Fuel characteristics of agropellets fabricated with rice straw and husk. *KoreanJ. Chem. Eng.* **2016**, *33*, 851–857. [CrossRef]
65. Ríos-Badrán, I.M.; Luzardo-Ocampo, I.; García-Trejo, J.F.; Santos-Cruz, J.; Gutiérrez-Antonio, C. Production and characterization of fuel pellets from rice husk and wheat straw. *Renew. Energy* **2020**, *145*, 500–507. [CrossRef]
66. Park, S.; Kim, S.J.; Oh, K.C.; Cho, L.; Kim, M.J.; Jeong, I.S.; Lee, C.G.; Kim, D.H. Investigation of agro-byproduct pellet properties and improvement in pellet quality through mixing. *Energy* **2019**, *190*, 116380. [CrossRef]
67. Djatkov, D.; Martinov, M.; Kaltschmitt, M. Influencing parameters on mechanical–physical properties of pellet fuel made from corn harvest residues. *Biomass Bioenergy* **2018**, *119*, 418–428. [CrossRef]
68. Miranda, M.T.; Sepúlveda, F.J.; Arranz, J.I.; Montero, I.; Rojas, C.V. Physical-energy characterization of microalgae Scenedesmus and experimental pellets. *Fuel* **2018**, *226*, 121–126. [CrossRef]
69. Wattana, W.; Phetklung, S.; Jakaew, W.; Chumuthai, S.; Sriam, P.; Chanurai, N. Characterization of Mixed Biomass Pellet Made from Oil Palm and Para-rubber Tree Residues. *Energy Procedia* **2017**, *138*, 1128–1133. [CrossRef]
70. Brunerová, A.; Müller, M.; Šleger, V.; Ambarita, H.; Valášek, P. Bio-pellet fuel from oil palm empty fruit bunches (EFB): Using European standards for quality testing. *Sustainability* **2018**, *10*, 4443.
71. Jeguirim, M.; Limousy, L.; Fossard, E. Characterization of coffee residues pellets and their performance in a residential combustor. *Int. J. Green Energy* **2016**, *13*, 608–615. [CrossRef]
72. Fernández-Puratich, H.; Hernández, D.; Lerma Arce, V. Characterization and cost savings of pellets fabricated from Zea mays waste from corn mills combined with Pinus radiata. *Renew. Energy* **2017**, *114*, 448–454. [CrossRef]
73. Toscano, G.; Alfano, V.; Scarfone, A.; Pari, L. Pelleting vineyard pruning at low cost with a mobile technology. *Energies* **2018**, *11*, 2477. [CrossRef]
74. Brand, M.A.; Jacinto, R.C. Apple pruning residues: Potential for burning in boiler systems and pellet production. *Renew. Energy* **2020**, *152*, 458–466. [CrossRef]
75. Suardi, A.; Latterini, F.; Alfano, V.; Palmieri, N.; Bergonzli, S.; Pari, L. Analysis of the Work Productivity and Costs of a Stationary Chipper Applied to the Harvesting of Olive Tree Pruning for Bio-Energy Production. *Energies* **2020**, *13*, 1359. [CrossRef]
76. Suardi, A.; Latterini, F.; Alfano, V.; Palmieri, N.; Bergonzoli, S.; Karampinis, E.; Kougioumtzis, M.A.; Grammelis, P.; Pari, L. Machine Performance and Hog Fuel Quality Evaluation in Olive Tree Pruning Harvesting Conducted Using a Towed Shredder on Flat and Hilly Fields. *Energies* **2020**, *13*, 1713–1728. [CrossRef]
77. Pradhan, P.; Mahajani, S.M.; Arora, A. Production and utilization of fuel pellets from biomass: A review. *Fuel Process. Technol.* **2018**, *181*, 215–232. [CrossRef]

78. Scatolino, M.V.; Neto, L.F.C.; Protásio, T.P.; Carneiro, A.C.O.; Andrade, C.R.; Guimarães Júnior, J.B.; Mendes, L.M. Options for Generation of Sustainable Energy: Production of Pellets Based on Combinations Between Lignocellulosic Biomasses. *Waste Biomass Valori* **2018**, *9*, 479–489. [CrossRef]
79. Dołżyńska, M.; Obidziński, S.; Kowczyk-Sadowy, M.; Krasowska, M. Densification and combustion of cherry stones. *Energies* **2019**, *12*, 1–15. [CrossRef]
80. González, W.A.; López, D.; Pérez, J.F. Biofuel quality analysis of fallen leaf pellets: Effect of moisture and glycerol contents as binders. *Renew. Energy* **2020**, *147*, 1139–1150. [CrossRef]
81. Peng, J.; Bi, X.T.; Lim, C.J.; Peng, H.; Kim, C.S.; Jia, D.; Zuo, H. Sawdust as an effective binder for making torrefied pellets. *Appl. Energy* **2015**, *157*, 491–498. [CrossRef]
82. Harun, N.Y.; Parvez, A.M.; Afzal, M.T. Process and energy analysis of pelleting agricultural and woody biomass blends. *Sustainability* **2018**, *10*, 1–9.
83. Liu, Z.; Mi, B.; Jiang, Z.; Fei, B.; Cai, Z.; Liu, X. Improved bulk density of bamboo pellets as biomass for energy production. *Renew. Energy* **2016**, *86*, 1–7. [CrossRef]
84. García, R.; Gil, M.V.; Rubiera, F.; Pevida, C. Pelletization of wood and alternative residual biomass blends for producing industrial quality pellets. *Fuel* **2019**, *251*, 739–753. [CrossRef]
85. Platace, R.; Adamovics, A.; Kalnacs, J. Engineering for Rural Development Production of Reed Canary Grass-Wood Pellets. *Eng. Rural Dev.* **2016**, 371–374.
86. Barbanera, M.; Lascaro, E.; Stanzione, V.; Esposito, A.; Altieri, R.; Bufacchi, M. Characterization of pellets from mixing olive pomace and olive tree pruning. *Renew. Energy* **2016**, *88*, 185–191. [CrossRef]
87. Hosseinizand, H.; Sokhansanj, S.; Lim, C.J. Co-pelletization of microalgae Chlorella vulgaris and pine sawdust to produce solid fuels. *Fuel Process. Technol.* **2018**, *177*, 129–139. [CrossRef]
88. Cui, X.; Yang, J.; Shi, X.; Lei, W.; Huang, T.; Bai, C. Experimental investigation on the energy consumption, physical, and thermal properties of a novel pellet fuel made from wood residues with microalgae as a binder. *Energies* **2019**, *12*, 3425. [CrossRef]
89. Wang, Y.; Wu, K.; Sun, Y. Pelletizing Properties of Wheat Straw Blending with Rice Straw. *Energy Fuels* **2017**, *31*, 5126–5134. [CrossRef]
90. Lisowski, A.; Matkowski, P.; Dąbrowska, M.; Piątek, M.; Świętochowski, A.; Klonowski, J.; Mieszkalski, L.; Reshetiuk, V. Particle Size Distribution and Physicochemical Properties of Pellets Made of Straw, Hay, and Their Blends. *Waste Biomass Valori* **2018**, *11*, 63–75. [CrossRef]
91. Jiang, L.; Yuan, X.; Xiao, Z.; Liang, J.; Li, H.; Cao, L.; Wang, H.; Chen, X.; Zeng, G. A comparative study of biomass pellet and biomass-sludge mixed pellet: Energy input and pellet properties. *Energy Convers. Manag.* **2016**, *126*, 509–515. [CrossRef]
92. Abedi, A.; Dalai, A.K. Study on the quality of oat hull fuel pellets using bio-additives. *Biomass Bioenergy* **2017**, *106*, 166–175. [CrossRef]
93. Abedi, A.; Cheng, H.; Dalai, A.K. Effects of Natural Additives on the Properties of Sawdust Fuel Pellets. *Energy Fuels* **2018**, *32*, 1863–1873. [CrossRef]
94. Zhang, T.; Qiu, L.; Wang, Y.; Zhang, C.; Kang, K. Comparison of Bio-Oil and Waste Cooking Oil as Binders during the Codensification of Biomass: Analysis of the Pellet Quality. *Bioenergy Res.* **2019**, *12*, 558–569. [CrossRef]
95. Obidziński, S.; Piekut, J.; Dec, D. The influence of potato pulp content on the properties of pellets from buckwheat hulls. *Renew. Energy* **2016**, *87*, 289–297. [CrossRef]
96. Obidziński, S.; Dołżyńska, M.; Kowczyk-Sadowy, M.; Jadwisieńczak, K.; Sobczak, P. Densification and Fuel Properties of Onion Husks. *Energies* **2019**, *12*, 4687.
97. Gageanu, I.; Cujbescu, D.; Persu, C.; Voicu, G. Impact of using additives on quality of agricultural biomass pellets. *Eng. Rural Dev.* **2018**, *17*, 1632–1638.
98. Gageanu, I.; Persu, C.; Cujbescu, D.; Gheorghe, G.; Voicu, G. Influence of using additives on quality of pelletized fodder. *Eng. Rural Dev.* **2019**, *18*, 362–367.
99. Mišljenović, N.; Čolović, R.; Vukmirović, D.; Brlek, T.; Bringas, C.S. The effects of sugar beet molasses on wheat straw pelleting and pellet quality. A comparative study of pelleting by using a single pellet press and a pilot-scale pellet press. *Fuel Process. Technol.* **2016**, *144*, 220–229.
100. Jamradloedluk, J.; Lertsatitthanakorn, C. Influences of Mixing Ratios and Binder Types on Properties of Biomass Pellets. *Energy Procedia* **2017**, *138*, 1147–1152. [CrossRef]

101. Si, Y.; Hu, J.; Wang, X.; Yang, H.; Chen, Y.; Shao, J.; Chen, H. Effect of Carboxymethyl Cellulose Binder on the Quality of Biomass Pellets. *Energy Fuels* **2016**, *30*, 5799–5808. [CrossRef]
102. Dąbrowska-Salwin, M.; Lisowski, A.; Kostrubiec, M.; Swietochowski, A. Pressure agglomeration of biomass with addition of calcium carbonate. *Eng. Rural Dev.* **2016**, 542–546.
103. Iftikhar, M.; Asghar, A.; Ramzan, N.; Sajjadi, B.; Chen, W. Biomass densification: Effect of cow dung on the physicochemical properties of wheat straw and rice husk based biomass pellets. *Biomass Bioenergy* **2019**, *122*, 1–16. [CrossRef]
104. Gehrig, M.; Wöhler, M.; Pelz, S.; Steinbrink, J.; Thorwarth, H. Kaolin as additive in wood pellet combustion with several mixtures of spruce and short-rotation-coppice willow and its influence on emissions and ashes. *Fuel* **2019**, *235*, 610–616. [CrossRef]
105. García, R.; González-Vázquez, M.P.; Martín, A.J.; Pevida, C.; Rubiera, F. Pelletization of torrefied biomass with solid and liquid bio-additives. *Renew. Energy* **2020**, *151*, 175–183. [CrossRef]
106. Tang, Y.; Chandra, R.P.; Sokhansanj, S.; Saddler, J.N. The Role of Biomass Composition and Steam Treatment on Durability of Pellets. *Bioenergy Res.* **2018**, *11*, 341–350. [CrossRef]
107. Pirraglia, A.; Gonzalez, R.; Saloni, D.; Denig, J. Technical and economic assessment for the production of torrefied ligno-cellulosic biomass pellets in the US. *Energy Convers. Manag.* **2013**, *66*, 153–164. [CrossRef]
108. Kizuka, R.; Ishii, K.; Sato, M.; Fujiyama, A. Characteristics of wood pellets mixed with torrefied rice straw as a biomass fuel. *Int. J. Energy Env. Eng.* **2019**, *10*, 357–365. [CrossRef]
109. Gaitán-Alvarez, J.; Moya, R.; Puente-Urbina, A.; Rodriguez-Zuñiga, A. Physical and compression properties of pellets manufactured with the biomass of five woody tropical species of Costa Rica torrefied at different temperatures and times. *Energies* **2017**, *10*, 1205. [CrossRef]
110. Azócar, L.; Hermosilla, N.; Gay, A.; Rocha, S.; Díaz, J.; Jara, P. Brown pellet production using wheat straw from southern cities in Chile. *Fuel* **2019**, *237*, 823–832. [CrossRef]
111. Puy, N.; Alier, S.; Bartrolí, J. Production of torrefied pellets from agroforestry biomass for local and regional use. In Proceedings of the 24th European Biomass Conference and Exhibition, Amsterdam, The Netherlands, 6–9 June 2016.
112. Manouchehrinejad, M.; Mani, S. Torrefaction after pelletization (TAP): Analysis of torrefied pellet quality and co-products. *Biomass Bioenergy* **2018**, *118*, 93–104. [CrossRef]
113. Song, X.; Yang, Y.; Zhang, M.; Zhang, K.; Wang, D. Ultrasonic pelleting of torrefied lignocellulosic biomass for bioenergy production. *Renew. Energy* **2018**, *129*, 56–62. [CrossRef]
114. Kudo, S.; Okada, J.; Ikeda, S.; Yoshida, T.; Asano, S.; Hayashi, J.I. Improvement of Pelletability of Woody Biomass by Torrefaction under Pressurized Steam. *Energy Fuels* **2019**, *33*, 11253–11262. [CrossRef]
115. Gong, S.H.; Im, H.S.; Um, M.; Lee, H.W.; Lee, J.W. Enhancement of waste biomass fuel properties by sequential leaching and wet torrefaction. *Fuel* **2019**, *239*, 693–700. [CrossRef]
116. Rejdak, M.; Czardybon, A.; Ignasiak, K.; Robak, J. Utilization of waste forest biomass: Pelletization studies of torrefied sawmill wood chips. In *Proceedings of the 11th Conference on Interdisciplinary Problems in Environmental Protection and Engineering EKO-DOK 2019, Polanica-Zdroj, Poland, 8–10 April 2019*; E3S Web Conferences: Les Ulis, France, 2019; Volume 100.
117. Shahrukh, H.; Oyedun, A.O.; Kumar, A.; Ghiasi, B.; Kumar, L.; Sokhansanj, S. Techno-economic assessment of pellets produced from steam pretreated biomass feedstock. *Biomass Bioenergy* **2016**, *87*, 131–143. [CrossRef]
118. Obernberger, I.; Thek, G. *The Pellet Handbook: The Production and Thermal Utilisation of Pellets*; Routledge: Abingdon, UK, 2010; ISBN 1844076318.
119. Mobini, M.; Sowlati, T.; Sokhansanj, S. A simulation model for the design and analysis of wood pellet supply chains. *Appl. Energy* **2013**, *111*, 1239–1249. [CrossRef]
120. Mobini Dehkordi, M.M. On the design and analysis of forest biomass to biofuel and bioenergy supply chains. Ph.D. Thesis, University of British Columbia, Vancouver, BC, Canada, April 2015.
121. Trømborg, E.; Ranta, T.; Schweinle, J.; Solberg, B.; Skjevrak, G.; Tiffany, D.G. Economic sustainability for wood pellets production–A comparative study between Finland, Germany, Norway, Sweden and the US. *Biomass Bioenergy* **2013**, *57*, 68–77. [CrossRef]
122. Koppejan, J.; Sokhansanj, S.; Melin, S.; Madrali, S. Status overview of torrefaction technologies. *IEA Bioenergy Task* **2012**, *32*, 1–54.
123. Radics, R.I.; Gonzalez, R.; Bilek, E.T.M.; Kelley, S.S. Systematic review of torrefied wood economics. *BioResources* **2017**, *12*, 6868–6884. [CrossRef]

124. Beets, M. A Torrefied Wood Pellet Supply Chain. A detailed cost analysis of the comptetitiveness of torrefied wood pellets compared to white wood pellets. Master's Thesis, Utrecht University, Utrecht, The Netherlands, 2017.
125. Yun, H.; Clift, R.; Bi, X. Process simulation, techno-economic evaluation and market analysis of supply chains for torrefied wood pellets from British Columbia: Impacts of plant configuration and distance to market. *Renew. Sustain. Energy Rev.* **2020**, *127*, 109745. [CrossRef]

Review

Revolutionizing Towards Sustainable Agricultural Systems: The Role of Energy

Ilaria Zambon [1,*], Massimo Cecchini [1], Enrico Maria Mosconi [2] and Andrea Colantoni [1]

[1] Department of Agricultural and Forestry Sciences (DAFNE), Tuscia University, Via San Camillo de Lellis, I-01100 Viterbo, Italy; cecchini@unitus.it (M.C.); colantoni@unitus.it (A.C.)

[2] Department of Economics Engineering, Society and Business Organization, Tuscia University, Via del Paradiso 47, I-01100 Viterbo, Italy; enrico.mosconi@unitus.it

* Correspondence: ilaria.zambon@unitus.it

Received: 29 July 2019; Accepted: 23 September 2019; Published: 25 September 2019

Abstract: Innovations play a significant role in the primary sector (i.e., agriculture, fisheries and forestry), ensuring a greater performance towards bioeconomy and sustainability. Innovation is being progressively applied to examining the organization of joint technological, social, and institutional modernizations in the primary sector. Exploring the governance of actor relations, potential policies, and support structures is crucial in the phase of innovation, e.g., during research activities, often applied at the national or sectorial scale. However, when opposing normative guidelines for alternative systems of agriculture arise (e.g., the industrial agriculture paradigm), modernizations in agricultural and forestry may contribute to outlining more sustainable systems. To date, innovations in the primary sector do not seem as advanced as in other sectors, apart from industrial agriculture, which sometimes appears to be the most encouraged. The present review aims to shed light on innovations that have been identified and promoted in recent years in the primary sector, including agriculture and forestry. The need to pursue sustainable development in this sector requires the inclusion of a fourth dimension, namely energy. In fact, energy sustainability is an issue that has been much discussed in recent years. However, the need for progressive technological progress is indispensable to ensure long-lasting energy efficiency. The aim is to understand what innovations have been implemented recently, highlighting opportunities and limitations for the primary sector.

Keywords: innovation; agriculture; forestry; energy; sustainability

---

1. Introduction

Innovations play a significant role in the primary sector (i.e., agriculture, fisheries and forestry), ensuring a greater performance towards bioeconomy and sustainability [1–3]. Innovation has been progressively applied to examine the organization of joint technological, social and institutional modernizations in the primary sector [4–6]. Exploring the governance of actor relations, potential policies, and support structures is crucial in the phase of innovation, e.g., during research activities [7], often applied at the national or sectorial scale [5,7]. However, when opposing normative guidelines for alternative systems of agriculture arise (e.g., the industrial agriculture paradigm), modernizations in agriculture and forestry may contribute to outlining more sustainable systems [5,8–12]. To date, innovations in the primary sector do not seem as advanced as in other sectors, apart from industrial agriculture, which sometimes appears to be the most encouraged nowadays [10]. This leads to the need (i) to adopt a multifunctional approach to agriculture, (ii) to focus explicitly on ecological aspects, and (iii) to take on technological advances that involve different disciplines [5,13,14].

Innovations in the primary sector can enable a cooperative achievement emerging from different actors (frequently in new mixtures), providing innovative modes of production and new organizational structures and activities to better support widespread learning, adapting, and adjusting [5,15].

By permitting relations across different stakeholders (e.g., professional, sectoral, organizational, or cultural), the involved actors can (i) experiment together, (ii) propose new technologies, practices, and institutions, and (iii) support evolution towards a more sustainable agriculture [5,15]. Thanks to the interactions among different actors, changes to more sustainable agriculture comprise the development of modernization [15,16].

Complex challenges facing an evolution towards more sustainable agricultural and forestry systems are often connected to resource competition (e.g., energy, water, land, biodiversity), socio-economic apprehensions (e.g., community development, rural livelihoods, emerging markets), and environmental integrity (e.g., climate change) [17–20]. Such challenges characteristically span numerous natural resource management systems (e.g., agriculture and forestry, but also water, conservation and energy) and related ecosystem services (including the provision, regulatory, cultural and support services) [5,21,22]. Agriculture-related fields identify the requirement for better adoptive relationships across scales and sectors to report multifaceted sustainability trials [13,23]. Consequently, more networked methods are required to move toward innovation and sustainable agriculture, with the aim to simplify the boundary crossing and to coordinate actors across different spatial scales [5]. With reference to crossing scales, changes toward sustainability require the skill to move innovation processes ahead by working across scales [14,24]. Correspondingly, landscape approaches have the latent character to mature more combined approaches for sustainable agriculture and to enable the compulsory connections among the systems, services and, sectors interested in agricultural and forestry areas [16,25,26]. For instance, multifunctionality and circular economies are some of the main ideas in sustainable agricultural and forestry systems [5]. Multifunctionality includes production, environmental, and human features of agriculture, which are crucial to sustainability [19,27,28]. Sectoral divisions (e.g., agriculture versus energy) or sub-sectoral divides within agriculture owing to specialization (e.g., separation of crop and livestock sectors) have long delayed endorsing multifunctional methods in agriculture and practical crossovers with non-agricultural sectors [5,27,29–31].

The present review aims to shed light on innovations that have been identified and promoted in recent years in the primary sector, including agriculture and forestry. The need to pursue sustainable development in this sector perceives the requirement to include a fourth dimension, namely energy [32]. Examining one of the recent notions of 'sustainability', energy can be assumed as the fourth dimension of sustainability. With the four sustainable dimensions (energy, economy, society and environment), rural districts can provide greater maintainable growth, also focusing on a sustainable (energy) future, which in literature is perceived by the "agro-energy districts" [32–36]. However, very often such innovative realities present their intrinsic limits. For this reason, today, the need for progressive technological progress is indispensable to ensure long-lasting energy efficiency. The general aim of the present review is to understand what innovations have been carried out recently, highlighting opportunities and limitations for the primary sector.

2. Energy for a Sustainable Development in the Primary Sector

Agriculture is one of the most significant sectors, which is characterized by the greatest potential for sustainable economic development [5,7,9,16,28–39]. Specifically, renewable sources represent good alternatives to fossil resources in the primary sector, which are inadequate in quantity and are prone to exhaustion [31,38,40–43]. Development of renewable energy as a main global resource of clean energy is one of the main global purposes of current policies, which, in the overall outline of sustainable development, is intended to reduce energy consumption and increase the security of supply, environmental protection, and maintainable technology development [38].

Forest and agricultural resources are conspicuous sources of energy that may be indispensable for greater local and technological development [23,41,44–47]. For this reason, recent scientific and technical research has sought to mature innovations that allow the larger reuse of resources with clean energy and low environmental impact [2,3,6,12,14,17].

To become economically sustainable, undeveloped renewable resource-based technologies must progress along the learning curve, finally influencing competitiveness with fossil resource-based options [8,48,49]. Furthermore, innovations are fundamental for ensuring environmental sustainability that decreases the potential impacts of renewable resource use on ecosystems [49,50]. The employment of a circular flow economy needs original technological, organizational and product solutions, and improved resource efficiency [1,47]. On the supply side, innovations should focus on technologies and products, while from the user side, main concern should be given to consumption and waste generation patterns [49,51].

The innovation lies in the fact that the advancement of energy efficiency is achieved through a gradual move from fossil energy sources to more environmentally friendly and renewable energy sources, endorsing energy efficiency through improved performance technologies and schemes [8,52,53] and affording a combined method reliable with the expansion of the liberalized market. For this purpose, today, a scheme that can assimilate advanced and innovative essentials of tariff control and direct control of the market should be implemented [43].

With the regards a prospective future changeover towards a sustainable bioeconomy, today, policy interventions, technology and product markets results are inadequate in offering suitable encouragements [49,54,55]. Furthermore, compulsory fossil resource-based technologies profit from past learning properties, increasing returns, and network externalities, all of which interrelate with the dedicated nature of investments to produce a technological path reliance [49]. Such a framework can be strengthened by a co-evolutionary growth of fossil resource-based organizations, codependent industries, consumption patterns, and private and public institutions ensuing in a "carbon lock-in" [4,7,9].

Research on the relationship among innovation systems, sustainability, and policies focused on energy system transitions [56–59]. Literature offers significant insights concerning appropriate policies, innovation economics, and systems [49,58]. Therefore, an inclusive and well-coordinated policy asset is compulsory to support a well-developed innovation system, which is in turn helpful in providing a sociotechnical path change [56,59].

In the last decade, climate and energy strategy was emphasized by the European declaration towards the achievement of definite environmental goals of energy strategy and advancement of agriculture-based bioenergy and biofuels [41,60–62]. Several global challenges require innovative approaches to knowledge discussion, e.g., those predicted in the European Innovation Partnership "Agricultural Productivity and Sustainability", which are of greatest importance to foster the application of explanations [52]. The change toward agriculture-based bioenergy has also been supported in reply to structural variations in agriculture that involve an examination of new endurance approaches [60].

2.1. Conjugating Innovation with Biomass

Climate change, scarcity of resources and materials, increasing population, and environmental pressures have encouraged a revaluation and assessment of the current fossil-based economy [3,63]. The bioeconomy concept has been given numerous descriptions and its conceptualization is still developing [64–66]. However, two main features can be shared. Firstly, bioeconomy will depend on renewable biomass in place of finite fossil inputs to produce an extensive variety of value-added products, e.g., bio-based products and bioenergy [3,32,64,67–70]. Secondly, these products will be formed in biorefineries succeeding a cascade principle with the intention of extremely valorizing the obtainable biomass [71,72]. Biomass is originally processed into high value products and the relative residues are used for lower value applications until the smallest amount of waste remains at the conclusion of the process [73–75]. Bioeconomy can therefore be considered an assortment of sectors and subsectors (e.g., agriculture and energy), employed in combination to derive products from renewable biological resources from the primary sector [3,71,72].

According to [3], innovation can develop following specific processes within the bioeconomy based on a four-staged literature research: innovation process, network management, stakeholder groups, and bioeconomy contextual factors. Idea development, invention, and commercialization are the three fundamental phases for activating an iterative process. During the latter process, the three fundamental phases are interconnected via learning cycles, enabling repetition of confident process steps to adjust to unexpected progresses and errors (Figure 1).

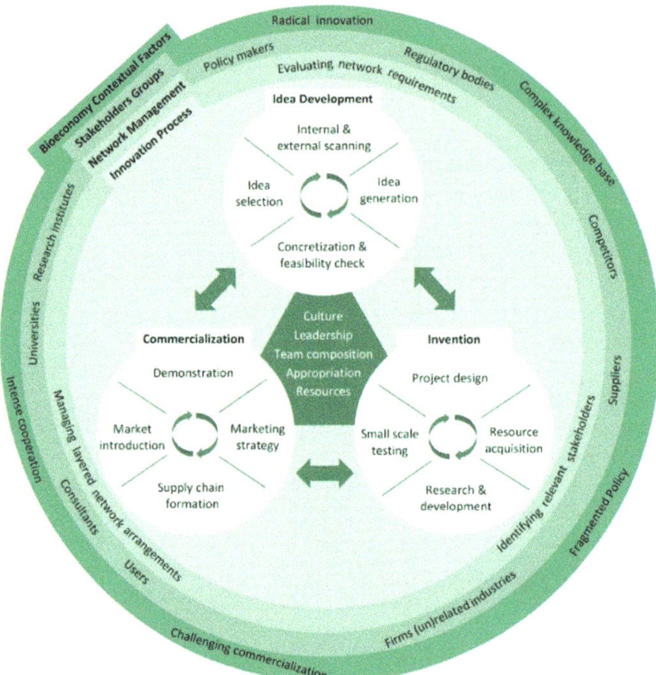

Figure 1. A schematic depiction of innovation in the bioeconomy framework (source: our elaboration with a reproduction with permission from [3], 2016).

Biomass feedstock are energy sources in the primary sector, which are derived from wastes associated with their processing (e.g., agricultural residues) [6,67,68,70,76]. However, natural resources for energy purpose expose contrasting opinions. Following some researchers, land-use change for the cultivation of energy crops can have important consequences on global food security, employment, the income of regional populations, and the biodiversity of ecological societies [37,76,77]. Land accessibility, land-use practices, and water availability are some of the main key factors for the large-scale production of biofuels [76,78–80]. A negative perspective has exposed that the future availability of arable land, using existing farming areas for energy crop cultivation, might result in food deficiency [76]. However, it may be practicable to convert abandoned and marginal areas into cultivated land for energy crop cultivation [32,55,67,69,77]. Global biofuel programs will usually subsidize the sustainable livelihood of agricultural employees by growing employment rates in most rural contexts since a large portion of feedstock cultivation and plant processing includes manual work [76].

Following a more optimistic perspective, biomass supply may be protected and extended using advanced management procedures and strategies [40,81]. For instance, new sources of cheap biomass for energy production can be selected by using lignocellulosic material that is inappropriate for common forestry and agricultural usage [82].

Investigation into innovations and technologies that can decrease land use and decrease accidents from renewable energy sources and the risk of resource competition have been carried out [83,84]. For instance, with bioenergy, food for consumption competes in the same areas as those for energy production [42].

2.2. Conjugating Innovation with Solar Energy

Innovation in the primary sector also benefits from solar energy [85–89]. Many studies have deepened understanding of this issue since the installation of solar systems on greenhouse structures became one of the recent strategies to allow both agricultural production and energy production [39,46,86,89–91]. However, the installation of photovoltaic panels on greenhouses involves certain limitations, which have been solved through specific studies [39,92]. Shading represents one of the main limitations for photovoltaic installation on the rooves of greenhouses [39,93–96]. The spatial shading variations in a greenhouse during a year can be supposed an appropriate factor for picking out the greatest combination of photovoltaic panels and crops present, guaranteeing both energy and production activities [52,88]. The choice of the covering shading on the spectral distribution of solar radiation incoming to greenhouses, the kind of material, and further properties of the greenhouse are crucial for defining pertinent technical structural details, environmental influences, and internal climatic settings [87,97–99].

The question of shade due to the presence of photovoltaic panels on greenhouse rooves, though, is related to the possible reduction of cultivation that is carried out within the greenhouse [52,100]. Nevertheless, several studies have already promoted different methods and strategies and projects of innovative greenhouses with the intention of limiting the shadow percentage and ensuring optimal agricultural production [39,46,85].

3. Discussion

Innovation is indispensable to reply to the critical apprehensions and challenges related to, for example, climate change, environmental trials, energy scarcity, and food security [1,13,21,22,27,37,50,76,77,99]. Numerous novelties will be in the form of products, processes, and services which can improve the success and efficiency of answering to these socioeconomic challenges and dealing with the measurement and mitigation of negative externalities [63].

Greater sustainability in the primary sector can be guaranteed through (i) the use of agricultural residues, i.e., biomass from agricultural and forestry wastes [16,39,67,68,70,82,101], and (ii) renewable energy, e.g., solar energy [39]. The importance of having adequate structures as solar systems installed on greenhouses helping to produce agricultural products is fundamental to maximize space and optimize both agricultural and energy production [32,39].

Agricultural innovation can be considered as a co-evolutionary process where technological, socioeconomic, and institutional changes are combined [4,7,102]. With production and technical knowledge, numerous factors play a crucial role as prerequisites for innovation, for instance, policy, legislation, funding, infrastructure, and market progresses [7]. However, a limited quantity of research is available concerning current technologies and innovations within the primary sector, such as in developed countries [102–104]. The latter countries are mainly vulnerable to environmental influences, e.g., related to climate change [34,42]. Such current challenges usually have an explicit influence on the agricultural and forestry sector [103].

Addressing the current complex and uncontrolled sustainability matters facing the primary sector, innovations must ensure continuous progress by (i) facilitating the development of rural and forestry economic realities, (ii) pursuing greater environmental mitigation, and (iii) bringing the political sphere into line with the more practical one [5,15,16]. However, dealing with such current challenges is progressively significant (i) bearing in mind mission-oriented innovation strategies and (ii) underscoring the requirement for innovation systems to achieve specific determinations for a sustainability changeover [5,10,105].

Agricultural innovation is no longer just about approving new technologies, technical practices, and alternative customs of organization [7]. Additionally, agricultural innovation is usually driven by different visions exposing different development directions, e.g., sustainable developments and energy efficiency [7,8,16,52,53,104]. However, soil should play a major role in food production in view of another fundamental issue today, namely the increase of both food requirements and the reduction in soil availability and fertility due to population growth and the growing phenomena of desertification and soil degradation [106–111]. In this requirement, the sustainability of agricultural practices can make sense of the reduction of waste and the agronomic energy reuse of by-products [75,82,101]. The role of energy is decisive in this framework also in view of a sustainable development in which the needs of fossil fuels (to produce energy biomass) must assess and reflect on the proper use of agricultural areas and in the recovery of abandoned and marginal areas, even where energy crops are grown [8,43].

4. Conclusion

Energy can assume a key role in current innovations, highlighting chances and limitations for sustainable development [32,33,38,39,76,82,83,112–115]. In this review, the primary sector is investigated due to different issues, e.g., alternative systems of agriculture, more sustainable modernizations in agricultural and forestry [116], reuse of rural land [75,77,82,101,107,116], environmental respect avoiding stress on natural resources, e.g., soil, which can lead to degenerative phenomena, e.g., soil degradation, which are especially severe if they relate to the loss of high-quality soils [37,80,106–111]. Innovation has become increasingly applied to survey the organization of joint technological, social, and institutional modernizations in the primary sector [6,99,103,106,113]. Sightseeing the governance of actor relations, latent policies, and support structures is critical to the stage of innovation, e.g., during research activities, at different spatial scales [117,118]. Modernizations in agriculture and forestry may contribute to planning more sustainable systems, even including energy as a fourth dimension together with economy, society, and environment [32,33,116]. In fact, energy sustainability has become a crucial question in recent years, especially where progressive technological progress is necessary to ensure long-lasting energy efficiency.

Author Contributions: Conceptualization, I.Z. and A.C.; methodology, M.C.; validation, E.M.M. and A.C.; formal analysis, I.Z.; investigation, I.Z., M.C., and A.C.; writing—original draft preparation, I.Z.; writing—review and editing, I.Z., M.C., E.M.M., and A.C.

Funding: The research was partially supported by the Italian Ministry for Education, University and Research (MIUR) according to the Italian Law 232/2016 within the fund for the financing of universities "Departments of Excellence".

Conflicts of Interest: The authors declare no conflict of interest.

References

1. Carus, M.; Dammer, L. Food or Non-Food: Which Agricultural Feedstocks Are Best for Industrial Uses? *Ind. Biotechnol.* **2013**, *9*, 171–176. [CrossRef]
2. Hellsmark, H.; Mossberg, J.; Söderholm, P.; Frishammar, J. Innovation system strengths and weaknesses in progressing sustainable technology: The case of Swedish biorefinery development. *J. Clean. Prod.* **2016**, *131*, 702–715. [CrossRef]
3. Van Lancker, J.; Wauters, E.; Van Huylenbroeck, G. Managing innovation in the bioeconomy: An open innovation perspective. *Biomass Bioenergy* **2016**, *90*, 60–69. [CrossRef]
4. Kilelu, C.W.; Klerkx, L.; Leeuwis, C. Unravelling the role of innovation platforms in supporting co-evolution of innovation: Contributions and tensions in a smallholder dairy development programme. *Agric. Syst.* **2013**, *118*, 65–77. [CrossRef]
5. Pigford, A.-A.E.; Hickey, G.M.; Klerkx, L. Beyond agricultural innovation systems? Exploring an agricultural innovation ecosystems approach for niche design and development in sustainability transitions. *Agric. Syst.* **2018**, *164*, 116–121. [CrossRef]
6. Turner, J.A.; Klerkx, L.; Rijswijk, K.; Williams, T.; Barnard, T. Systemic problems affecting co-innovation in the New Zealand Agricultural Innovation System: Identification of blocking mechanisms and underlying institutional logics. *NJAS Wagening. J. Life Sci.* **2016**, *76*, 99–112. [CrossRef]

7. Klerkx, L.; Van Mierlo, B.; Leeuwis, C. *Evolution of Systems Approaches to Agricultural Innovation: Concepts, Analysis and Interventions, Farming Systems Research into the 21st Century: The New Dynamic*; Springer: Berlin/Heidelberg, Germany, 2012; pp. 457–483.
8. Ingrao, C.; Bacenetti, J.; Bezama, A.; Blok, V.; Goglio, P.; Koukios, E.G.; Lindner, M.; Nemecek, T.; Siracusa, V.; Zabaniotou, A.; et al. The potential roles of bio-economy in the transition to equitable, sustainable, post fossil-carbon societies: Findings from this virtual special issue. *J. Clean. Prod.* **2018**, *204*, 471–488. [CrossRef]
9. Plumecocq, G.; Debril, T.; Duru, M.; Magrini, M.-B.; Sarthou, J.P.; Therond, O. The plurality of values in sustainable agriculture models: Diverse lock-in and coevolution patterns. *Ecol. Soc.* **2018**, *23*, 21. [CrossRef]
10. Schlaile, M.P.; Urmetzer, S.; Blok, V.; Andersen, A.D.; Timmermans, J.; Mueller, M.; Fagerberg, J.; Pyka, A. Innovation Systems for Transformations towards Sustainability? Taking the Normative Dimension Seriously. *Sustainability* **2017**, *9*, 2253. [CrossRef]
11. Stirling, A. Pluralising progress: From integrative transitions to transformative diversity. *Environ. Innov. Soc. Transit.* **2011**, *1*, 82–88. [CrossRef]
12. Touzard, J.-M.; Temple, L.; Faure, G.; Triomphe, B. Innovation systems and knowledge communities in the agriculture and agrifood sector: A literature review. *J. Innov. Econ.* **2015**, *17*, 117. [CrossRef]
13. Foran, T.; Butler, J.R.; Williams, L.J.; Wanjura, W.J.; Hall, A.; Carter, L.; Carberry, P.S. Taking Complexity in Food Systems Seriously: An Interdisciplinary Analysis. *World Dev.* **2014**, *61*, 85–101. [CrossRef]
14. Wigboldus, S.; Klerkx, L.; Leeuwis, C.; Schut, M.; Muilerman, S.; Jochemsen, H. Systemic perspectives on scaling agricultural innovations. A review. *Agron. Sustain. Dev.* **2016**, *36*, 46. [CrossRef]
15. Meynard, J.-M.; Jeuffroy, M.-H.; Le Bail, M.; Lefèvre, A.; Magrini, M.-B.; Michon, C. Designing coupled innovations for the sustainability transition of agrifood systems. *Agric. Syst.* **2017**, *157*, 330–339. [CrossRef]
16. Prost, L.; Berthet, E.T.; Cerf, M.; Jeuffroy, M.H.; Labatut, J.; Meynard, J.M. Innovative design for agriculture in the move towards sustainability: Scientific challenges. *Res. Eng. Des.* **2017**, *28*, 119–129. [CrossRef]
17. Bennett, E.; Carpenter, S.; Gordon, L.; Ramankutty, N.; Balvanera, P.; Campbell, B.; Cramer, W.; Foley, J.; Folke, C.; Karlberg, L. Toward a more resilient agriculture. *Solutions* **2014**, *5*, 65–75.
18. FAO (Food and Agriculture Organization). *Building a Common Vision for Sustainable Food and Agriculture: Principles and Approaches*; Food and Agriculture Organization of the United Nations: Rome, Italy, 2014.
19. FAO (Food and Agriculture Organization). *The State of Food and Agriculture 2016 (SOFA): Climate Change, Agriculture and Food Security*; Food and Agriculture Organization of the United Nations: Rome, Italy, 2016.
20. Hall, A.; Clark, N. What do complex adaptive systems look like and what are the implications for innovation policy? *J. Int. Dev.* **2010**, *22*, 308–324. [CrossRef]
21. Bommarco, R.; Kleijn, D.; Potts, S.G. Ecological intensification: Harnessing ecosystem services for food security. *Trends Ecol. Evol.* **2013**, *28*, 230–238. [CrossRef]
22. Ville, A.S.S.; Hickey, G.M.; Phillip, L.E. Addressing food and nutrition insecurity in the Caribbean through domestic smallholder farming system innovation. *Reg. Environ. Chang.* **2015**, *15*, 1325–1339. [CrossRef]
23. Berthet, E.T.; Segrestin, B.; Hickey, G.M. Considering agro-ecosystems as ecological funds for collective design: New perspectives for environmental policy. *Environ. Sci. Policy* **2016**, *61*, 108–115. [CrossRef]
24. Svensson, O.; Nikoleris, A. Structure reconsidered: Towards new foundations of explanatory transitions theory. *Res. Policy* **2018**, *47*, 462–473. [CrossRef]
25. Arts, B.; Buizer, M.; Horlings, L.; Ingram, V.; Van Oosten, C.; Opdam, P. Landscape Approaches: A State-of-the-Art Review. *Annu. Rev. Environ. Resour.* **2017**, *42*, 439–463. [CrossRef]
26. Sayer, J.; Sunderland, T.; Ghazoul, J.; Pfund, J.-L.; Sheil, D.; Meijaard, E.; Venter, M.; Boedhihartono, A.K.; Day, M.; Garcia, C.; et al. Ten principles for a landscape approach to reconciling agriculture, conservation, and other competing land uses. *Proc. Natl. Acad. Sci. USA* **2013**, *110*, 8349–8356. [CrossRef] [PubMed]
27. Brooks, S.; Loevinsohn, M. Shaping agricultural innovation systems responsive to food insecurity and climate change. *Nat. Resour. Forum* **2011**, *35*, 185–200. [CrossRef]
28. Rockström, J.; Williams, J.; Daily, G.; Noble, A.; Matthews, N.; Gordon, L.; Wetterstrand, H.; DeClerck, F.; Shah, M.; Steduto, P. Sustainable intensification of agriculture for human prosperity and global sustainability. *Ambio* **2017**, *46*, 4–17. [CrossRef]
29. Hassink, J.; Grin, J.; Hulsink, W. Multifunctional Agriculture Meets Health Care: Applying the Multi-Level Transition Sciences Perspective to Care Farming in the Netherlands. *Sociol. Rural.* **2013**, *53*, 223–245. [CrossRef]

30. Hassink, J.; Grin, J.; Hulsink, W. Enriching the multi-level perspective by better understanding agency and challenges associated with interactions across system boundaries. The case of care farming in the Netherlands: Multifunctional agriculture meets health care. *J. Rural. Stud.* **2018**, *57*, 186–196. [CrossRef]
31. Sutherland, L.-A.; Peter, S.; Zagata, L. Conceptualising multi-regime interactions: The role of the agriculture sector in renewable energy transitions. *Res. Policy* **2015**, *44*, 1543–1554. [CrossRef]
32. Zambon, I.; Colantoni, A.; Cecchini, M.; Mosconi, E.M. Rethinking Sustainability within the Viticulture Realities Integrating Economy, Landscape and Energy. *Sustainability* **2018**, *10*, 320. [CrossRef]
33. Bonazzi, F.A.; Cividino, S.R.; Zambon, I.; Mosconi, E.M.; Poponi, S. Building Energy Opportunity with a Supply Chain Based on the Local Fuel-Producing Capacity. *Sustainability* **2018**, *10*, 2140. [CrossRef]
34. Jeong, J.S. Biomass Feedstock and Climate Change in Agroforestry Systems: Participatory Location and Integration Scenario Analysis of Biomass Power Facilities. *Energies* **2018**, *11*, 1404. [CrossRef]
35. Moulogianni, C.; Bournaris, T. Biomass Production from Crops Residues: Ranking of Agro-Energy Regions. *Energies* **2017**, *10*, 1061. [CrossRef]
36. Sharara, M.A.; Sadaka, S.S. Opportunities and Barriers to Bioenergy Conversion Techniques and Their Potential Implementation on Swine Manure. *Energies* **2018**, *11*, 957. [CrossRef]
37. Altieri, M.A.; Nicholls, C.I. The adaptation and mitigation potential of traditional agriculture in a changing climate. *Clim. Chang.* **2017**, *140*, 33–45. [CrossRef]
38. Aschilean, I.; Rasoi, G.; Raboaca, M.S.; Filote, C.; Culcer, M. Design and Concept of an Energy System Based on Renewable Sources for Greenhouse Sustainable Agriculture. *Energies* **2018**, *11*, 1201. [CrossRef]
39. Marucci, A.; Zambon, I.; Colantoni, A.; Monarca, D. A combination of agricultural and energy purposes: Evaluation of a prototype of photovoltaic greenhouse tunnel. *Renew. Sustain. Energy Rev.* **2018**, *82*, 1178–1186. [CrossRef]
40. Fontes, C.H.D.O.; Freires, F.G.M. Sustainable and renewable energy supply chain: A system dynamics overview. *Renew. Sustain. Energy Rev.* **2018**, *82*, 247–259.
41. Lyytimäki, J. Renewable energy in the news: Environmental, economic, policy and technology discussion of biogas. *Sustain. Prod. Consum.* **2018**, *15*, 65–73. [CrossRef]
42. Owusu, P.A.; Asumadu-Sarkodie, S. A review of renewable energy sources, sustainability issues and climate change mitigation. *Cogent Eng.* **2016**, *3*, 1167990. [CrossRef]
43. Romano, S.; Cozzi, M.; Di Napoli, F.; Viccaro, M. Building Agro-Energy Supply Chains in the Basilicata Region: Technical and Economic Evaluation of Interchangeability between Fossil and Renewable Energy Sources. *Energies* **2013**, *6*, 5259–5282. [CrossRef]
44. Agrawal, A.; Nepstad, D.; Chhatre, A. Reducing Emissions from Deforestation and Forest Degradation. *Annu. Rev. Environ. Resour.* **2011**, *36*, 373–396. [CrossRef]
45. Bentsen, N.S.; Felby, C. Biomass for energy in the European Union—A review of bioenergy resource assessments. *Biotechnol. Biofuels* **2012**, *5*, 25. [CrossRef]
46. Carreño-Ortega, A.; Galdeano-Gómez, E.; Pérez-Mesa, J.C.; Galera-Quiles, M.D.C. Policy and Environmental Implications of Photovoltaic Systems in Farming in Southeast Spain: Can Greenhouses Reduce the Greenhouse Effect? *Energies* **2017**, *10*, 761. [CrossRef]
47. Ghisellini, P.; Cialani, C.; Ulgiati, S. A review on circular economy: The expected transition to a balanced interplay of environmental and economic systems. *J. Clean. Prod.* **2016**, *114*, 11–32. [CrossRef]
48. Osmani, A.; Zhang, J.; Gonela, V.; Awudu, I. Electricity generation from renewables in the United States: Resource potential, current usage, technical status, challenges, strategies, policies, and future directions. *Renew. Sustain. Energy Rev.* **2013**, *24*, 454–472. [CrossRef]
49. Purkus, A.; Hagemann, N.; Bedtke, N.; Gawel, E. Towards a sustainable innovation system for the German wood-based bioeconomy: Implications for policy design. *J. Clean. Prod.* **2018**, *172*, 3955–3968. [CrossRef]
50. Sala, S.; Anton, A.; McLaren, S.J.; Notarnicola, B.; Saouter, E.; Sonesson, U. In quest of reducing the environmental impacts of food production and consumption. *J. Clean. Prod.* **2017**, *140*, 387–398. [CrossRef]
51. Miceli, R. Energy management and smart grids. *Energies* **2013**, *6*, 2262–2290. [CrossRef]
52. Becerril, H.; Rios, I.D.L. Energy Efficiency Strategies for Ecological Greenhouses: Experiences from Murcia (Spain). *Energies* **2016**, *9*, 866. [CrossRef]
53. Gadenne, D.; Sharma, B.; Kerr, D.; Smith, T.; Kerr, D.; Smith, T. The influence of consumers' environmental beliefs and attitudes on energy saving behaviours. *Energy Policy* **2011**, *39*, 7684–7694. [CrossRef]

54. Hagemann, N.; Gawel, E.; Purkus, A.; Pannicke, N.; Hauck, J. Possible Futures towards a Wood-Based Bioeconomy: A Scenario Analysis for Germany. *Sustainability* **2016**, *8*, 98. [CrossRef]
55. Horlings, L.; Kanemasu, Y. Sustainable development and policies in rural regions; insights from the Shetland Islands. *Land Use Policy* **2015**, *49*, 310–321. [CrossRef]
56. Gallagher, K.S.; Grubler, A.; Kuhl, L.; Nemet, G.; Wilson, C. The Energy Technology Innovation System. *Annu. Rev. Environ. Resour.* **2012**, *37*, 137–162. [CrossRef]
57. Markard, J.; Raven, R.; Truffer, B. Sustainability transitions: An emerging field of research and its prospects. *Res. Policy* **2012**, *41*, 955–967. [CrossRef]
58. Ahorsu, R.; Medina, F.; Constantí, M. Significance and Challenges of Biomass as a Suitable Feedstock for Bioenergy and Biochemical Production: A Review. *Energies* **2018**, *11*, 3366. [CrossRef]
59. Edquist, C.; Zabala-Iturriagagoitia, J.M. Public Procurement for Innovation as mission-oriented innovation policy. *Res. Policy* **2012**, *41*, 1757–1769. [CrossRef]
60. Isoaho, K.; Karhunmaa, K. A critical review of discursive approaches in energy transitions. *Energy Policy* **2019**, *128*, 930–942. [CrossRef]
61. Kivimaa, P.; Hildén, M.; Huitema, D.; Jordan, A.; Newig, J. Experiments in climate governance—A systematic review of research on energy and built environment transitions. *J. Clean. Prod.* **2017**, *169*, 17–29. [CrossRef]
62. Kortelainen, J.; Rytteri, T. EU policy on the move–mobility and domestic translation of the European Union's renewable energy policy. *J. Environ. Policy Plan.* **2017**, *19*, 360–373. [CrossRef]
63. Boehlje, M.; Bröring, S. The increasing multifunctionality of agricultural raw materials: Three dilemmas for innovation and adoption. *Int. Food Agribus. Manag. Rev.* **2011**, *14*, 1–16.
64. Pfau, S.F.; Hagens, J.E.; Dankbaar, B.; Smits, A.J.M. Visions of Sustainability in Bioeconomy Research. *Sustainability* **2014**, *6*, 1222–1249. [CrossRef]
65. Pülzl, H.; Kleinschmit, D.; Arts, B. Bioeconomy—An emerging meta-discourse affecting forest discourses? *Scand. J. For. Res.* **2014**, *29*, 386–393. [CrossRef]
66. Vandermeulen, V.; Van Der Steen, M.; Stevens, C.V.; Van Huylenbroeck, G. Industry expectations regarding the transition toward a biobased economy. *Biofuels Bioprod. Biorefin.* **2012**, *6*, 453–464. [CrossRef]
67. Colantoni, A.; Delfanti, L.; Recanatesi, F.; Tolli, M.; Lord, R. Land use planning for utilizing biomass residues in Tuscia Romana (central Italy): Preliminary results of a multi criteria analysis to create an agro-energy district. *Land Use Policy* **2016**, *50*, 125–133. [CrossRef]
68. Colantoni, A.; Evic, N.; Lord, R.; Retschitzegger, S.; Proto, A.R.; Gallucci, F.; Monarca, D. Characterization of biochars produced from pyrolysis of pelletized agricultural residues. *Renew. Sustain. Energy Rev.* **2016**, *64*, 187–194. [CrossRef]
69. Johnson, T.G.; Altman, I. Rural development opportunities in the bioeconomy. *Biomass Bioenergy* **2014**, *63*, 341–344. [CrossRef]
70. Zambon, I.; Colosimo, F.; Monarca, D.; Cecchini, M.; Gallucci, F.; Proto, A.R.; Lord, R.; Colantoni, A. An Innovative Agro-Forestry Supply Chain for Residual Biomass: Physicochemical Characterisation of Biochar from Olive and Hazelnut Pellets. *Energies* **2016**, *9*, 526. [CrossRef]
71. De Besi, M.; McCormick, K. Towards a Bioeconomy in Europe: National, Regional and Industrial Strategies. *Sustainability* **2015**, *7*, 10461–10478. [CrossRef]
72. McCormick, K.; Kautto, N. The Bioeconomy in Europe: An Overview. *Sustainability* **2013**, *5*, 2589–2608. [CrossRef]
73. Fritsche, U.R.; Iriarte, L. Sustainability Criteria and Indicators for the Bio-Based Economy in Europe: State of Discussion and Way Forward. *Energies* **2014**, *7*, 6825–6836. [CrossRef]
74. Keegan, D.; Kretschmer, B.; Elbersen, B.; Panoutsou, C. Cascading use: A systematic approach to biomass beyond the energy sector. *Biofuels Bioprod. Biorefin.* **2013**, *7*, 193–206. [CrossRef]
75. Zwier, J.; Blok, V.; Lemmens, P.; Geerts, R.-J. The Ideal of a Zero-Waste Humanity: Philosophical Reflections on the Demand for a Bio-Based Economy. *J. Agric. Environ. Ethic* **2015**, *28*, 353–374. [CrossRef]
76. Srirangan, K.; Akawi, L.; Moo-Young, M.; Chou, C.P. Towards sustainable production of clean energy carriers from biomass resources. *Appl. Energy* **2012**, *100*, 172–186. [CrossRef]
77. Selvaggi, R.; Valenti, F.; Pappalardo, G.; Rossi, L.; Bozzetto, S.; Pecorino, B.; Dale, B.E. Sequential crops for food, energy, and economic development in rural areas: The case of Sicily. *Biofuels Bioprod. Biorefin.* **2018**, *12*, 22–28. [CrossRef]

78. Creutzig, F.; Ravindranath, N.H.; Berndes, G.; Bolwig, S.; Bright, R.; Cherubini, F.; Fargione, J. Bioenergy and climate change mitigation: An assessment. *GCB Bioenergy* **2015**, *7*, 916–944. [CrossRef]
79. Erb, K.H.; Luyssaert, S.; Meyfroidt, P.; Pongratz, J.; Don, A.; Kloster, S.; Haberl, H. Land management: Data availability and process understanding for global change studies. *Glob. Chang. Biol.* **2017**, *23*, 512–533. [CrossRef] [PubMed]
80. Keesstra, S.; Nunes, J.P.; Novara, A.; Finger, D.; Avelar, D.; Kalantari, Z.; Cerdà, A. The superior effect of nature based solutions in land management for enhancing ecosystem services. *Sci. Total Environ.* **2018**, *610*, 997–1009. [CrossRef] [PubMed]
81. Gold, S.; Seuring, S. Supply chain and logistics issues of bio-energy production. *J. Clean. Prod.* **2011**, *19*, 32–42. [CrossRef]
82. Al-Hamamre, Z.; Saidan, M.; Hararah, M.; Rawajfeh, K.; Alkhasawneh, H.E.; Al-Shannag, M. Wastes and biomass materials as sustainable-renewable energy resources for Jordan. *Renew. Sustain. Energy Rev.* **2017**, *67*, 295–314. [CrossRef]
83. Dincer, I.; Acar, C. Smart energy systems for a sustainable future. *Appl. Energy* **2017**, *194*, 225–235. [CrossRef]
84. Mafakheri, F.; Nasiri, F. Modeling of biomass-to-energy supply chain operations: Applications, challenges and research directions. *Energy Policy* **2014**, *67*, 116–126. [CrossRef]
85. Colantoni, A.; Monarca, D.; Marucci, A.; Cecchini, M.; Zambon, I.; Di Battista, F.; Maccario, D.; Saporito, M.G.; Beruto, M. Solar Radiation Distribution inside a Greenhouse Prototypal with Photovoltaic Mobile Plant and Effects on Flower Growth. *Sustainability* **2018**, *10*, 855. [CrossRef]
86. Hassanien, R.H.E.; Li, M.; Lin, W.D. Advanced applications of solar energy in agricultural greenhouses. *Renew. Sustain. Energy Rev.* **2016**, *54*, 989–1001. [CrossRef]
87. Kavga, A.; Souliotis, M.; Koumoulos, E.P.; Fokaides, P.A.; Charitidis, C.A. Environmental and nanomechanical testing of an alternative polymer nanocomposite greenhouse covering material. *Sol. Energy* **2018**, *159*, 1–9. [CrossRef]
88. Ureña-Sánchez, R.; Callejón-Ferre, Á.J.; Pérez-Alonso, J.; Carreño-Ortega, Á. Greenhouse tomato production with electricity generation by roof-mounted flexible solar panels. *Sci. Agric.* **2012**, *69*, 233–239. [CrossRef]
89. Sahu, B.K. A study on global solar PV energy developments and policies with special focus on the top ten solar PV power producing countries. *Renew. Sustain. Energy Rev.* **2015**, *43*, 621–634. [CrossRef]
90. Castillo, C.P.; Silva, F.B.E.; LaValle, C. An assessment of the regional potential for solar power generation in EU-28. *Energy Policy* **2016**, *88*, 86–99. [CrossRef]
91. De Luca, G.; Fabozzi, S.; Massarotti, N.; Vanoli, L. A renewable energy system for a nearly zero greenhouse city: Case study of a small city in southern Italy. *Energy* **2018**, *143*, 347–362. [CrossRef]
92. Simeoni, P.; Nardin, G.; Ciotti, G. Planning and design of sustainable smart multi energy systems. The case of a food industrial district in Italy. *Energy* **2018**, *163*, 443–456. [CrossRef]
93. Trypanagnostopoulos, G.; Kavga, A.; Souliotis, M.; Tripanagnostopoulos, Y. Greenhouse performance results for roof installed photovoltaics. *Renew. Energy* **2017**, *111*, 724–731. [CrossRef]
94. Pérez-Alonso, J.; Pérez-García, M.; Pasamontes-Romera, M.; Callejon-Ferre, A.J. Performance analysis and neural modelling of a greenhouse integrated photovoltaic system. *Renew. Sustain. Energy Rev.* **2012**, *16*, 4675–4685. [CrossRef]
95. Tudisca, S.; Di Trapani, A.M.; Sgroi, F.; Testa, R.; Squatrito, R. Economic analysis of PV systems on buildings in Sicilian farms. *Renew. Sustain. Energy Rev.* **2013**, *28*, 691–701. [CrossRef]
96. Tudisca, S.; Di Trapani, A.M.; Sgroi, F.; Testa, R.; Squatrito, R. Assessment of Italian energy policy through the study of a photovoltaic investment on greenhouse. *Afr. J. Agric. Res.* **2013**, *8*, 3089–3096.
97. Both, A.J.; Benjamin, L.; Franklin, J.; Holroyd, G.; Incoll, L.D.; Lefsrud, M.G.; Pitkin, G. Guidelines for measuring and reporting environmental parameters for experiments in greenhouses. *Plant Methods* **2015**, *11*, 43. [CrossRef] [PubMed]
98. Cuce, E.; Harjunowibowo, D. Renewable and sustainable energy saving strategies for greenhouse systems: A comprehensive review. *Renew. Sustain. Energy Rev.* **2016**, *64*, 34–59. [CrossRef]
99. Ghoulem, M.; El Moueddeb, K.; Nehdi, E.; Boukhanouf, R.; Calautit, J.K. Greenhouse design and cooling technologies for sustainable food cultivation in hot climates: Review of current practice and future status. *Biosyst. Eng.* **2019**, *183*, 121–150. [CrossRef]

100. Marucci, A.; Monarca, D.; Cecchini, M.; Colantoni, A.; Cappuccini, A. Analysis of internal shading degree to a prototype of dynamics photovoltaic greenhouse through simulation software. *J. Agric. Eng.* **2015**, *46*, 144. [CrossRef]
101. Carlini, M.; Mosconi, E.M.; Castellucci, S.; Villarini, M.; Colantoni, A. An Economical Evaluation of Anaerobic Digestion Plants Fed with Organic Agro-Industrial Waste. *Energies* **2017**, *10*, 1165. [CrossRef]
102. Leeuwis, C.; Aarts, N. Rethinking Communication in Innovation Processes: Creating Space for Change in Complex Systems. *J. Agric. Educ. Ext.* **2011**, *17*, 21–36. [CrossRef]
103. Long, T.B.; Blok, V.; Coninx, I. Barriers to the adoption and diffusion of technological innovations for climate-smart agriculture in Europe: Evidence from the Netherlands, France, Switzerland and Italy. *J. Clean. Prod.* **2016**, *112*, 9–21. [CrossRef]
104. Smith, A.; Voß, J.-P.; Grin, J. Innovation studies and sustainability transitions: The allure of the multi-level perspective and its challenges. *Res. Policy* **2010**, *39*, 435–448. [CrossRef]
105. Mazzucato, M. From market fixing to market-creating: A new framework for innovation policy. *Ind. Innov.* **2016**, *23*, 140–156. [CrossRef]
106. Zdruli, P.; Lal, R.; Cherlet, M.; Kapur, S. New world atlas of desertification and issues of carbon sequestration, organic carbon stocks, nutrient depletion and implications for food security. In *Carbon Management, Technologies, and Trends in Mediterranean Ecosystems*; Springer: Cham, Switzerland, 2017; pp. 13–25.
107. Zambon, I.; Sabbi, A.; Schuetze, T.; Salvati, L. Exploring forest 'fringescapes': urban growth, society and swimming pools as a sprawl landmark in coastal Rome. *Rend. Lincei* **2015**, *26*, 159–168. [CrossRef]
108. Islam, M.S.; Wong, A.T. Climate Change and Food In/Security: A Critical Nexus. *Environments* **2017**, *4*, 38. [CrossRef]
109. Keesstra, S.D.; Bouma, J.; Wallinga, J.; Tittonell, P.; Smith, P.; Cerdà, A.; Montanarella, L.; Quinton, J.N.; Pachepsky, Y.; Van Der Putten, W.H.; et al. The significance of soils and soil science towards realization of the United Nations Sustainable Development Goals. *Soil* **2016**, *2*, 111–128. [CrossRef]
110. Tóth, G.; Hermann, T.; Da Silva, M.R.; Montanarella, L. Monitoring soil for sustainable development and land degradation neutrality. *Environ. Monit. Assess.* **2018**, *190*, 57. [CrossRef]
111. Gomiero, T. Soil Degradation, Land Scarcity and Food Security: Reviewing a Complex Challenge. *Sustainability* **2016**, *8*, 281. [CrossRef]
112. Geissdoerfer, M.; Savaget, P.; Bocken, N.M.; Hultink, E.J. The Circular Economy—A new sustainability paradigm? *J. Clean. Prod.* **2017**, *143*, 757–768. [CrossRef]
113. Anadon, L.D.; Chan, G.; Harley, A.G.; Matus, K.; Moon, S.; Murthy, S.L.; Clark, W.C. Making technological innovation work for sustainable development. *Proc. Natl. Acad. Sci. USA* **2016**, *113*, 9682–9690. [CrossRef]
114. Shahbaz, M.; Mallick, H.; Mahalik, M.K.; Sadorsky, P. The role of globalization on the recent evolution of energy demand in India: Implications for sustainable development. *Energy Econ.* **2016**, *55*, 52–68. [CrossRef]
115. Stock, T.; Obenaus, M.; Kunz, S.; Kohl, H. Industry 4.0 as enabler for a sustainable development: A qualitative assessment of its ecological and social potential. *Process. Saf. Environ. Prot.* **2018**, *118*, 254–267. [CrossRef]
116. Knickel, K.; Redman, M.; Darnhofer, I.; Ashkenazy, A.; Chebach, T.C.; Šūmane, S.; Tisenkopfs, T.; Zemeckis, R.; Atkociuniene, V.; Rivera, M.; et al. Between aspirations and reality: Making farming, food systems and rural areas more resilient, sustainable and equitable. *J. Rural. Stud.* **2018**, *59*, 197–210. [CrossRef]
117. Secco, L.; Favero, M.; Masiero, M.; Pettenella, D.M. Failures of political decentralization in promoting network governance in the forest sector: Observations from Italy. *Land Use Policy* **2017**, *62*, 79–100. [CrossRef]
118. Bosworth, G.; Rizzo, F.; Marquardt, D.; Strijker, D.; Haartsen, T.; Thuesen, A.A. Identifying social innovations in European local rural development initiatives. *Innov. Eur. J. Soc. Sci. Res.* **2016**, *29*, 442–461. [CrossRef]

MDPI
St. Alban-Anlage 66
4052 Basel
Switzerland
Tel. +41 61 683 77 34
Fax +41 61 302 89 18
www.mdpi.com

Energies Editorial Office
E-mail: energies@mdpi.com
www.mdpi.com/journal/energies